Packet Switching

TOMORROW'S COMMUNICATIONS TODAY

Roy D. Rosner

LIFETIME LEARNING PUBLICATIONS
A division of Wadsworth, Inc.
Belmont, California

Designer: Rick Chafian
Developmental Editor: Susan Weisberg
Illustrator: John Foster
Composition: Syntax International

Printed in the United States of America

3 4 5 6 7 8 9 10—85 84 83

Library of Congress Cataloging in Publication Data

Rosner, Roy D., 1943–
Packet switching, tomorrow's communications today.

Includes bibliographies.
1. Packet switching (Data transmission)
I. Title.
TK5105.R67 621.38 81-8120
ISBN 0-534-97965-3 AACR2

Packet Switching

TOMORROW'S COMMUNICATIONS TODAY

To my darling Judith,
and our three "packets"—
Stuart, Matthew, and Rachel.

Contents

Preface

Packet switching has emerged as a telecommunications technique with unlimited potential. Because it permits communications resources to be used at utmost efficiency, packet switching can adapt to a wide range of user services and user demands. Public packet switching networks have been built or are planned in more than 20 countries around the world, and numerous private and experimental networks are currently using packet switching. Presently packet switching is used primarily in connection with computer and data communication. However, its effectiveness for voice, video, and other wideband telecommunications services has been demonstrated, and as advanced data processing techniques improve the computer processors that form the heart of the packet switches, this application of the technique will undoubtedly become widespread.

This book explains how packet networks operate, how they compare to alternative techniques, and how the many options in the design and use of packet networks are interrelated. The needs of both suppliers and users of packet networks are addressed throughout the book, particularly because one's viewpoint can have a significant impact on design choices.

INTENDED AUDIENCE

The book is intended as both a learning tool and a reference for systems analysts, computer scientists, engineers, teleprocessing system planners, and managers of organizations that supply services to the public at large or to their own organizations. It can also aid systems analysts, communications staff officers, and managers of network users in making a knowledgable evaluation of services to be obtained. In other words, it is meant to assist anyone who needs a working knowledge of the design considerations, alternatives, and techniques for meeting present and future communications requirements.

STRUCTURE AND ORGANIZATION OF THE BOOK

In order to cover this broad range of viewpoints, the book is organized into six major parts.

In Part One the basic concept of packet switching is introduced and described against the backdrop of the ubiquitous worldwide circuit switched telephone network. Both qualitative and quantitative comparisons are drawn among packet switching, circuit switching, and message switching. Case examples highlight various operational principles of packet switching.

Part Two presents the details of packet network operation, in terms of both internal network operations and protocols, and the user-network interface. Of particular significance in this part is the discussion of the international network standard known as X.25. Through nearly unprecedented international cooperation this standard has been a key catalyst in the widespread acceptance and utilization of packet switching. Part Two concludes with a discussion on the control, monitoring, and management of packet networks.

In Part Three the discussion turns to various topological considerations in the design of packet networks. Initial packet networks had exclusively distributed, nonhierarchical structures. Later generations of packet networks, however, have been moving toward more hierarchical structures. These aspects are compared and illustrated in Part Three. This part concludes with a discussion of the routing techniques that can be employed in packet switched networks.

Part Four introduces the important concepts of packet broadcast and packet contention networks. Although the discussion initially focuses on satellite-based channels, contention operation on terrestrial radio or wire facilities is considered later in this part. Part Four also describes the utilization of packet broadcast and packet contention to result in packet networks without packet switches, particularly the ALOHA technique and some variations on it. Because of the extremely flexible and efficient operations that this mode of packet switching produces, it will expand significantly over the next decade. Part Four concludes with a discussion of a unique estimating technique for the early design phases of a packet network that combines satellite access with terrestrial interconnects.

Part Five discusses the offerings of commercial communications carriers, some of which use packet switching, and many of which do not. The cost and tariff structures of such services illustrate both user and supplier costs. Part Five concludes with a discussion of some major new communications systems being developed by the biggest organizations in the information services field. These systems may have substantial impact on both users and suppliers of packet switched network services.

In Part Six the future of packet switching (which may already be the present by the time this book is published) is discussed. Technological improvements in the hardware are expected to bring packet switching and other switching methods closer together in both operational technique and performance. Hybrid techniques will lead to the integration of voice, video, graphics, and computer data into single, multifunction networks. Part Six concludes with a brief summary of the trends and future directions of packet switching, with some suggestions for dealing with the rapidly changing technology.

This book is intended to be largely self-contained. An annotated list of carefully chosen supplementary readings at the end of each chapter provides detailed information that can lead interested readers beyond the principles of packet switching into the specifics of hardware design or software coding. However, the emphasis of the book is on the principles of operation and the relative advantages of packet switching compared to other techniques. As you will see through the course of the book, neither the hardware of packet switches nor their operational software is unique. What makes this telecommunications technique unique are the functional characteristics that combine the flexibility of modern data processing with the efficiencies of buffered and queued communications systems. This book aims to provide insight into aspects of packet switching that have unlimited potential for a broad range of applications throughout the 1980s.

A ROADMAP TO THE BOOK

Because this book covers a broad range of topics relevant to the general subject of packet switching, it may not be necessary for each reader to read every section. The figure below breaks down the material according to particular reader interests. Beyond the general background and overview, the basic division separates the interests of network users from network designers and suppliers. An overview of the field is provided for management personnel within each of these two areas of interest. At the designer, analyst, and engineering levels interests have been separately addressed for:

Designers of user facilities (both hardware and software) that will interface with packet switched networks

Systems analysts and applications software designers for facilities that will use packet switched networks

Designers of packet switched network facilities and services to be supplied to other end users

Systems analysts and designers of applications packages and interface equipment and software

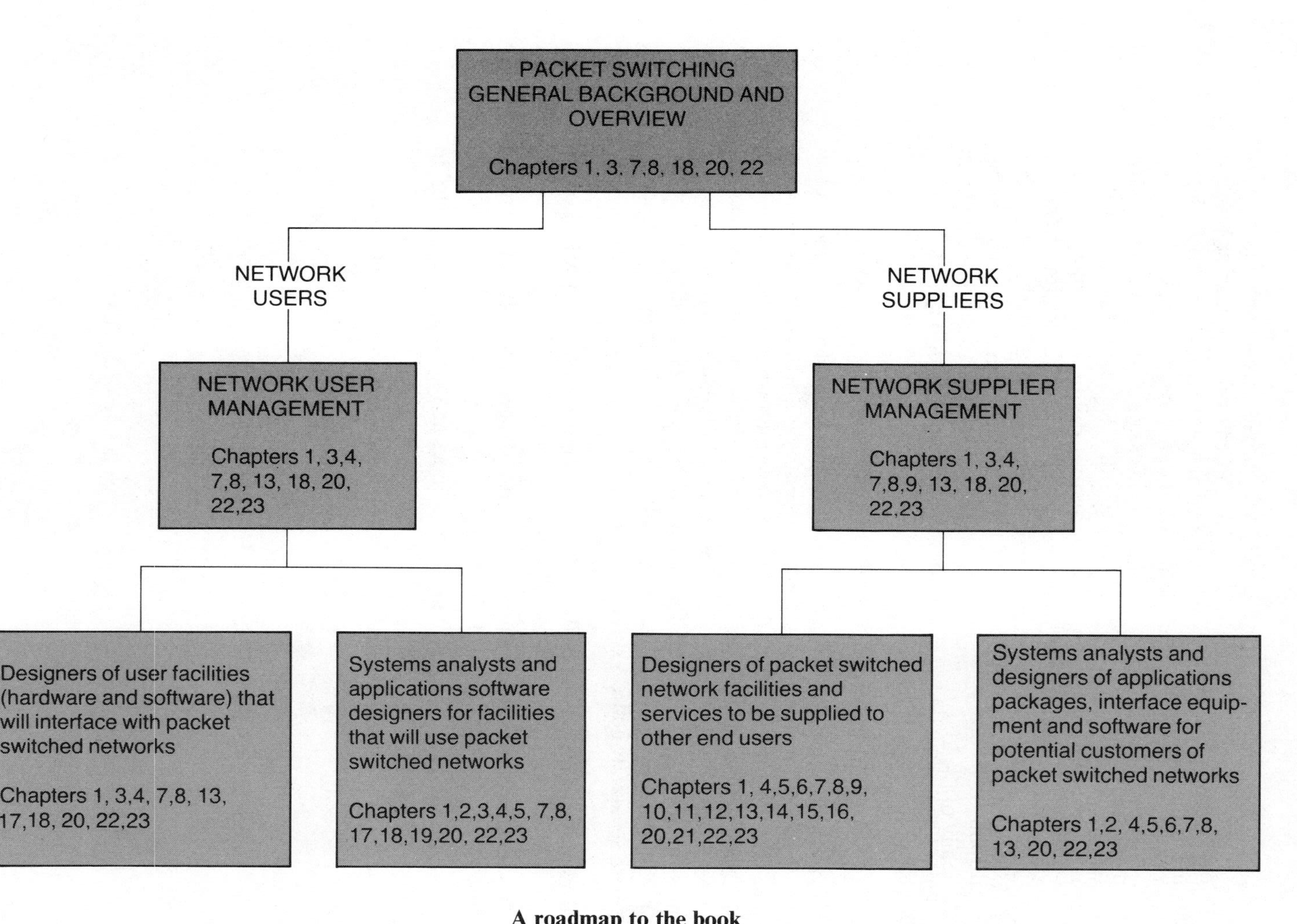

A roadmap to the book

ACKNOWLEDGMENTS

Most of my personal involvement with packet switching took place during my tenure at the Defense Communications Engineering Center, Reston, Virginia. I would like to gratefully acknowledge the far-sighted leadership and support of the many people involved with the early efforts to apply this fledgling technology to important military communications problems. In particular, Brigadier General Kirby LaMar, Brigadier General Herbert Schulke, and Dr. Harry A. Helm were instrumental in providing the encouragement and freedom to explore, innovate, and develop new ideas and concepts. In addition, then Colonel, now Major, General John H. Jacobsmeyer made major contributions to the practical application of many of the ideas in this book.

Roy D. Rosner

PART ONE

What Is a Packet?

1

To Switch or Not to Switch

THIS CHAPTER:

will introduce the concept of switched communications networks.

will discuss the technology that makes switching highly advantageous and cost effective.

WHAT IS PACKET SWITCHING?

Let us begin looking at communications networks by imagining a way of moving information—be it voices speaking, computers rapidly emitting ones and zeros, or picture images—from one physical location to another. The information would be carefully copied onto huge rolls of paper hung up in homes and offices. The rolls would be strung from each home and office to the nearest post office, where the information would be transferred to even larger rolls of paper strung through space from city to city. The paper would slowly be unrolled as information was added, and would move slowly through space, to have the information removed and transferred at the distant end.

This picture may seem far-fetched, but it is exactly analogous to what happens every time we make a telephone call. An actual, physical path is established between each pair of callers for the entire duration of their call. Even though they speak in separate words and sentences, and even though one person generally listens while the other is talking, a continuous two-way path in space, time, or frequency spectrum—a continuous "roll" or circuit—is provided between each pair of users.

With voice communication, where the information content (in an information theoretical sense) is high, use of such a circuit is fairly efficient. However, when we begin to use such a connection to pass information between a keyboard terminal and a computer, with appreciable amounts of "think time" between information bursts, maintaining that physical connection through space is very inefficient.

Contrary to **circuit switching, packet switching** involves moving information from place to place on an as-needed basis, where the amount of information and the end points change with time. In many situations, transferring information through communications networks in packets, or bursts, like envelopes full of letters or bills, can result in vastly increased efficiency, reduced costs to users, and, as we shall see, a wide variety of new communication services.

Much of the following discussion will be based on the fact that packet switching technology has been developed largely with computer communications users in mind. This assumes either a person at some kind of data terminal, entering information to and receiving information from a computer located at some distance, or pairs or groups of computers transferring information among themselves. However, the basic ideas are also applicable to information transfer for voice, video, graphic, or any other electronic communication.

WHY USE A SWITCHED NETWORK?

The primary function of switches in a communications network is to allocate the network resources to a particular information transfer at a given time. There are three fundamental reasons for using switching in a communications network:

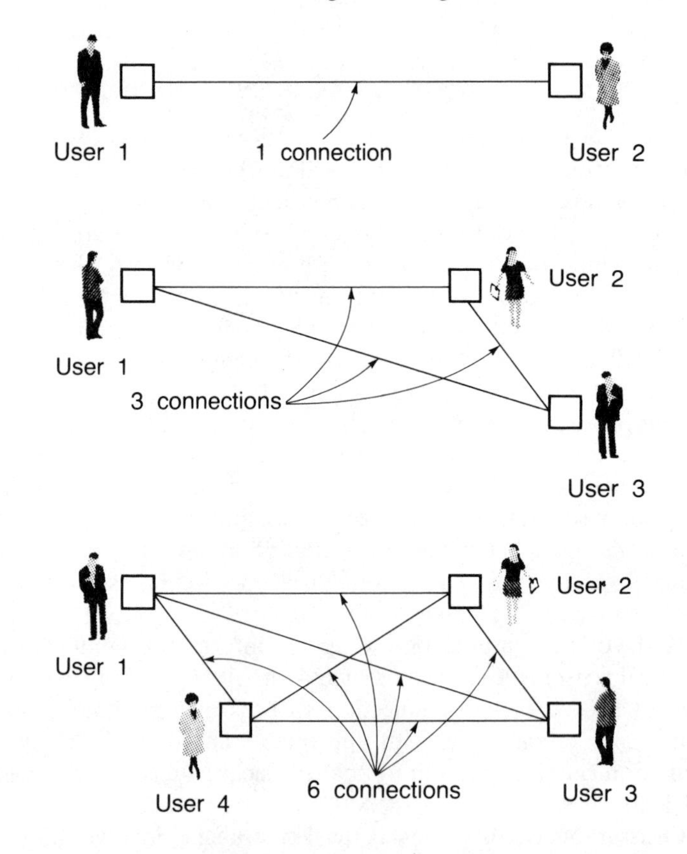

Figure 1-1. The relation of users and number of connections.

1. Switching reduces the number of connections needed to link all the users of the network.
2. Switching allows varying combinations of users to communicate at different times.
3. Switching uses "cheap" data processing to efficiently share the total resources of the network, particularly the "expensive" transmission capacity.

Reducing the Number of Connections

Consider two communications users who want to converse with each other. Only a single connection (which can carry the communications in both directions) between them is necessary (Figure 1-1). If there are three users in the network, three connections will be necessary, even if only one connection is used at a time. If four users are involved, six connections are required; and the number of connections increases in a geometric progression as additional users are added. In fact, as Figure 1-2 shows, for an arbitrary number of users—N—we will need a very large number of connections, given as $C = N(N - 1)/2$. This is because from each of the N users will be $N - 1$ connections going to each of the other

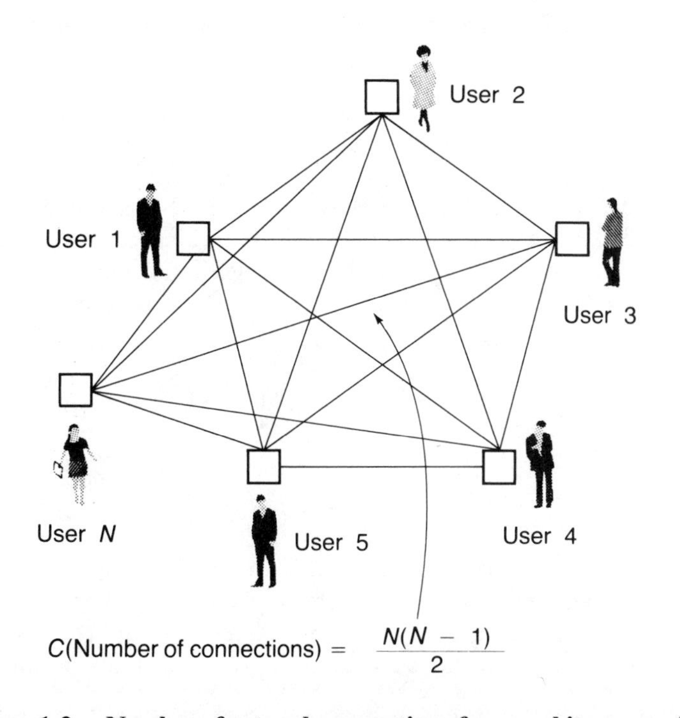

Figure 1-2. Number of network connections for an arbitrary number of users (N).

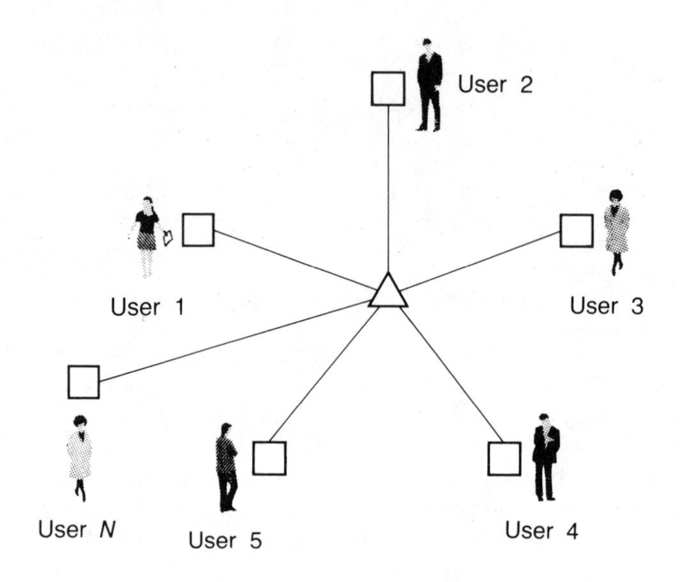

Figure 1-3. Number of lines needed with a central switch.

users. Since each connection has two ends and can be used for communications in either direction, the total number of connections is N users times $N - 1$ connections, all divided by 2.

However, if we add a switch, as shown in Figure 1-3, and connect each user to the switch, all N users can then communicate through a total of N lines. The addition of the switch results in a very large savings in the number of connections needed in the network. In fact that difference (C_d) can be simply expressed as:

$$C_d = C - N$$

$$= \frac{N(N-1)}{2} - N = \frac{N(N-1) - 2N}{2}$$

$$C_d = \frac{N(N-3)}{2}$$

Figure 1-4 shows a plot of the values of N, the number of users; C, the number of connections required without a switch; and C_d, the number of connections saved if a switch is added. As N becomes larger than approximately 10, the potential number of connections saved starts reaching into the hundreds and thousands.

Of course, switches are not free, so we would have to determine if the savings in connection cost are greater than the cost of the switch.

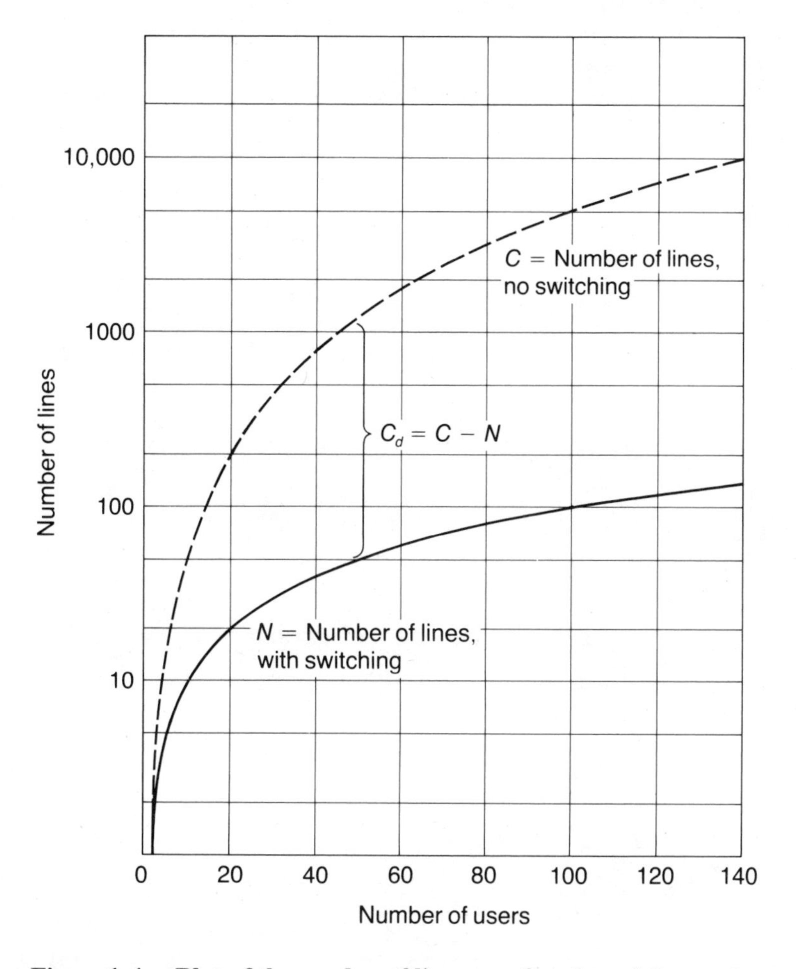

Figure 1-4. Plot of the number of lines as a function of the number of users.

Saving Resources

As users are added to a network, it is highly unlikely that each new user will want to communicate with every other user already in the network. The huge number of connections implied by C is just a fiction. As long as we know which other users each user will ever want to talk to, we can simply put in the required connections. Figure 1-5 illustrates a group of users spread out across the United States. The small squares represent computers, and the small circles represent people at terminals with a need to communicate with or access information in the various computers. As we just pointed out, not every user in this network requires connectivity to every other user. Figure 1-6 shows the actual connectivity pattern required by each user's "community of interest."

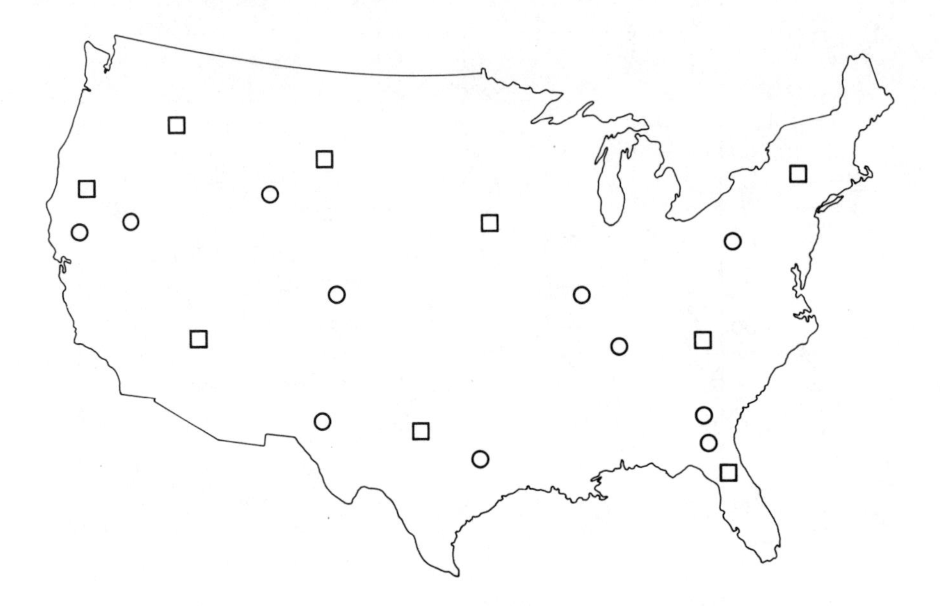

Figure 1-5. Locations of hypothetical network users.

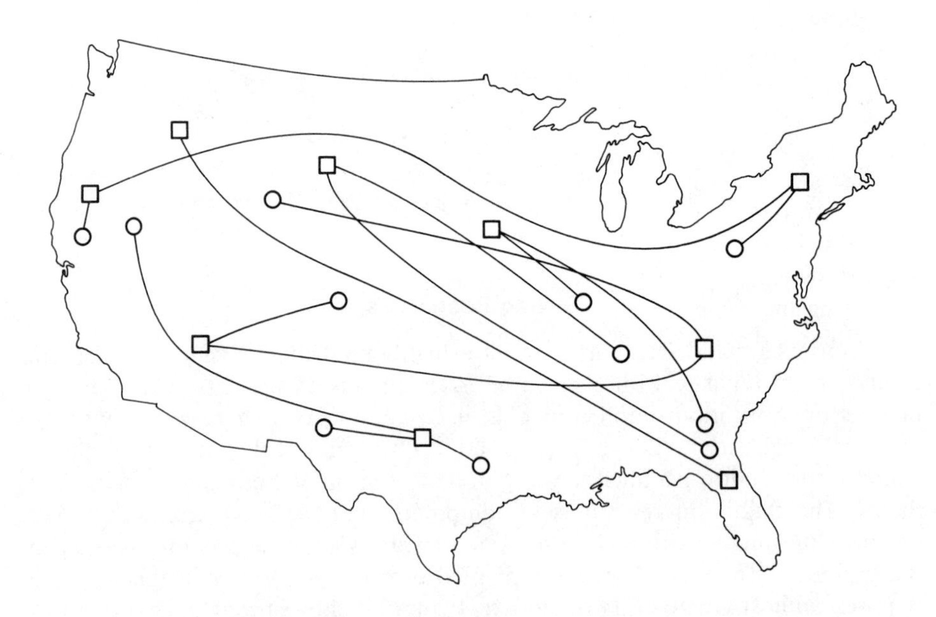

Figure 1-6. Required connectivity for hypothetical network users.

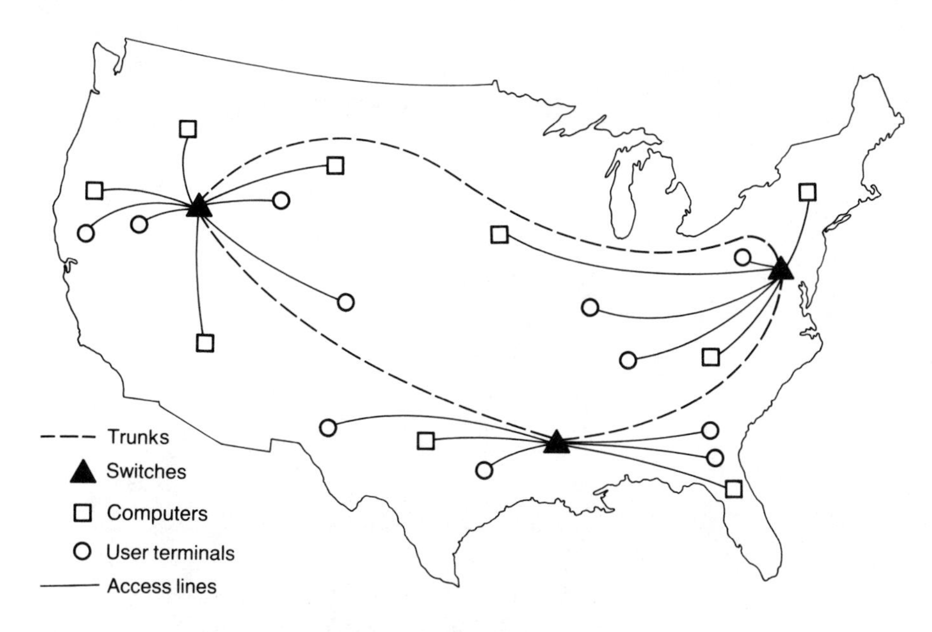

Figure 1-7. Network connectivity using three switches.

Channel-Mile Cost. The cost of a communications network depends not only on the number of connections in the network but also on the length of those connections (known as **line-haul cost** or **channel-mile cost**). A quick approximation reveals that about 16,000 channel-miles of connectivity are required to meet the pattern shown in Figure 1-6.

Even though this connectivity requires many fewer than $N(N - 1)/2$ connections, because of the limited communities of interest switching may still be advantageous. Figure 1-7 adds to the pattern three small triangles, representing network switches. Two different kinds of connections replace the connections going from one user to another. There are connections going between the users and the switches, called **access lines,** and there are connections going between the network switches, called interswitch **trunks** (or backbone trunks, or simply trunks).

Using the same scale as on Figure 1-7 we would find about 7800 miles of access lines and about 4500 miles of interswitch trunks—or a total of about 12,300 channel-miles of connectivity. Assuming for the moment that the **channels** or connections used for trunks are technically similar to those used for access lines, this represents a savings of about one-fourth of the connectivity required by the nonswitched network in Figure 1-6. Are these savings substantial? Are the savings, in fact, sufficient to offset the cost of the switches?

Trends in Processing and Transmission Costs

In a wide spectrum of cases, particularly where computer and data communications are concerned, the technology has made the tradeoff of switching for

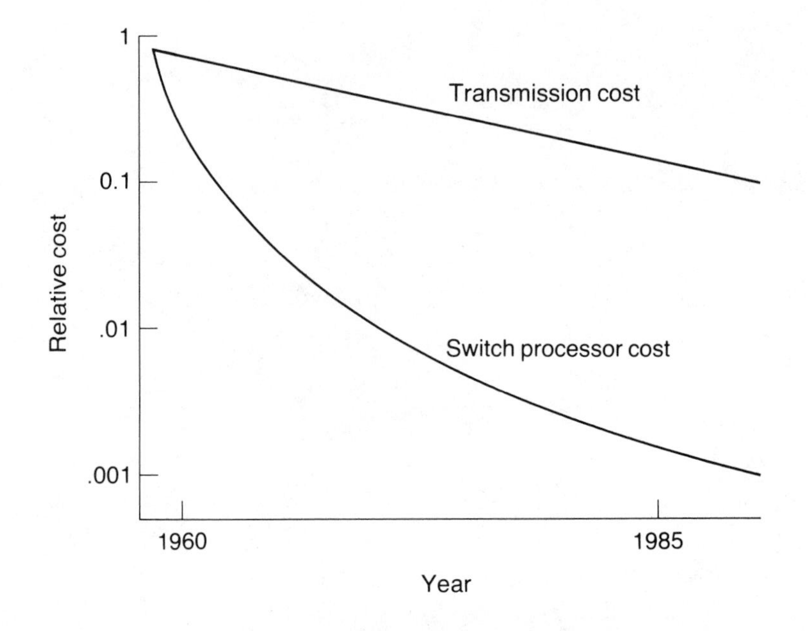

Figure 1-8. Relative cost trends for communications transmission and switch processing.

connectivity highly advantageous. Both processing and transmission have benefited from advances in the state of the art of electronic technology. However, the cost of transmission, especially as seen by the end user, is also affected by the regulatory environment, the large physical plant required, only modest competition, and many other factors. As a result, processing costs have decreased much more dramatically than transmission costs over the past two or three decades (see Figure 1-8).

Lawrence Roberts (1974)* estimated the components of cost associated with transmitting 1 million **bits** of data across the country, a nominal distance of 3000 miles. This information was based on Roberts's experience with the **ARPANET,**** the first packet switching network to serve a large group of distributed users. According to Roberts's data (see Figure 1-9), by 1971 the cost of

* References will be identified by the author's last name and year of publication. The complete bibliographical information will be found at the end of each chapter, along with other related and suggested references.

** Frequent reference will be made to the ARPANET throughout this book because it was the foundation of many of the present principles of packet switching, because there is a very large body of published information about it, and because it provided a research and test environment for new techniques and concepts of packet switching. While we use these examples because they represent in many cases, the state of the art in packet switching, it is important to keep in mind that these are *only* examples. The exact implementation of the ARPANET is a particular case of packet switching, with a single

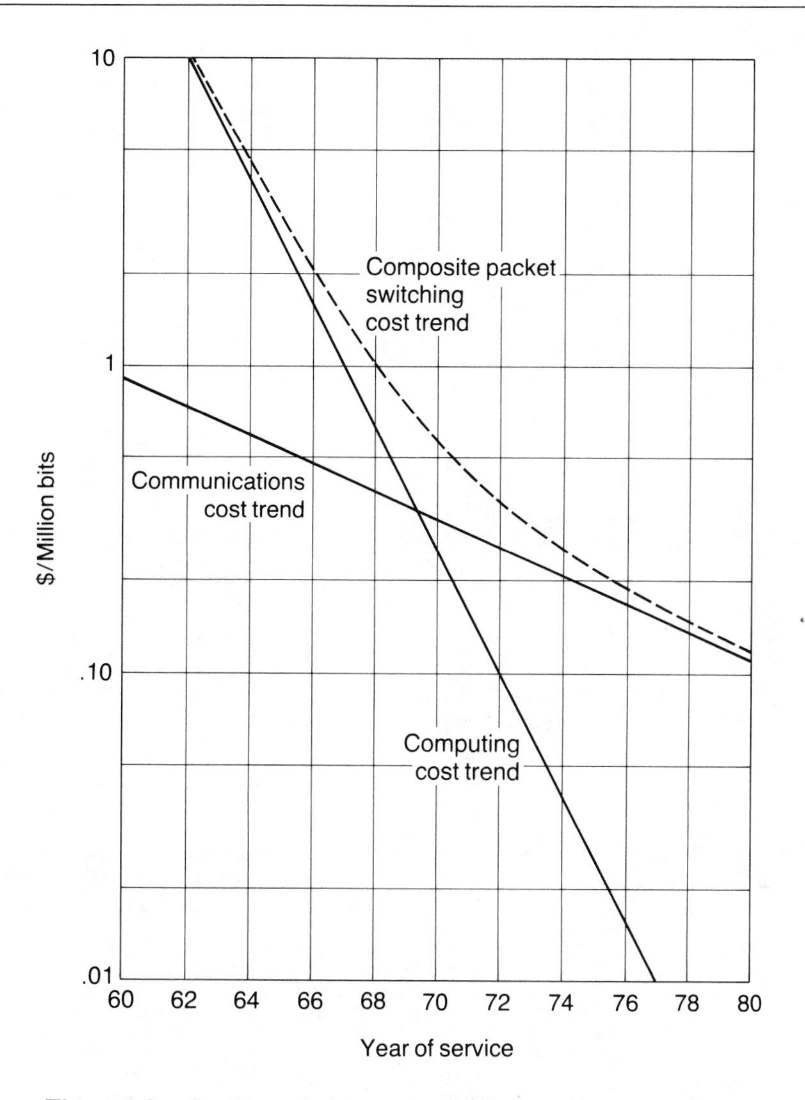

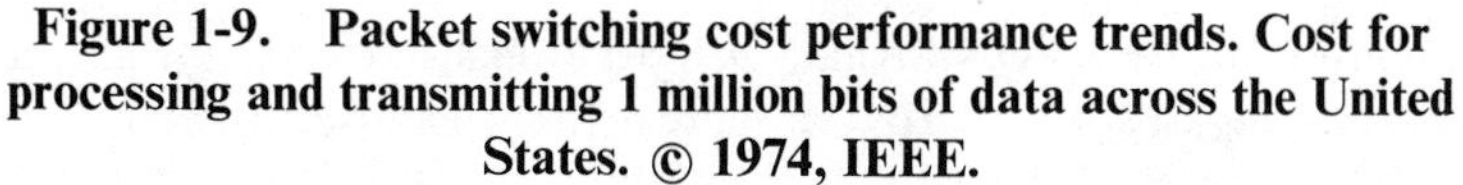

Figure 1-9. Packet switching cost performance trends. Cost for processing and transmitting 1 million bits of data across the United States. © 1974, IEEE.

Footnote continued from page 10.

set of design parameters. For example, the ARPANET uses packets that are nominally 1000 bits in length. Through the early years of packet switching this packet length was accepted as being the only allowable one for a packet network. In fact, for various reasons we will elaborate upon in later sections, packet length of anywhere between 100 and 10,000 bits makes sense, depending on implementation details. More descriptive and historical information on the ARPANET will be found in Chapter 7. ARPA is an acronym for the Advanced Research Projects Agency of the United States Department of Defense; it is sometimes referred to as DARPA.

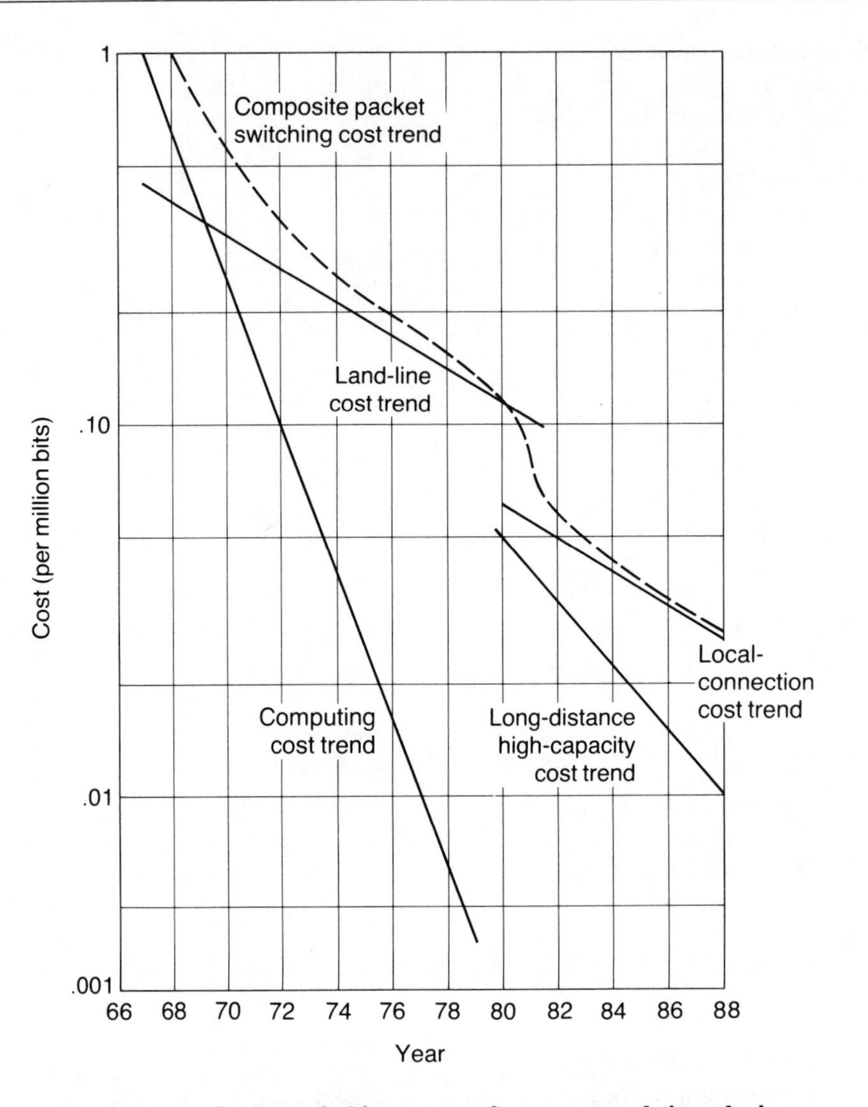

Figure 1-10. Packet switching cost performance trends introducing low-cost high-capacity carriers. Incremental cost of transmitting and processing 1 million bits of data. © 1974, IEEE.

processing a million bits of data and the cost of transmitting a million bits of data were about equal at approximately 35¢. However, by the late 1970s the cost of transmission was at least ten times that of computation on an incremental cost basis. Without some major breakthrough in transmission costs, this differential will be about 100 times by the mid-1980s.

Factors Influencing Costs. The new specialized domestic and international carriers as well as high-capacity satellite carriers entering the marketplace with a wide range of new and innovative services could reduce the cost of transmission by a quantum jump so that it would again become nominally equal to the cost of processing (Figure 1-10). However, even if such a breakthrough occurred, it would likely be limited to transmission where long distances and large economies of scale could be achieved. Despite such a substantive reduction in part of the communications costs the general network user would see total costs composed of three major cost components: the cost of (1) processing; (2) "cheap" large-capacity, nationwide transmission; and (3) the local or regional communications to the nearest point where economies of scale were feasible. The composite cost, as shown in Figure 1-10 still shows a heavy dependence upon the cost of the communications services.

Thus whether we look at the composite line in Figure 1-9 or in Figure 1-10 the conclusion is the same: With the cost of processing data so much less than the cost of transmitting it, any application of processing that reduces the amount of data to be transmitted will probably result in a net cost reduction. Since switching—especially for digital information—is a form of processing, there is a further reason to use switched common user-shared connectivity.

SUMMARY

We have now seen the explanation for the three major reasons to use switching in a communications network—that switching reduces the number of connections needed to link all the users; that it provides the ability to connect different users at different times; and that it permits us to apply "cheap" data processing in order to share "expensive" transmission.

In addition, switched networks provide many other advantages. They are more reliable than single point-to-point connections and can more easily protect against transmission errors. They can permit the expansion of capacity upon demand. They can provide for the matching of dissimilar users. We will look at these factors further as we discuss the details of packet switching and compare it to other techniques. We'll begin to highlight these comparisons in the next chapter, Packet Switching in a Telephone World.

SUGGESTED READINGS

ROBERTS, LAWRENCE G. "Data by the Packet." *IEEE Spectrum*, vol. 11, no. 2 (February 1974), pp. 46–51.

This article traces the cost trends in both computing and communications. The early usage trends of the ARPANET are discussed, as are trends in satellite communications that have an impact on packet switching.

ROBERTS, LAWRENCE G. "The Evolution of Packet Switching." *Proceedings of the IEEE*, vol. 66, no. 11 (November 1978), pp. 1307–1313.

This article traces the evolution of packet switching beginning with the early works by Paul Baran at RAND Corporation in the early 1960s. A number of different public and private packet networks are briefly described, and major issues in the path of further development of worldwide packet networks are considered.

2

Packet Switching in a Telephone World

THIS CHAPTER:

will contrast data communications with voice communications.

will describe the difference between transparent networks and transactional networks.

will introduce the applicability of packet switching.

SWITCHING THROUGH THE TELEPHONE

Every reader of this book deals with the world's premier switched network nearly every day—the telephone. Because the worldwide telephone network has the potential of providing universal access, why can't we simply tie all communications users together by the telephone circuits?

Applications of telephone networks to all communications requirements is essentially a matter of economy. The determining factor is the difference in the continuity of communications usage that we looked at briefly in Chapter 1. Long, continuous streams of information can profitably be transmitted through a system like that of the telephone, whereas for short, relatively infrequent bursts of information such a system is inefficient.

TRANSPARENT AND TRANSACTIONAL SWITCHING

The concepts of information transmission can be divided into two basic systems—(1) transparent networks, and (2) transactional networks.

Transparent Networks

In a **transparent** network the switches are designed to act as a bridge between communications connections such that the end-to-end connection formed within the network functions as a pair of wires between the users. Information that goes into the connection at one end comes out at the other end in real time, without intentional change or modification. Of course, as with any physical medium, there are slight unintentional modifications, such as delays due to the physical speed of propagation through the medium; amplitude, phase, or frequency distortions of the transmitted information; and occasional introduced errors or interference. Other than the processing of the call set-up information needed to establish the connection, no processing is done within the network. The common "plain old telephone service," or **POTS,** is the best example of a transparent network.

Transactional Networks

In a **transactional** approach the users introduce information into the network in the form of a complete entity. The network guarantees acceptance of the information and to some extent guarantees delivery of the information. Much like a letter dropped into a mailbox, information that comes into the network is physically disconnected from the input that originated it. The information is fairly certain to come out of the system at the destination address—you just are not sure exactly when and in what condition. When the information does come out of the system, however, it represents a complete transaction. The word *transaction* may be somewhat misleading here, as in business it generally implies a two-way information exchange. In a communications transaction, the contents of the information package arriving through the network have usefulness at the destination end, without the need for any additional information. The best examples of transactional networks are message networks, such as telex and telegram. Packet switching is also a transactional network.

Overlapping Approaches

The distinction between transparent and transactional switching tends to blur as other forms of communications processing are introduced to a network. For example, circuit switching in a normal telephone network provides a full-time path through the network for the duration of a call. However, other techniques can provide the appearance of a point-to-point connection for a definable interval of time. For example, with terminal polling in a computer network a large number of computer terminals are linked to a common connection. As each terminal takes a turn in sending information over the shared circuit, the network looks like a transparent connection. However, since the terminal may have to "wait its turn" before each transmission, the network is not truly transparent. Switched networks that employ multidrop connections, multiplexing of users onto a common connection, or the concentration of users where there are fewer connections than there are users, attempt to give the functional impression of a transparent connection, although the realities of the network operation mean that the network is moving toward transactional processing. As we shall see later (particularly in Chapter 7), however, even packet switching can be implemented in a "virtual circuit" mode, which attempts to approximate a transparent connection.

APPROACHES TO SWITCHING COMPARED

Beyond the functional distinctions of transactional and transparent switching approaches for network users, there are some very important technical distinctions between these approaches. Table 2-1 summarizes the system attributes of these two extremes in switched networks. (Keep in mind that the transparent technique is best represented by the everyday telephone network and the transactional approach by telegrams.) Refer to this table as you read the following comparison of the technical attributes of the two basic switching approaches.

User Interface

This refers to the way users interface the switch in the network and the way the switch looks at the users' requests for service.

In a transparent system calls are presumed to last for a matter of many seconds or minutes. The time interval between each user's request for service from the switch is sufficiently long that, on the average, the switch has enough time to scan each line in sequence, looking for a change of state, which is interpreted as a request for service from the switch. Thus transparent systems are generally dependent upon a *scanning* form of user/switch interface.

In a transactional system calls or transactions arrive from the users more frequently and have a much shorter average duration. In the normal mode of operation when all, or some predetermined part, of the user transaction has arrived at the switch, the processor is *interrupted* to accept and begin processing this new transaction through the network.

User Timing

This refers to the continuity of the communication flow through the switched network. Since a transparent system attempts to provide or emulate a real physical connection between the users, the time

Table 2-1. Attributes of Two Switching Approaches

Transparent	*System Attribute*	*Transactional*
Scanning	←— INTERFACE —→	Interrupt
Synchronous	←— TIMING —→	Asynchronous
Fixed	←— BANDWIDTH/BIT RATE —→	Adaptable
Dedicated	←— RESOURCE ALLOCATION —→	Protocol
Fixed	←— ROUTING/OVERHEAD —→	Variable
Compatible	←— TERMINAL DESIGN —→	Separable

relationships of the input information through the network are preserved. If element B occurs 1.7 milliseconds after element A at the input of the system, a transparent system must insure that the time between A and B remains 1.7 milliseconds, with A occurring first, at the system output. User timing, preserved through the communications network, is said to be **synchronous.**

In a transactional network it is entirely possible to have two sequential transactions come out of the network in the reverse order from the way they went into the network. The time spacing and temporal relationships of the entered information need not be preserved precisely. The user timing is thus said to be **asynchronous.**

(Note that terms *synchronous* and *asynchronous* here refer to the information flow across the network on a user-to-user basis. These terms are often also used in relation to the way data sets interface to the communication circuit.)

Connection Bandwidth or Bit Rate

This refers to the amount of **capacity** made available to a user when a connection is established through the network. In a transparent network this capacity is generally a fixed amount, which is identical from call to call and identical among all users. A transparent network may be composed of 4000 Hertz wide-voice channels, all 4800 bit/second data channels, or some other basic channel capacity. To maintain transparency, all users must use the same capacity. The bandwidth or bit rate is thus said to be *fixed* and, moreover, is constant throughout the network.

In a transactional network capacity is assignable on an as-needed basis. The bandwidth or bit rate is *adaptable*. Large transactions can temporarily be given greater capacity, while smaller transactions may use smaller amounts of capacity on an economical basis.

Resource Allocation

The primary function of switches in a network is to allocate the network resources to a particular information transfer at a given time. In a transparent approach the resource allocation is done through a **dedicated line** to a user pair over the duration of the information transfer.

In a transactional approach, *protocol* is used as the fundamental means by which the resources are allocated. (A **protocol** is the rule or procedure by which the information is transported through the network.) Elements of the information transfer, handled as complete transactions, consist of a set of protocol elements to allocate network resources only during useful parts of the information transfer.

Routing/Overhead

In a transparent network approach the **routing** of the information through the network is *fixed* at the beginning of the information flow and remains the same for the duration of the transparent connection. Consequently, the **overhead** information required to establish communication is also fixed and is the same whether the information exchange is of long or short duration.

In a transactional approach each succeeding transaction need not follow exactly the same route; routing is easily made highly *variable* in order to achieve more uniform and more efficient utilization of all available resources. However, since each transaction must employ a set of protocols, the overhead required is also variable. In general, the amount of overhead increases as the length of the information transfer increases.

Terminal Design

By definition a transparent switching approach requires that the end terminals of the network be *compatible* with each other; that is, they must have common formats, codes, languages, signaling, and data transfer rates.

In the transactional approach the terminal design is *separable*, and different user terminal characteristics can be used at either end. This is made possible (and, as we shall see later, desirable) because the network processing elements have time to do conversion between one set of terminal characteristics and another.

DATA COMMUNICATIONS VERSUS VOICE COMMUNICATIONS

In this section we will look at the greatly different characteristics between typical voice communications information flows and the typical information flows associated with data communications. These applications characteristics are closely related to the suitability and efficiency of different approaches to switched network design.

Statistical Characteristics of Voice Communication

Over the long history of telephony, a well-defined profile of the voice communications user (the telephone talker) has evolved. Electrical performance characteristics (e.g., earpiece volume, background noise, spectral bandwidth transmitted) has a direct relation to the level of user satisfaction with the service. Calling patterns as a function of time of day, day of the year, and geographic location have, in a statistical sense, become quite predictable. Finally, the nature of usage by individual callers, in terms of the average frequency of new call originations and the average duration of calls successfully completed is also accurately characterized statistically.

Call arrival rates and average holding times are of the most interest in the application of various approaches to network switching. Depending upon the type of user (such as business or residential), time of day (such as during the business day or in the evening), and day of the year (such as a weekday, weekend, or special holiday), the average call holding time for a voice call is typically between 3 and 7 minutes. On a global average holding times of 5 minutes provide a good approximation.

The Data Communications Environment

Table 2-2 summarizes the important characteristics of the data communications environment. Five different types of transactions are noted, ranging from short, interactive transactions to long bulk data transfers of millions of bits.

Associated with each type of transaction is a nominal message length and a nominal target or desired delivery delay. The nominal message length is based on flow between users viewed as a set of discrete transactions. These transactions can be handled as complete entities for delivery through the switched network. The permitted time for that delivery is characterized by delivery delay.

Interactive Transactions. This type of transaction is most typically the result of a person working at a terminal device (such as a keyboard-printer or a **keyboard CRT**). The work is being supported or assisted on a continuous basis by a computer at a location remote from the user, which is supplying information on demand from the user, performing computations, modifying software, and so on. Typical inputs from the person to the computer consist of a few characters or possibly a few lines of typed data (anywhere from 20 to 1000 bits). Typical responses from the computer to the data terminal user range from a few lines to possibly a full **CRT** screen full of data (anywhere from 600 to 10,000 bits). In either case in order to maintain reasonable continuity of the interactive thought process, the individual transactions would have to proceed through the network with delivery delays of less than 1 second.

Table 2-2. The Data Communications Environment

Transaction Type	*Nominal Message Length*	*Delivery Delay*
Interactive (person/computer continuous thought process)	600/6000 bits	Less than 1 second
Inquiry/response	600/6000 bits	1 to 30 seconds
Data base update	600 bits	Seconds to minutes
Bulk (1) data transfer	10^4 to 10^6 bits	Tens of seconds to minutes
Bulk (2) data transfer	10^6 to 10^8 bits	Tens of minutes to hours

Inquiry Response Transactions. The message or transaction length characteristics of inquiry response systems are similar to those of interactive systems except that there is no continuous interactive thought between the user and the computer. In a request to an inventory data base for spare part availability or a credit card validity check, for example, a single inquiry into the data system generates a single response. Information received by the user as a result of this interchange is a part of the normal work being performed by the user, in contrast to the interactive case, where continuous interchange with the computer is central. It is therefore permissible in the inquiry response case to have network-imposed delays amounting to a few seconds without destroying the relationship between the computer and the user.

Data Base Updates. Data base updates and bulk transfers differ from the first two cases in that, by and large, the transaction or message is a one-way transmission of information. Except possibly for an acknowledgement of receipt by the destination, there is no immediate dependence on the return of information based on transmitted information. For example, in a point-of-sale terminal, where credit card purchases are entered to a master billing data base, the information related to the purchase needs only to be entered into the data base. The sales clerk is not dependent on information transmitted from that data base in order to complete his actions with the customer. Of course, in a practical sense, the clerk (and the customer) get a more confident feeling if the terminal "beeps" or a light comes on indicating that the information was indeed accepted. However, the time relationships in such a situation permit a relatively loose constraint on the timeliness of the data transfer.

Bulk Data Transfers Bulk data transfers have been grouped into two different categories. Bulk (1) is typical of movement of several pages of data to a remote printer, transfer of a major file of a data base, or the loading of code in a remote computer. It consists of as many as a million bits of data. Since the actual transmission time of such a long transfer is many seconds, the additional delay of tens of seconds, or possibly a minute or two, is not detrimental to the information transfer.

Bulk (2) represents very large data base transfers and the remote transfer of extremely large amounts of data. For example, several periodicals are composed at a central point but printed at regional printing plants around the country. The fully composed pages are digitized and transmitted as very long data transfers (often overnight) to the regional printing plants. Here again, the actual transmission times are on the order of hours, and imposed network delays of many minutes or even hours are generally not harmful to the utility of the information. In fact, if such information can be saved in the network until the network is lightly loaded, and then used to give better average utilization of the network resources, it is possible to capitalize on the delay to provide an economical means of moving these very large bulk (2) type of data transfers.

DATA CALL STATISTICS

The most important difference between voice telephone calls and data exchanges lies in the message length and delay characteristics. As voice calls have a holding time in the range of several minutes, they are generally unaffected if a few seconds (maybe up to 15 seconds) are necessary to get the call established initially. During the period that users are carrying on the voice call, the system delay has to be held small (a quarter-second or less) and constant so that the speech will be undistorted at the listener's end. Data communications transactions, on the other hand, often have a very short duration (a few seconds or less) and can tolerate some amount of delay; thus they can generally trade message length and delay for economy of operation.

Figure 2-1 plots the distribution of call holding times for about 27,000 potential and actual data users within the U.S. Department of Defense. The **holding time** is based on the actual transmission time of the various messages and transactions, where the transmission rate is the bit rate of operation for the data terminal involved. This means that the holding time, measured in seconds, depends not only on the length of the transaction (in bits), but also on the speed of the terminal (in bits/second, or **bps**). For example, a 600-bit inquiry will take only one-quarter of a second using a 2400-bit/second terminal. However, the same 600-bit inquiry will take 8 seconds using a 75-bit/second teleprinter terminal. Figure 2-1 thus reflects not only the distribution in user transaction lengths, but also the distribution in user terminal characteristics.

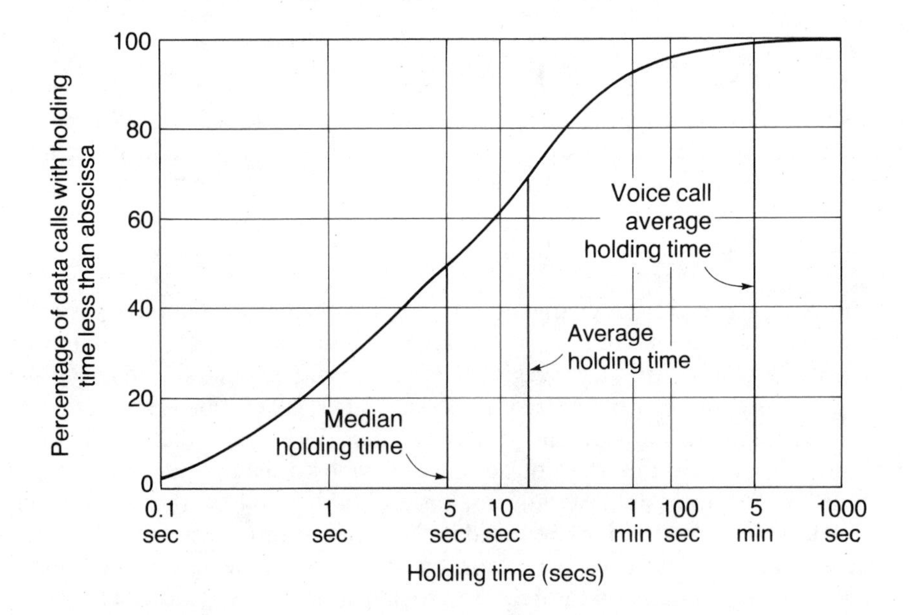

Figure 2-1. Distribution of data call holding times.

Call Holding Patterns

Figure 2-1 shows that about 25% of the data calls last for 1 second or less. About 50% of the calls last 5 seconds or less, making the median length data call nominally 5 seconds. Furthermore, 90% of the data calls are less than 50 seconds long. Using a figure of 300 seconds (5 minutes) as the average length of a voice call, we find that more than 99.5% of the data calls are shorter than the average voice call. We also find that the average holding time for the data transactions is much longer than the median holding time (about 15 seconds average versus 5 seconds median) because a very few very long calls pull up the overall average.

This information suggests several conclusions. First, voice and data communications are vastly different from each other. Secondly, trying to put data communications through a voice network that has been optimized over the years for the unique characteristics of person-to-person voice communication may not be very efficient or very economical. Finally, the very short duration of most data calls and data communications transactions implies that a network would have to act quickly in response to data-oriented calls.

THE WHY AND WHERE OF PACKET SWITCHING

Circuit switching, the fundamental basis for all of the worldwide telephone networks, has been in use for the better part of a century. Packet switching, to a large measure a product of the computer age, can trace its history back at most to the mid-1960s. Now that we have looked at various forms of communications transfer in some detail, we should be able to consider how and why packet switching can coexist with a worldwide network of millions of telephones.

Table 2-3 contrasts in brief the key features and basic technical aspects of circuit switching and packet switching. All are a product of the different ways the two approaches operate—that is, on a transparent or a transactional basis.

Table 2-3. Features of Two Types of Switched Network

Circuit	*Feature*	*Packet*
Call establishment and disestablishment	PROTOCOL	Interface and data flow
Fully transparent	TRANSPARENCY	Transaction oriented
Inherently full duplex	DUPLEX OPERATION	Independently bidirectional
Keyed to voice call statistics	HARDWARE	Data call oriented
Line dominated/ scan oriented	SYSTEM LIMITATIONS	Origination dominated/ interrupt driven

Circuit and Packet Switching Contrasted

Protocol

While both approaches employ protocols, circuit switching needs only a *call establishment* and *call disestablishment* protocol. The compatibility of the two end users and the transparency of the system keep an orderly, logical, and useful flow of information between the users.

Packet switching is protocol dependent throughout its operation. Well-defined protocols *interface* each of the users to the network and guide the *data flow* through the network.

Transparency

Circuit switching is a *fully transparent* network. Once a call has been established, the switched network is functionally equivalent to a solid pair of wires between the end users.

Packet switching is at the other end of the spectrum, providing a *transactional* form of communication.

Duplex Operation

Circuit switched connections are inherently **full duplex.** This means that both parties can simultaneously send and receive information. This capability is provided even though most of the time one person is talking and the other is listening. In long-distance communications, where amplifiers must be used along the communications circuit, two circuits, one going in each direction, must be assigned to a single call. (This is known as a four-wire circuit, since it functionally represents a pair of wires in each direction, or a total of four wires in the connection.)

Packet switching uses the principle of moving one packet, or transaction, at a time through the network. Two-way communication is the sequential exchange of transactions going in opposite directions. On long-distance circuits, which employ separate circuits in each direction, the use of those circuits is *independently bidirectional.* This means that transactions can be simultaneously flowing in each circuit, but at any time the information going in one direction is independent of the information flowing in the other direction.

Hardware

The hardware for circuit switched networks is *keyed to* the statistical nature of *voice calls.* For instance, since typical calls last several minutes or more, the hardware is permitted the "luxury" of taking 5 or more seconds to establish a new call. Because voice callers generally originate at most only a few calls per hour, and because once a call is established the line requires very little additional service from the switch, the hardware can look at the line only occasionally (like once every second or two) to see if the user wants any service from the switch.

Packet switching hardware, which is *data call oriented*, is primarily high-speed digital computer based. Interswitch lines are sized to capacities defined by the short, **bursty** nature of the transactional data communications, which requires quick response from the switches.

System Limitations

In any system there is an ultimate limit to the amount of capacity available. Because, as we saw, the cost of processing is increasingly cheap relative to the cost of transmission, it is important that the switches do not place the limit on system capacity, so that all available transmission can always be used to the fullest extent possible.

In circuit switching, the limitation in the switches is generally the number of lines that can be handled at any given time; the system is said to be *line dominated.* Circuit switches are *scan oriented*; that is, the switch call processor scans line by line among all of its customers, looking for new requests for service. This is possible primarily because of the low call origination rate associated with voice customers.

Packet switches are typically limited by the total amount of traffic flowing through the switch. This information flow is dependent upon the number of transactions originated in the network; thus the situation is said to be *origination dominated.* Because the switch has relatively little time in which to get a new data transaction under way (or else the delay will get too long), the processor is *interrupt driven.* There is not enough time to scan the customers line by line, so the initiation of a new transaction is permitted to interrupt the processor from one task to initiate action on the new service request.

Thus you can see from this brief survey that, despite the ubiquitous availability of telephone communications, there are several important reasons to develop packet switching networks rather than to tie in with the existing system. In the next chapter we will look in more detail at the various switched networks and their comparative advantages for data communications users.

SUMMARY

We have seen that communications networks can be said to be either transparent or transactional. Transparent techniques, such as circuit switching, pass information without changing or modifying it. Transactional techniques, such as packet switching, move complete entities, transactions, or messages through the network without a fixed or permanent path. There is a vast difference between the statistical nature of voice and data communications. Voice calls tend to be long, duplex, and highly interactive. Data calls or transactions tend to be short, bursty, often unidirectional, and much more frequent. Packet switching can match the unique operational requirements of data users, thereby applying relatively low-cost data-processing techniques to achieve efficient communications transmission.

SUGGESTED READINGS

DUDICK, A. L., FUCHS, E., and JACKSON, P. E. "Data Traffic Measurements for Inquiry-Response Computer Communications Systems." Proceedings of the Conference on Information Processing '71, Ljubljana, Yugoslavia, August 1971. Published for IFIPS by the North Holland Publishing Company (Amsterdam), 1972, pp. 634–641.

This paper derives usage statistics on two credit bureau systems, a banking system, and a production control system. All four systems studied used telephone or specially designed keypads for data entry. This type of operation is typical of a broad range of common data systems, particularly in the credit card verification and electronic funds transfer modes of operation.

JACKSON, P. E., and STUBBS, C. D. "A Study of Multiaccess Computer Communications." Proceedings of the Spring Joint Computer Conference, Atlantic City, N.J., 1969. Mountvale, N.J.: AFIPS Publications, pp. 491–504.

This paper reports on the operation of three time-sharing systems—two involved with scientific data processing and one with business processing. A wide variety of statistics are presented, including utilization and message length figures, user and computer response times, user think times, and the effects of various data line speeds on system operation.

3

Circuit Switching/Message Switching/Packet Switching— A Quick Comparison

THIS CHAPTER:

will compare the three major types of communications switching.

will look at the advantages and disadvantages of each technique for switching data communications.

As we have seen in Chapters 1 and 2, switching is an essential element in an economical and flexible communications network. The various characteristics of users of such a network have led to the development of several different switching techniques. Let us now compare these techniques particularly in terms of their satisfaction of user needs.

CIRCUIT SWITCHING

Circuit switching dedicates a network resource to a call or transaction on an exclusive use basis, and thus can handle calls on the basis of one line per call in progress. Physical resources in time, space, or frequency spectrum are dedicated to the exclusive use of a single call for the duration of that call. There is a call establishment delay associated with the allocation of the needed resources to a new call, and because of the competition for limited resources, blocking within the switches or communications lines can occur.

With a long technological history, circuit switching is well understood, extremely well developed, and widely deployed in the form of the worldwide **common user** telephone system. In addition, advances in solid-state technology and highly capable computerlike processors lead to constant improvement in its capabilities and, consequently, its cost-effectiveness. The technology also permits the replacement of the electromechanical portions (the actual switch contacts) of the circuit switches to be replaced by solid state devices. Often these are most economically implemented by converting the inputs and outputs to the switch to a digital format, and using an internal structure of the circuit switch closely approximating that of a digital computer.

Fundamental Characteristics of a Circuit Switch

Regardless of its particular implementation, the circuit switch must have two fundamental characteristics. (1) It must retain a high degree of transparency across the switch matrix. The minimum requirement is to approximate a pair of wires with a bandwidth of 4000 Hertz, compatible with the transmission characteristics of the worldwide telephone networks. (2) Since telephone service is so highly standardized, and the investment in the nationwide and worldwide telephone plant is so huge, changes in technology are of necessity very slow and evolutionary. New switches applied to the common user telephone networks, no matter how modern the implementation, must be "backward compatible" with the technology in existence. To some extent, this has been a restricting force in the technological advance of circuit switching techniques. On the other hand, advanced digital, solid-state circuit switches have demonstrated many of the characteristics needed to handle data communications efficiently, especially when deployed in specialized networks. However, if a new, specialized network were to be developed, direct, quantitative comparison between switching techniques would be required, using techniques we will discuss in Chapter 6.

There has been little motivation to make fundamental changes in the transparency and backward compatibility aspects of circuit switching since the user community tends to be quite constant in its use of circuit switched voice systems. New features and new capabilities of the circuit switching technology have been used primarily to improve the utility to the millions of voice users. We have seen features such as call forwarding, user-initiated conferencing, repertory and abbreviated dialers, and the like. New switching and call processing techniques and high-speed signaling between switching centers (something called *Common Channel Interoffice Signaling*, CCIS) have reduced the average time required to set up a new call. However, because voice calls tend to be long, call initiation times shorter than about 5 seconds are not likely to be achieved in the universal common user circuit switched telephone networks.

MESSAGE SWITCHING

For **message switching,** each switch within the network stores the network messages in their entirety, giving very reliable service. Competition occurs between messages, but rather than traffic being blocked when all resources are occupied, traffic is delayed (stored) until capacity is available to deliver currently stored traffic.

Storage

Though most people have never had first-hand experience with a message switch or any terminal connected to one, they have received messages as a result of the message switching process—telegrams, for instance. Message switching has been applied primarily to the service of writer-to-reader communications, where the final output of the communications process is a tangible product in a readable format, providing a permanent record of the information transfer (in fact, message switching is often referred to as *record communications*). Because message switching can use much narrower transmission bandwidths than voice communications, accurate transmission is possible on poor-quality communications circuits. Thus it is commonly used under adverse conditions and to remote locations. In message communications the sender and receiver do not have to be simultaneously available to the communications system. In fact, the very transactional nature of the message switching process makes it generally difficult, if not impossible, to conduct communications between two users on a real-time basis.

Message switching is particularly useful in government, military, or business communications where legal or historical reasons demand written records. One of the largest and most complex message systems, known as **AUTODIN,** enables the U.S. government to carry worldwide military and diplomatic communications.

Message switching systems generally keep a record of all communications using electronic storage (usually on magnetic tape) at the switching centers. Storage times, often between 1 and 12 months, make the information retrievable long after the written copy is delivered. The message storage feature is a direct consequence of the processing technique employed by the switches. Switch processing is on a **store-and-forward** basis; that is, each message, in its entirety, is stored at each switch before it is forwarded to the next.

Message Switching in Data Communications

As data and computer communications requirements have evolved, they have used message switching networks because of the digital nature of communications flow and processing through the message switching networks. Many of the early applications involved the transfer of large quantities of digital data (like a data base on a reel of tape), so they could tolerate the long delays encountered in a message switched network. However, the presence of long computer transfers within the message network acts to the disadvantage of other users. Since each message has to be handled in its entirety, the presence of a long message in the system delays the processing and transmission of shorter messages. As the technology matured, some computers could enter information into the network much more rapidly than the computer or terminal at the destination could receive it, creating further problems.

PACKET SWITCHING

Packet switching, which in large measure is a special case of message switching, wherein the maximum message length is severely restricted, also is based upon the principles of queueing and blocked messages delayed. However, since individual packets have a short lifetime in the system, both average delay and delay variance can be kept quite small.

Complete Information Transfer

In discussing communications and the differing kinds of communication requirements, we have alternatively talked about *calls*, *messages*, and *transactions*. In one sense these terms are interchangeable. They all represent a communications entity—a meaningful exchange of information that starts at the beginning and is complete at the end. The distinction among them is based on the form that the communications would take. In referring to voice communications, we generally speak of **calls.** In dealing with telegrams or writer-to-reader forms of communication, we generally refer to *messages*. In dealing with communications involving computers, we tend to refer to **transactions.** The word *calls* may also be used for computer-based communications using the readily available telephone network with special adapters (known as **modems**). Although in this book the three terms may be used interchangeably, the important point is that they all refer to the notion of a complete information transfer entity.

Packet Switching in Operation

The concept of packet switching is based on the ability of modern, high-speed digital computers to act on transmitted information so as to divide the calls, messages, or transactions into pieces called *packets*. Depending on the implementation and the form of the information, there may be more than one level of subdivision. For example, in one well-known implementation of packet switching the user messages are divided into "segments," and the segments are further divided into packets.

Packets move around the network, from switching center to switching center, on a **hold-and-forward** basis. That is, each switching center, after receiving a packet, "holds" a copy of it in temporary storage until the switch is sure that it has been received properly by the next switch or by the end user. Unlike message switching, which uses the principle of store-and-forward switching, the copies of the packets are destroyed (actually, written over in memory) when the switch is confident that the packet has been successfully relayed.

This form of operation permits the network to achieve low overhead for short messages and eliminates the set-up time for calls going through circuit switched networks. Because all communications are broken down into similar component pieces—the packets—long messages and short messages can move through the network with a minimum of interference with each other. By moving the packets through the network in (nearly) real time, the switches can adapt their operation quickly in response to changing traffic patterns or failure of part of some network facility.

The Origins and Early History of Packet Switching

Communications Security. The origins of packet switching are more strongly based in voice communications than in data communications. This fact is not often recognized since all current applications of packet switching deal primarily (if not exclusively) with data- or computer-based communications. Paul Baran and his associates at RAND Corporation in the early 1960s are generally credited with "inventing" packet switching as it is described in this book. They were working on the problem of making military voice communication circuits safe from wiretapping (an application known as *communications security*).

Baran's ideas started with the notion of breaking a voice conversation between two parties into short separate pieces (packets), as depicted in Figure 3-1. Further, at each switch the pieces of a call could even be mixed with pieces of other calls and sent, piece by piece, over several different routes to the destination. Only at the destination would it be possible to collect all of the pieces and, after reassembling them in proper order, make the voice intelligible. If the wires were tapped anywhere in the network, or if communications between microwave relay points were intercepted, all that would be heard would be the garble of dozens of interleaved bits and pieces of many conversations. Although these ideas were

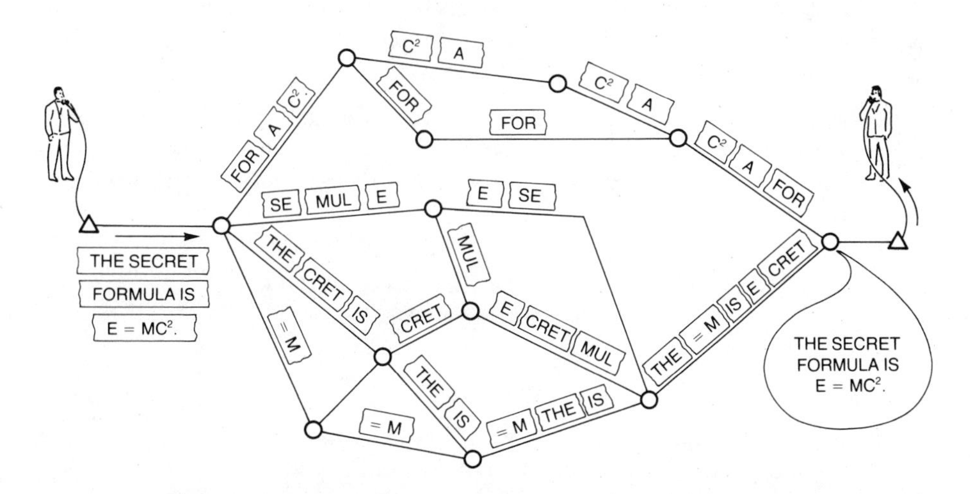

Figure 3-1. A packet switched network used for voice communications.

published in 1964, the technology was not really available to perform the complex processing, routing, and control functions required to implement this concept in a large-scale network.

ARPANET and Beyond. In the meanwhile, the Advanced Research Projects Agency (ARPA) of the U.S. Department of Defense was supporting numerous large computer installations at various universities and laboratories doing basic and applied research throughout the United States. Because of differences in time zone, computer center workload, and specialized hardware and software at the various installations, it seemed desirable to find a way of sharing resources that would enable the facilities to operate more efficiently. As no appropriate networking capabilities were available, ARPA embarked on developing them.

In order to achieve a usable degree of resource sharing, high-speed, high-capacity, low-delay communications paths between the computers are required. The cost of such lines is so large, however, that switching techniques must be applied to insure efficient utilization. Given these conditions, the ARPA researchers hit upon the potential of applying packet switching to this computer-based communication requirement. The evolution from this point to the state of the art in packet switching as it presently exists has been heavily guided by the explosion in data-processing capacity and the reductions in computational costs seen through the late 1960s and 1970s.

At the time of the ARPANET's inception, the trend in computers was toward extremely large computing facilities, following the principle that the cost of a computer facility went up much more slowly than its size and processing power. However, the introduction of minicomputers, followed by microcomputers and then home computers, led to a much wider distribution of computer resources. This wide distribution of computer resources has in turn led to a rapid growth in the need for communications among these resources. Packet switching technology has been able to adapt and meet this need as well.

Now that we have seen generally how circuit switching, message switching, and packet switching work, let us look at their relative advantages and disadvantages for data communications, and some of their potential applications to other forms of communications usage.

CIRCUIT SWITCHING OF DATA—ADVANTAGES AND DISADVANTAGES

Advantages

Figure 3-2 summarizes the relative advantages and disadvantages of using circuit switching techniques for data communications.

Compatible with Voice. The biggest advantages are gained when circuit switching is used in the readily available voice-based telephone networks. The data is transformed to a mode of operation that is compatible with the voice communi-

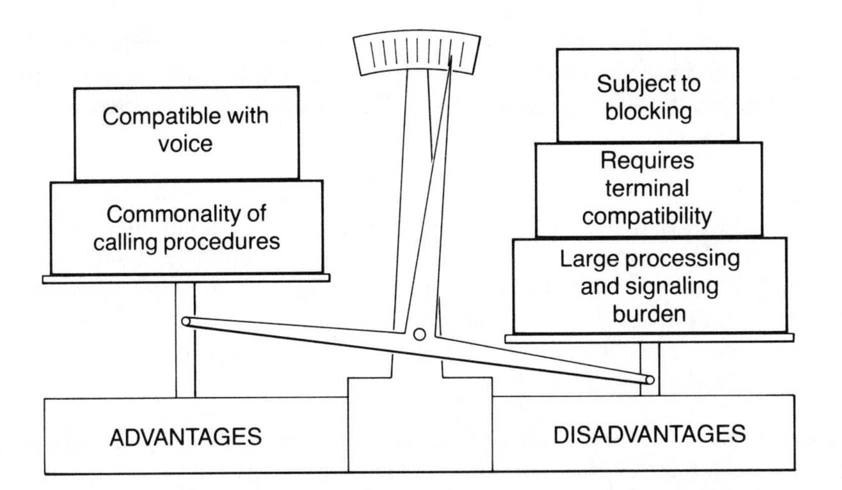

Figure 3-2. Advantages and disadvantages of circuit switching of data.

cations being carried in the network. Economies can thus be realized, since, in terms of total communications, much more voice communication than data communication flows through the networks at any given time. However, to a great extent these economies are lost because of the data transformation process and the differences among voice and data users.

Common Calling Procedures. A second advantage of circuit switching is the fact that the calling procedures are the same for data as for voice calls. Consequently, for the data user, establishing calls through the network would not require special training. Information exchange between user devices would be limited to actual user information, with the establishment of the call (and its disconnection at the end) done "off line" by the operator or an external device. In other words, no protocols would have to be included in the data exchange to account for the network switching function.

Disadvantages

System Burdens. Circuit switching of data communications has several important disadvantages. If the communications facilities are to be used efficiently, calls have to be set up rapidly and can only last for a very short time, on the average (see Figure 2-1). In this case, there would be a very large processing and signaling burden placed on the circuit switching facilities. The physical implementation of circuit switches themselves has evolved with the statistical characteristics of the voice user in mind. The internal processing and switch-to-switch signaling speeds would have to be drastically increased to accommodate data communications. This is not to say that the technology is not available to achieve the required capabilities. It is only to point out that most of the circuit switching plant that exists today, in the form of telephone company equipment and user-owned private branch exchange equipment, is not capable of the high call processing rates that would be required for efficient usage of the networks by data communications messages and transactions.

Incompatibility. As we saw earlier, a disadvantage or circuit switching of data is the requirement that the terminal devices at each end of a circuit switched connection must be compatible. Furthermore, if a single computer is supporting the processing of a large number of separate terminals simultaneously, the computer needs a separate line for each terminal through the circuit switched network. But at the computer end each line must have the characteristics of the terminal at the distant end of the circuit (see Figure 3-3).

Blocking. New data messages or data transactions entering a circuit switched network are subject to **blocking.** This simply means that at a given instant of time all possible paths between one user and another are in use by some other subscribers. The most common form of blocking occurs in telephone communications, when the destination line is being used and the new caller gets the familiar busy signal. Blocking can also occur when the network is heavily loaded, and the switches can find no path through the network. (This condition is often indicated in the phone networks by a double rate busy signal.)

The problem of blocking is illustrated in Figure 3-3, where the computer has four lines into the network, and four terminals are currently in communication

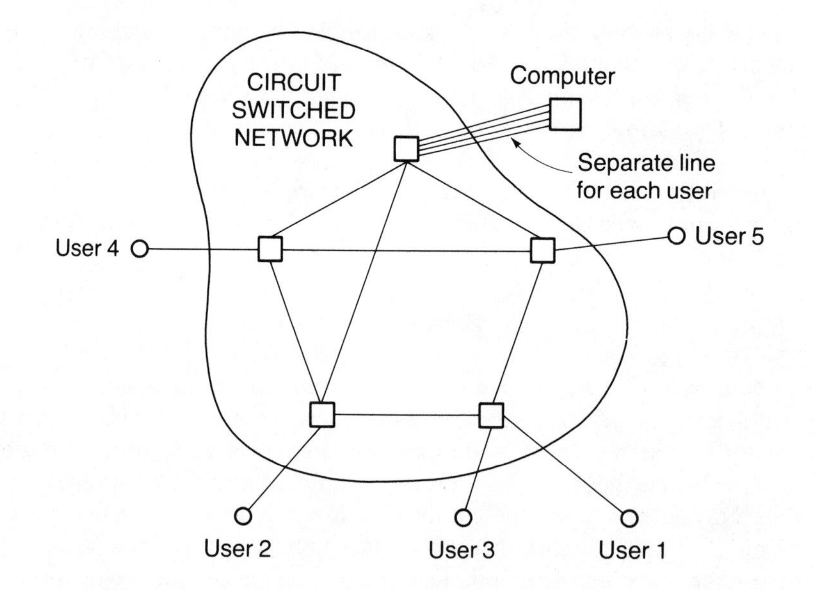

Figure 3-3. Connection of a computer to a number of different terminals, using circuit switching.

with the computer. If a fifth user terminal becomes active and attempts to establish a connection to the computer, it will be blocked since all lines from the terminating switch to the computer are busy. This condition is particularly frustrating for data communications users, since, as we saw in the last chapter, most data messages and transactions are very short. Everyone will recognize the situation of just knowing that, if we had tried to establish the circuit a second or two earlier, or a second or two later, one of those four lines probably would have been available. Instead, we have to wait and try again.

MESSAGE SWITCHING OF DATA—ADVANTAGES AND DISADVANTAGES

Advantages

The relative advantages and disadvantages of message switching data communications are summarized in Figure 3-4. Message switching of data eliminates most of the disadvantages of circuit switching.

Internal Conversion. Compatible terminals are not required at each end of the network, since the internal processing of the message switches can perform any required speed, code, or protocol conversions required between incompatible terminals.

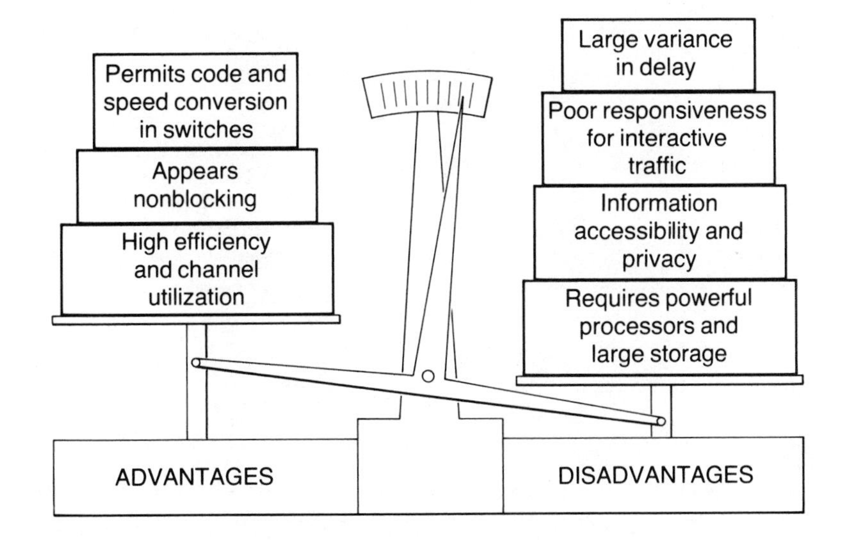

Figure 3-4. Advantages and disadvantages of message switching of data.

Nonblocking. The message storage capability at the switches makes the network appear to be nonblocking since, if the path to the destination is busy, the message switch will accept and store for future delivery any messages or transactions originated by a new user, although the initiator of the call is not sure how long the new information will be stored in the network before it can be delivered. What is actually happening is that potential blocking is translated into potential delay. However, if the average delay can be kept low for data communications, a small delay is far more convenient than being blocked, which requires the full reinitiation of the message.

Efficiency. The storage of messages in the switches of a message switched network leads to extremely high efficiency of network resources and very high utilization of the interswitch communications channels. As new information enters the network for future delivery, the message storage process creates a backlog, or reservoir, of traffic. This reservoir tends to fill the gaps between new message arrivals into the network and smooths the peaks and valleys of demand upon the network, allowing the switches and communications links to be used efficiently.

Disadvantages

Several disadvantages result from the way message switching must be implemented. The storage of entire messages at each switch requires that the switches be equipped with large and powerful processors, with a large complement of peripheral equipment in the form of tape and disk storage media. Because messages may be stored indefinitely in the network, and a short message may fall in line just

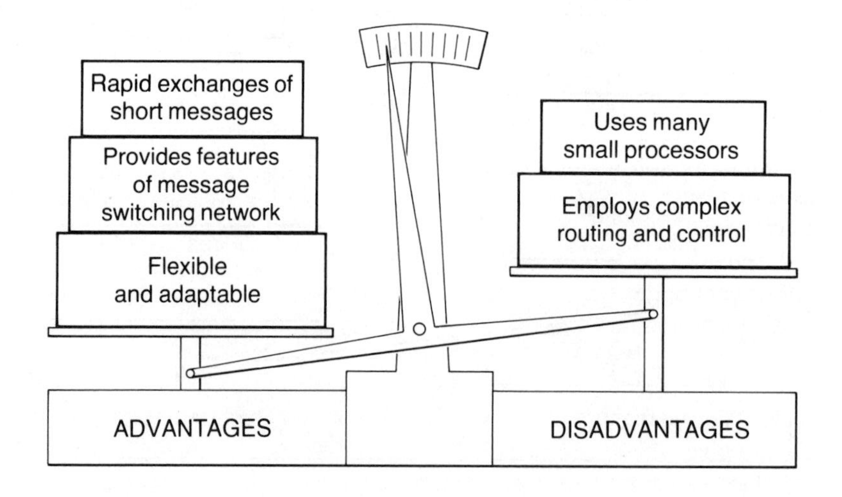

Figure 3-5. Advantages and disadvantages of packet switching of data.

behind a very long one, there is a very large variance in the delay seen by individual users; that is, some messages may get through the network in just a second or two, while others may take minutes or hours. This makes the operation of message switched networks cumbersome for interactive computer users, where the thought processes of the user would be greatly disturbed by widely varying delay with each inquiry and response through the network. A final point is that, since each message passes through and is stored in full at the switches, personnel associated with the switching operation have access to the users' information. Some users may see this as a security or privacy concern.

PACKET SWITCHING OF DATA—ADVANTAGES AND DISADVANTAGES

It is not surprising that a network switching technique developed especially with the needs of computer and data communications users in mind has several unique advantages for that community. The comparative advantages and disadvantages of packet switching data are illustrated in Figure 3-5.

Advantages

Packet switching, because of the processing capabilities of the switching facilities in the network, retains most of the features and advantages of message switching techniques. That is, the network can provide format, code, and speed conversions between unlike terminal devices. It appears essentially nonblocking and can achieve very high network efficiency and line utilization. In addition, by use of **logical multiplexing** on a single line, large computers can simultaneously converse with a number of lower speed devices, using a single high-speed access line into the network. This is illustrated in Figure 3-6. In this example a computer using a single 1200-bit/second line can maintain communication with eight terminal devices, each operating at 150 bits/second.

Timing. Since long messages are broken down into packets, and since most short messages or transactions will fit into a single packet (or less), long messages and short messages do not interfere with one another. As a result, the packet switched network provides for rapid exchange of short messages, with consistent delay patterns, except under extreme overload circumstances. The switches are designed to operate in nearly real time, with most of the capacity limitations due to relatively more expensive transmission lines.

Flexibility. To the extent that the operation is nearly real time, and since all the packets of a given message or transaction do not have to follow precisely the same route to the destination, the network is very flexible and adaptable. Failures that may occur during a particular transaction can be routed around, totally unrecognized by the user.

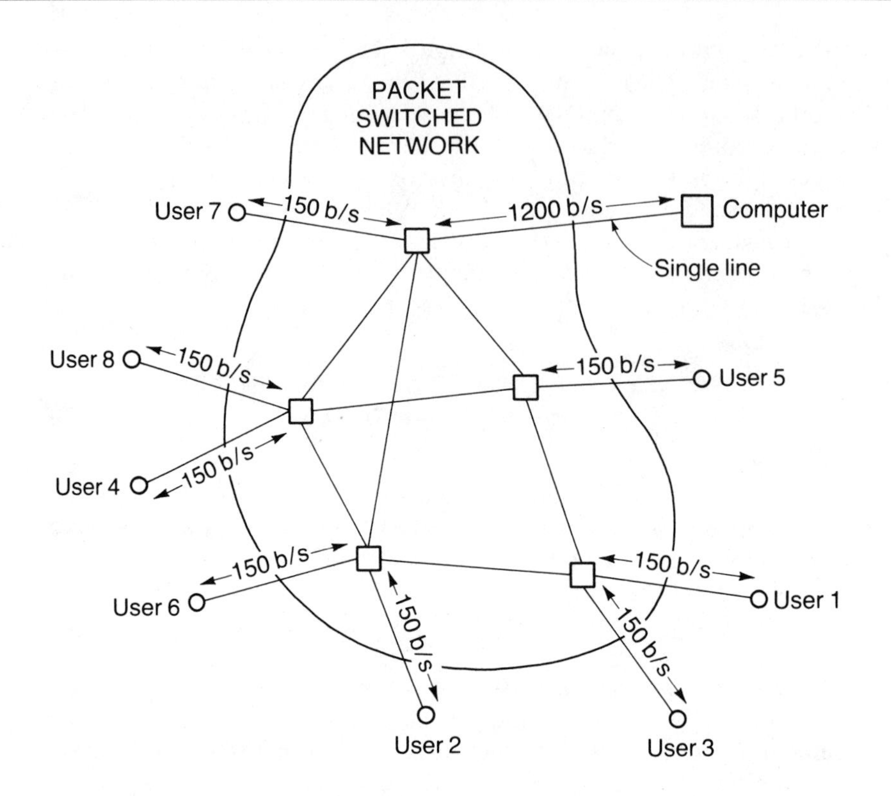

Figure 3-6. Connection of a computer to a number of different terminals, using packet switching.

Disadvantages

The principal disadvantages of packet switching lie in the way it must be implemented in order to achieve its advantages. To achieve flexibility and adaptability, many small switches and processors must be employed, and therefore complex routing and control procedures are necessary. An entire technology of protocol design and protocol layering, which we shall discuss in Part 2 of this book, has evolved to deal with the issue of user-to-switch interfaces and the operation of networks in failure and congestion modes.

SUMMARY

As we examined the concepts and techniques of circuit, message, and packet switching, emphasizing the applicability of each technique to computer-based information systems, we saw that each has various advantages and disadvantages. As in any area of endeavor, choosing the best switching technique involves a set of tradeoffs between the attributes and shortcomings of any alternative. Those tradeoffs tend to favor packet switching for the movement of data communications through large networks.

In the next few chapters we shall further explore the network environment and develop some tools and techniques that we can use to better quantify the advantages and disadvantages introduced in this chapter.

SUGGESTED READING

KLEINROCK, LEONARD. "On Resource Sharing in a Distributed Communication Environment." *IEEE Communications Magazine*, vol. 17, no. 1 (January 1979), pp. 27–34.

This article considers the nature and cost savings of resource sharing in a teleprocessing network. It shows how packet switching is a natural outgrowth of the pooling of communications resources in a large teleprocessing network.

SANDERS, RAY W., and MCLAUGHLIN, R. A. "Networks at Last?" *Datamation*, vol. 26, no. 3 (March 1980), pp. 122–128.

This article traces the evolution of computer networking over the decade of the 1970s, from the first appearances of packet switching, and competition among U.S. common carriers. The authors define their expectations for the continued evolution of various networking and switching techniques, including circuit switching and hybrid networks.

4

Getting Started with Network Basics

THIS CHAPTER:

will examine the network environment and its components.

will look at the public telephone network as an example of a network environment.

will explore the dichotomy between the user's view and the supplier's view of a network.

will introduce design considerations for communications networks.

THE NETWORK ENVIRONMENT

The architect of a building must work in an environment of numerous constraints. The design must be both pleasing and functional when viewed from without. The internal components, most of which are not seen by the building occupants, must insure that the structure is both strong and efficient. Finally, since the budget of the owner is generally limited, there are finite resources available to meet the first two objectives.

Such is the case with the design of networks. The users of the network have an external view of the network, with little concern for its internal structure. The owner and operator—whom we will call the supplier—of the network is concerned with maximizing the efficiency and flexibility of the internal structure to meet the varying demands of the many users. The constraints imposed by these two views of the same environment require some interesting and difficult design compromises in the architecture, design, and implementation of communications networks.

The External Environment. Figure 4-1 shows a generalized view of a network as seen by the users—in other words, the external environment. The network is seen as a "transport mechanism," a vehicle by which the information of the users

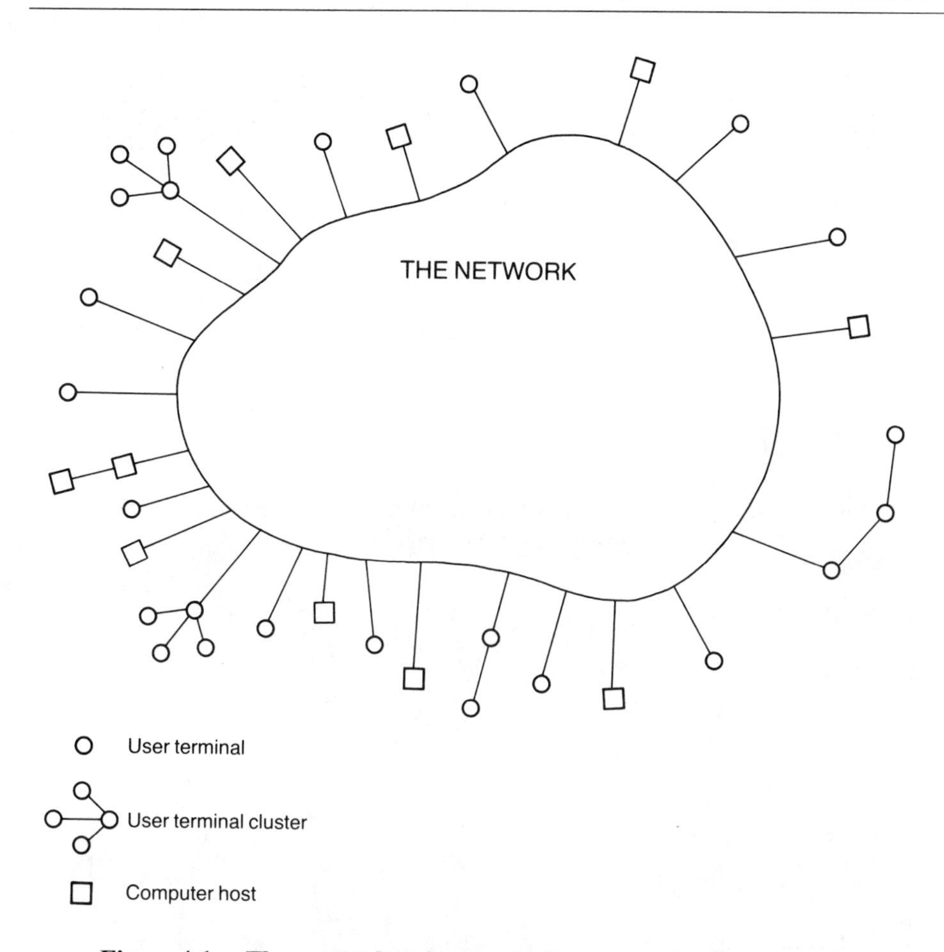

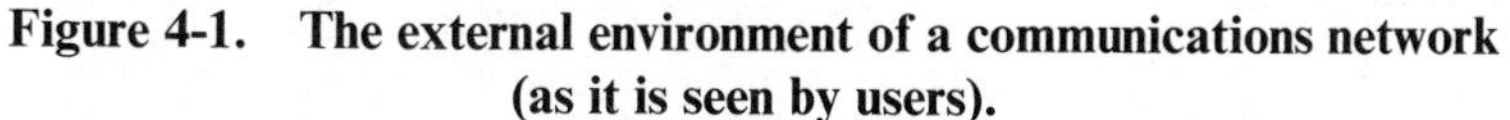

Figure 4-1. The external environment of a communications network (as it is seen by users).

passes from one user device to another. Each device connected to the network should be as accessible as any other, unless, of course, a user has a particular reason for excluding access to certain terminals (for example, restricting access to electronic funds transfer terminals).

The Internal Environment. An entirely different view of the same network is represented by Figure 4-2. Here we see the network as the supplier sees it—the internal environment. The network includes switches and transmission lines among the switches. The number and capacity of these costly resources have to be limited by the budget of the supplier, the current demand of the users, and a realistic projection of possible future demand.

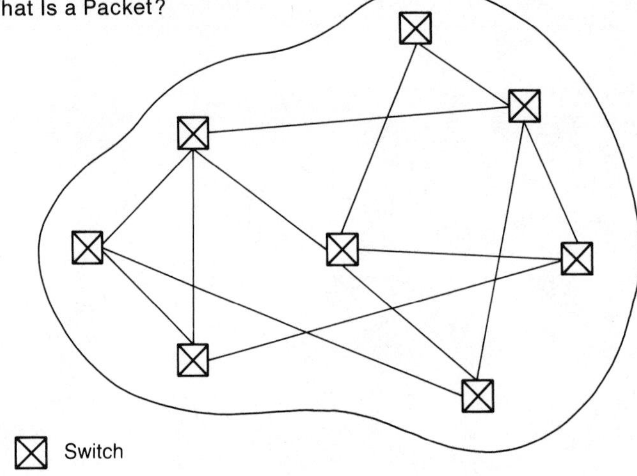

Figure 4-2. The internal environment of a communications network (as it is seen by suppliers).

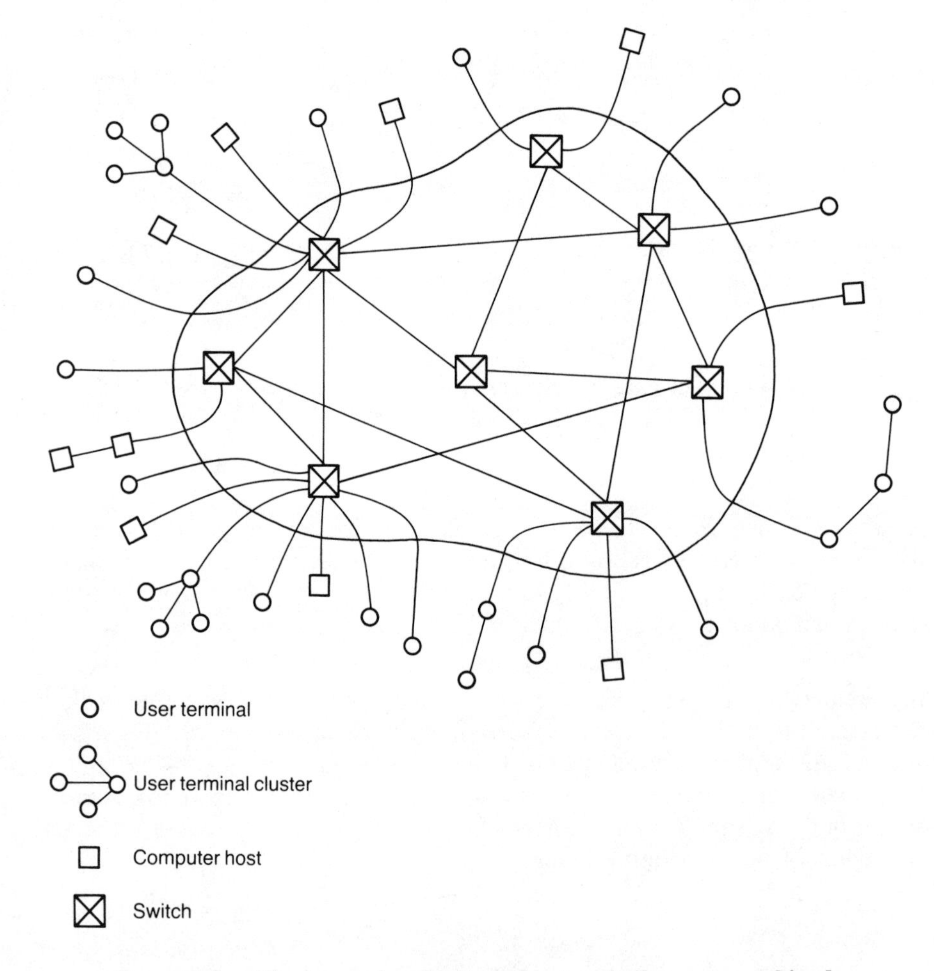

Figure 4-3. The internal and external network elements combined.

The Composite Network. When the lines from the user terminal devices to the first switch in the network to which they are connected are added, the result is the composite picture shown in Figure 4-3. This figure illustrates some of the key components: terminals, switches, access lines, and trunks. The switches may be circuit switches, message switches, packet switches, or combinations of these. In the future they may even be single devices that employ the principles of several of the techniques at the same time.

THE EXAMPLE OF THE PUBLIC TELEPHONE NETWORK

We can understand the principles of internal and external network environment by looking at the nationwide telephone network as an example.

Elements of the Telephone Network

Figure 4-4 is a schematic representation of the telephone network. Only three types of network switches are shown in this representation: (1) local central office switches, (2) toll switches, and (3) tandem switches. **Tandem switches** is a general

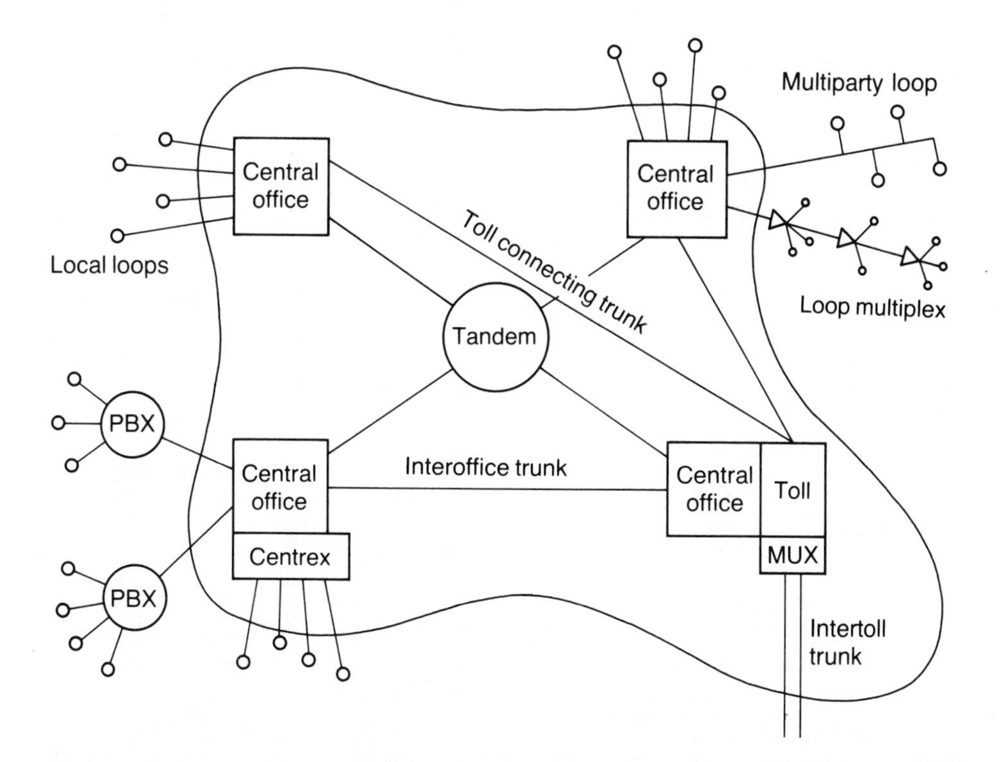

Figure 4-4. The elements of the public telephone network.

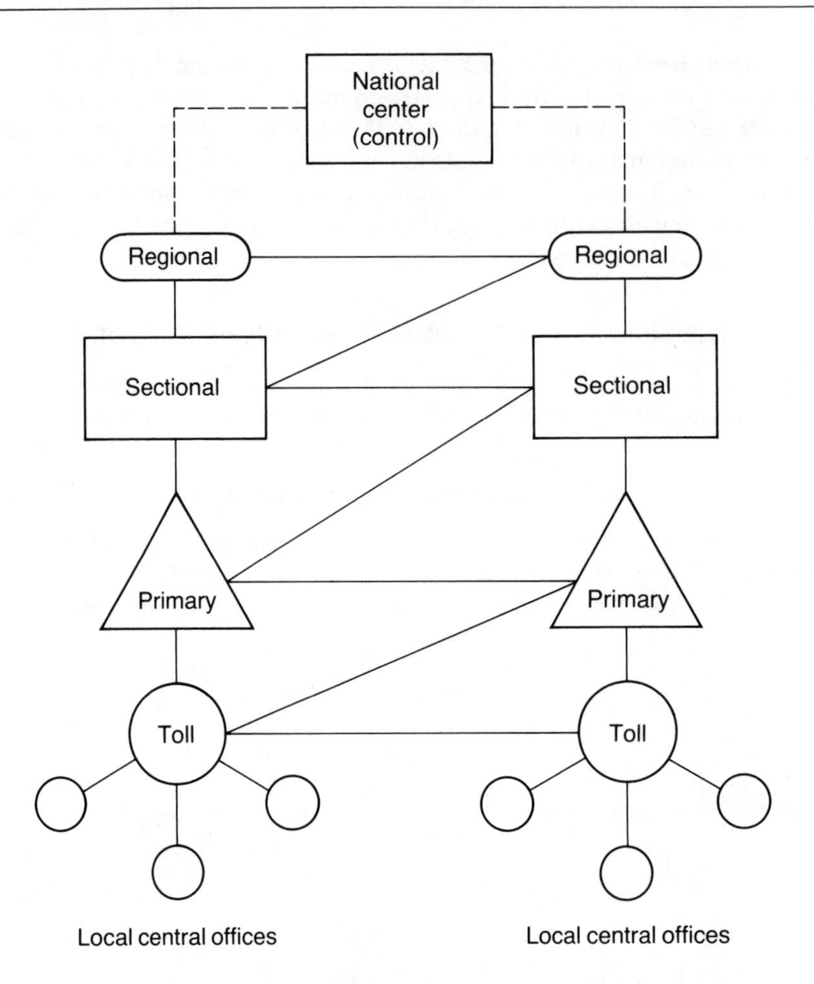

Figure 4-5. The switching and routing hierarchy of the public telephone network.

term, which, in the telephone network, encompasses primary, sectional, and regional switching centers. The full switching hierarchy is shown in Figure 4-5.

Local Central Office Switches. Users are connected to the network through local central offices. Individual subscribers, such as residential telephones, are in general connected on an individual basis to the central office, often using a separate pair of wires from each instrument to the central office. Groups of users who share a common interest, such as all the users in a single company or office building, are often handled using concentrating devices, commonly called *private branch exchanges* (*PBXs*). These devices allow many fewer lines than the total number of users to connect the users to the central office. This system is based on the statistical principle that not all users will be making calls at the same time. In addition, application of PBXs permits special services to be supplied, such as the ability to call anyone else within the company by dialing only two or three digits. In some cases PBX services can be supplied through special arrangements at the central office. This is known as *Centrex* service.

Toll Switches. Central offices can complete calls between subscribers connected to that office or subscribers connected to offices within the same local calling area by direct routing of the call over interoffice trunks. In cases where the call is outside the local call area, and a "toll," or "long-distance," charge is applicable, the call must be routed through a toll office. Most toll offices are located within the same physical building as one or more central office switches, but they are functionally separated within the hierarchical routing and call processing structure.

Tandem Switches. In Figure 4-4 all switching offices above the toll office fall into the general category of tandem switching offices. This tandem switch does not connect directly to any subscribers but only functions to tie together other switches in the end-to-end route between two subscribers.

Connections between Switching Centers

Switching centers are generally connected by high-capacity interswitch trunks, made up of dozens, hundreds, or often thousands of individual channels. A number of different access arrangements are also possible ways to minimize the cost of connecting individual subscribers to the switching centers. The familiar multiparty **loop,** or "party line," for example, though nearly extinct for voice users, is quite efficient for data communications. The capacity of a single line can be divided in a number of ways to serve quite a large number of data users simultaneously. However, all the users would have to be routed between the same two endpoints in the network since the network is providing a transparent, circuit switched connection between the endpoints. Such an arrangement is useful to connect, for example, 24 airline reservation clerks in Washington, D.C., to a central reservation computer in Atlanta, since all of the clerks in the same local area are communicating to the same destination through the network.

Polling and multidrop circuits are often used in practice to achieve such shared line arrangements. **Multidrop** circuits refer to the parallel connection of a number of user terminals on a common communication circuit. **Polling** is the most commonly used discipline on such a connection. A central control station (generally the main computer serving the community of terminals) periodically interrogates (polls) each terminal on the multidrop circuit. When a terminal is polled, it can reply with a certain amount of data, or it can indicate that it has no information to transmit at this time.

Hierarchical Structure

Above the local central offices, which directly interconnect with the network users, lies a carefully structured hierarchy of high-capacity tandem switching offices. The structure of the hierarchy is integrally related to the network routing plan, which attempts to find a path through the network, using any available capacity, even if the most direct path through the network is not available.

Routing. When a subscriber initiates a call, the routing of the call follows a series of sequential routing decisions through the hierarchy. The first decision depends on whether the destination of the call is connected to the same switch as the initiator. If it is, then the routing is complete. If the destination is not at the same switch, then the availability of a circuit directly to the destination office is checked. If no such circuit is available, the call is routed upward to the toll switch.

The toll switch will look for an available channel directly to the terminating office and will route the call directly if such a circuit is available. If there is no direct circuit from the toll office to the destination office, the call will be passed horizontally to the toll office closest to the destination office. If there is no circuit available to the nearest toll office, the call is passed to the primary office immediately above the toll office serving the destination office. If such a circuit is not available, then the call will again be passed up, to the hierarchy primary center, immediately above the originating toll office.

The process again repeats itself—first searching for a direct path from the primary office directly to the toll office serving the destination office, then trying for a channel horizontally connected to a primary office serving the destination office, and, if that route is not available, routing the call to the sectional office, which lies above the primary office serving the ultimate destination. Following this basic pattern, literally hundreds of possible routes between every pair of subscribers is searched in order to complete a connection.

Understanding the Routing Concept. Understanding the basic routing concept for the common phone network is important because one of the unique aspects of packet switching is the dynamic nature of the routing through the network. The hierarchy shown in Figure 4-5 has a well-defined pattern and structure. In order to perform its function, each switch must have basic information about the network structure, the pattern of connectivity, and the availability of capacity toward the destination. Furthermore, the routing decisions are sequential, which means that the longer the path between the two endpoints of the call, the longer the call will take to set up because of the multiple, sequential decisions that are necessary. After a call is successfully established, it remains connected for the duration of the call. If some part of the connection is disrupted by the failure of a connecting channel or the disruption of a switch, the call connection is lost, and the process must be started all over again.

Any changes in the network structure, such as the addition of new switching offices or the connection of new transmission routes, have to be reflected by changes in the routing tables used by the switches throughout the network.

THE NETWORK USER AND THE NETWORK SUPPLIER—TWO SIDES OF THE COIN

The "Marketplace" Situation

There often appears to be a dichotomy between the supplier's view of network services and the user's view. In a general sense, this dichotomy is no different from any other marketplace situation, where the buyer is trying to pay the lowest price and the seller is trying to extract the highest price. However, there are many subtle issues that go beyond the normal marketplace give-and-take. These issues have an impact on the way the network is designed, no matter which networking techniques are employed.

In addition, many specialized networks, particularly those designed to satisfy the internal data communications needs of a company, agency, or organization, are constrained by the needs of both the user and the supplier at the same time. When both the supplier and the users of the network are contained within the same organization, the supplier tends to see the network as a revenue producer, whereas the user sees it as a revenue consumer.

The User

Let us start by looking at the network from the user's viewpoint. Figure 4-6 presents a complete picture of the objectives of a switched data network from the user's viewpoint. What do you see? Nothing! Nothing is what the network user

Figure 4-6. Objectives of a switched data network from the user's viewpoint.

would really prefer to see. If cost were no object, the user would really like to have a full-period, point-to-point, high-capacity connection to every point on earth with which he might want to communicate. A network with no switches, and unlimited point-to-point connectivity would yield a communications environment that posed no restrictions, no delays, no errors, no constraints, and, above all, no costs to the user.

A More Balanced Picture

Given that this view is not very practical, Table 4-1 presents a more balanced view of the objectives of a switched data network. Let's look at these counter-balancing viewpoints one pair at a time.

Cost/Revenue

The user, seeing the network as a revenue consumer, tries to minimize cost. This means using resources efficiently, sharing capacity, and always looking to more economical alternatives.

The supplier, seeing the network as a revenue producer, tries to maximize that revenue. This means accommodating the largest possible amount of traffic within the limitation of available resources.

Delay/Utilization

The network user wants to experience little or no delay in transmitting information through the network.

The network supplier wants to maximize the utilization of the resources in the network. The result, as we will see from queueing theory, is increased delay.

Capacity/Conflict

The network user wants to have capacity always available. A network that cannot frequently accept new user traffic is not very acceptable.

Table 4-1. Objectives of a Switched Data Network

User's View	*Supplier's View*
Minimize cost	Maximize revenue
Minimize delay	Maximize utilization
Have capacity always available	Share capacity among many users (conflict)
Change configuration rapidly	Develop through evolutionary changes
Use as little service as possible	Sell as much service as possible

From the supplier's view, the basis of network operation is making individual services more economical by allocating capacity among a large community. This invariably leads to conflict when all available capacity is filled.

Change/Evolution

The network user wants the potential to change the configuration rapidly. It is important for the network to be able to meet new requirements as rapidly as possible. In addition, as technology changes, the network services should be able to adapt rapidly.

The network supplier must make changes in an evolutionary fashion. New capabilities and services should be interoperable and technically compatible with the existing plant. Additions to the plant cannot be made very quickly because of the investment capital requirements, regulatory approval processes, and physical construction time. Spare capacity to meet changing demand is not desirable because it does not produce revenue.

Buying/Selling

As in any market situation, the user tries to minimize costs by using as little service as possible. However, communications services can often replace or supplement more costly services. For example, certain types of computer conferencing services have been used as replacements for business meetings. Data services can often substitute for more expensive voice calls. Simply minimizing actual network utilization does not necessarily reduce the user's total expenditure.

Similarly, the supplier is trying to sell as much service as possible. New services can often make existing services more useful and thus increase the amount of revenue they generate.

DESIGN CONSIDERATIONS

Any communications network entails a number of major design considerations for both user and supplier. The network supplier must balance flexibility with economy. The user must balance costs of service with network capabilities and reliability. Figure 4-7 illustrates a decision tree approach to such considerations for switched data networks. The issues that are raised here will be dealt with in more detail in later chapters.

Connectivity. The first consideration is connectivity. Is switching really necessary, or would a network of point-to-point connections, using concentrators and multiplexers for economy, be adequate? Once we decide on a switched network approach, we must choose which particular switching approach to use.

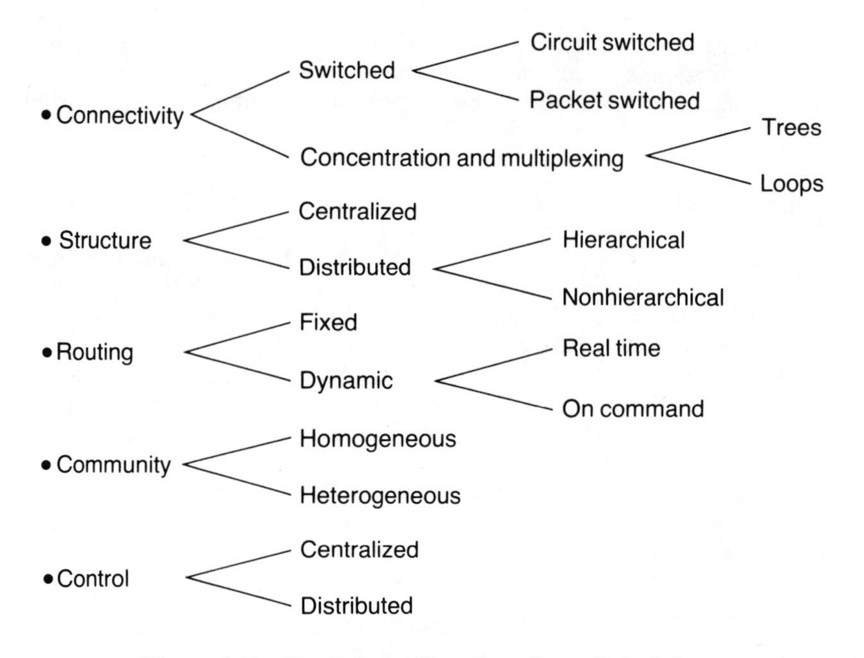

Figure 4-7. Design considerations for switched data network development.

Structure. What physical structure should our network take? Is it centralized, with all users trying to gain access and communication with a large central resource, such as a master data base or parts warehouse? Or is the network distributed, like the phone network, with pairs, or possibly groups, of users throughout the network communicating with each other? If the latter, should the network be configured in a logical hierarchy, like the phone network, or in a nonhierarchical structure?

Routing. How should the connections and information be routed through the network? Should the information always follow the same fixed route, with the attendant simplicity of operation? Or should the routing be dynamic, using excess capacity, where it exists, to find any available route through the network?

Community. What kind of community is our network trying to serve? Is it homogeneous, with similar user characteristics, similar statistics, and, more important, similar physical components and hardware? Or is the community of network users heterogeneous, with different requirements, different equipment, and different demands? This consideration will have a large impact on the complexity of the network design itself, as well as on the design of the interfaces between the users and the network.

Control. Finally, how is the network to be controlled? Is the control centralized at one point, where performance can be monitored or optimized? Or should the control be distributed, possibly down to the level where individual users can participate?

Figure 4-7 summarizes the interrelationships of the various considerations. Some are important largely for network suppliers. However, understanding how the supplier implements, manages, and controls the network can be important for users facing a wide choice of possible suppliers with competing features, different interface requirements, and very different cost and tariff structures.

SUMMARY

In this chapter we have taken a closer look at the environment in which communications networks are designed and used. The public telephone network helped to illustrate the different points of view—internal and external—of a network. The differences in these viewpoints entail design considerations, which lead to a multitude of choices to be made. Having introduced these choices, we will now begin to gather the information necessary for making choices intelligently.

SUGGESTED READINGS

JOEL, A. E., Jr. "What Is Telecommunications Circuit Switching?" *Proceedings of the IEEE*, vol. 65, no. 9 (September 1977), pp. 1237–1253.

A broad introduction to the entire field of telecommunications circuit switching, this article deals with the past and current technology and provides a foundation for understanding the potential impact of new technological advancements. A comprehensive bibliography, with references going back to the turn of the century, traces the history of circuit switching and public telephone networks.

PEARCE, J. G. "The New Possibilities of Telephone Switching." *Proceedings of the IEEE*, vol. 65, no. 9 (September 1977), pp. 1254–1263.

This article, continuing where the previous one leaves off, looks at the possibilities for circuit switched networks as the complexity of electronic switches increases while unit costs decrease. In addition, it explores a number of possibilities for much more flexible services to user terminals.

5

What Is an Erlang? An Introduction to Telecommunications Traffic Engineering

THIS CHAPTER:

will briefly introduce telecommunications traffic engineering.

will define *blocking* in a circuit switched network.

will look at the relationship between delay and capacity in a packet switched network.

will consider why, when we talk about communications systems, bigger is better.

STATISTICAL BEHAVIOR OF NETWORK USERS

We have seen that one of the most important aspects of switching in communications networks is *sharing*. The basic premise of sharing is that not all users of a network will want to communicate with each other at the same time. However, it is impossible to know in advance when every user will want to communicate with some other user. The behavior of the users, and therefore of the network, can only be defined statistically.

The Supplier's View

Network suppliers must consider a dual set of statistics—**call arrival rate** and average call length, or **holding time.**

Call Arrival Rate. When the entire population of network users is viewed from a vantage point inside the network, there is a random length of time—known as

interarrival time—between the initiation of one user's call and the initiation of the next user's request for a call. For example, when the network is very busy, there may typically be 2 seconds of time between the initiation of two successive calls, whereas the time between successive calls may be 2 minutes when the network is not busy.

Another way of looking at the same phenomenon is to take a common time interval as a reference and determine, on the average, how many new calls (or messages, or packets) enter the network during that interval. For example, we may find that, on the average, 25 new calls or messages enter the network per minute. We may choose whatever time basis seems most appropriate, but as long as the measurement base is specified, the call arrival rate can be related to the needed time base. A call arrival rate of 25 calls per minute is equivalent to $25 \times 60 = 1500$ calls per hour, or $25/60 = 0.433$ calls per second.

It is important to recognize that the call arrival rate and the interarrival time are inversely related to each other. If the call arrival rate is high, so that the network is experiencing a large number of calls per minute, then the average interarrival time is going to be small; that is, the time between successive new calls into the network will be short.

Holding Time. The other set of statistics important to the network supplier is the length of the calls users are transferring through the network. For voice calls this length is usually measured in minutes or seconds; for data transactions the number of bits is generally a more useful measure. For example, voice calls are approximately 5 minutes (300 seconds) long, on the average. Data communications, on the other hand, are very application dependent, and vary widely from a few bits to many millions of bits in length (see Table 2-2, Chapter 2).

TRAFFIC AND USER SATISFACTION

The principle of sharing in the switching function of networks is dependent on the random nature of the call arrival process and the call holding process. If there are a large number of users, and sufficient total resources available, the randomness will tend to average out, and most of the time users will be satisfied with the service they receive from the network. This is where the challenge of traffic analysis lies: we have to supply enough resources so that the users are satisfied but not so many resources that capacity is wasted. The essential factors can be estimated from statistical knowledge of the amount of network demand and the number of available network resources. Much of the mathematics that relates the amount of offered demand, the number of resources, and certain other statistical information to get an estimate of successfully completed traffic was developed by A. K. Erlang, who, in the early 1920s, began the application of statistical analysis to early telephone exchanges.

The Erlang

We will define network demand in terms of the call arrival rate and the average holding time.

Let λ = call arrival rate, in calls per hour

τ = average holding time, in hours

We will define the network demand, or *traffic intensity* (E) by the product of these two figures. Thus

$$E = \lambda\tau \quad \text{(measured in \textbf{Erlangs})}$$

Let us look more closely at the physical reality of the communications situation before we try to fully appreciate these figures.

Figure 5-1 shows a very simple network, which employs two switches with a single line between them, able to carry the communications in either direction. Each switch has a group of subscribers attached to it. At each switch most of the subscribers talk among themselves and only occasionally have to call one of the subscribers tied to the distant switch. Looking at the line for the busiest 1-hour period of the day (the **busy hour**) can demonstrate the meaning of traffic capacity and Erlangs.

A Traffic Scenario. We can postulate many possible scenarios during the busiest hour of the day (as measured by the amount of time that the line between the switches is in use). Let us assume that user Alpha calls user Beta and communicates with him for an hour. During that time nobody else can use the circuit, but since Alpha is paying for the call, the network supplier is happy; he has gotten a full hour's worth of utilization from the transmission line. In other words, during the busy hour, line utilization was 100%. According to our definition, λ is equal to one call per hour, and τ is also equal to 1 (that is a call length of 1 hour). By multiplying these two numbers together, we find that $E = \lambda\tau$, which is simply $E = 1 \times 1 = 1$, or 1 Erlang. This means that during the busy hour the line between the switches carried 1 Erlang of traffic, which was in fact 100% of the possible utilization of that line.

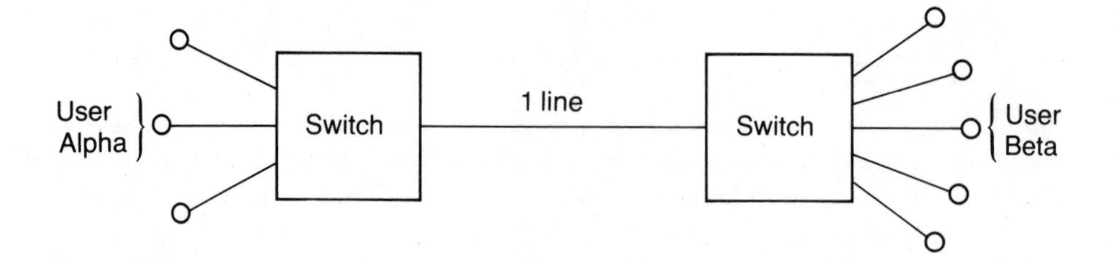

Figure 5-1. Utilization of a single line connecting two switches.

This was a good deal for the network supplier, because the line was operating at full capacity for the hour that Alpha was talking to Beta. But what about all of the other subscribers of the network? If any of them were trying to call across the network at any time during the busy hour, the call attempt would have been rejected because the connecting line was busy. The call attempt of any user other than Alpha or Beta is said to have been *blocked.* Not only that, if the line is in use 100% of the time—that is, if it carries 1 Erlang of traffic—the probability is unity that new calls arriving at either switch will be excluded from the network. And, as we will see, if the network facilities are used to near-full capacity, the probability of new calls being successfully completed is low.

A Fairer Scenario. Before we elaborate on this point further, let us look at Figure 5-1 again, but with a somewhat fairer scenario. In this case we will assume that all callers who complete a call on the interswitch circuit talk exactly the same length of time, that being the average holding time of 5 minutes. If each call lasts exactly 5 minutes, it is possible to accommodate 12 such calls during a 1-hour period, in theory reaching a total of 60 minutes, or 100%, utilization during the busy hour.

In order to reach 100% utilization, each call would have to begin at the exact instant that the preceding call ends. If all of the users at each switch are independent of each other (in a statistical sense), it is very unlikely that such would occur. A realistic situation would require at least a little idle time on the connecting channel between the end of one call and the beginning of the next.

However, let us assume for the moment that, by making reservations in advance or by some other clever mechanism, we do achieve this situation. In this scenario, therefore, λ is equal to 12 calls per hour, and the holding time of each call is τ, equal to 5/60 or 1/12 hours. The traffic intensity in this case is $E = \lambda\tau$, or $E = 12 \times \frac{1}{12} = 1$. The traffic intensity is again 1 Erlang, representing 100% utilization of the line.

Averages in Traffic Intensity

This scenario illustrates three things. First, the measure of traffic intensity, which is the product of the call arrival rate and average holding time, is independent of the values of individual parameters. Traffic intensity as a useful measure is sensitive to the product of its two components. Secondly, achievement of 100% utilization of a communications resource is essentially impossible in a realistic situation of statistically unrelated users. Finally, the performance of the system can be related to the amount of utilization of the resources and the probability that new attempts will be either satisfied or blocked.

In order to satisfy the statistically unrelated group of subscribers to a communications network, more resources must be provided than the number of Erlangs that have to be satisfied, with the absolute maximum amount of service that a given facility can provide being 1 Erlang per resource. On the other hand, if there is only one line between the switches, as in Figure 5-1, and if the probability of success for newly originated calls is to be reasonably high (say 90% or more), the amount of traffic carried has to be much less than 1 Erlang. This will assure that there is a high probability that the line is idle (and thus available to a new call) when a new call is originated from within the statistically uncorrelated population.

In order to move to more realistic scenarios than the two that we have just discussed, we can generalize to the concept of *average traffic intensity* carried on a facility. This is the sum of the individual users' traffic over the measurement period:

$$\text{average } E = \sum_{N=1}^{N=k} \tau_N$$

where k is the total number of calls completed over the measurement period, and τ_N is the length of each call.

Average Offered Traffic Intensity. In addition to the average traffic intensity carried, there is an analogous expression for the average *offered traffic intensity*. The offered traffic is the amount of traffic that would be carried if unlimited resources were available and all requests for service could always be accommodated. The offered traffic intensity is the sum of the carried traffic plus the blocked traffic.

Since, in general, it is impossible to know each individual user's future demand, which would be needed to calculate the average offered load from the formulation above, we generally try to deal in averages. Let's say that the system in Figure 5-1 has 500 subscribers at each switch. On the average, each subscriber originates one call per hour destined for someone at the other switch, and the average length of each completed call is 5 minutes. We can now go back to our definition and compute the offered load as $1000 \times 5/60 = 83.33$ Erlangs. That is, $\lambda = 1000$ calls per hour, and $\tau = 5$ minutes or 5/60 hrs. If the offered load were in fact 83.33 Erlangs, and only one line were available between the switches, at least 99% of the offered traffic would be blocked since the one line can carry at most 1 Erlang of traffic. On the other hand, if there were 100 lines available between the switches, there would be enough capacity available to accommodate 83.33 Erlangs. Not only that, if calls last an average of 5 minutes (300 seconds), and there are 100 lines available, calls typically would be beginning and ending on one of the lines every 3 seconds. If the switches could select any of the 100 lines to complete a new call, we would expect that somewhere in the group of 100 lines there would almost always be a line available to accommodate a new call, even if the offered traffic were 83.33 Erlangs. Every once in a while (less than 1% of the time) a new call would be attempted while all 100 lines were in use. This would result in about 1% of the calls being blocked, a far different situation from that of a single line, where more than 99% of the calls would be blocked.

Another sample calculation of offered traffic during a sample busy hour is shown below. In this case the calls are grouped into three traffic classes, each with its own characteristic average length. In addition, the basic calculation is made in seconds, with the total number of call-seconds (CS) divided by 3600 (the number of seconds in an hour) to arrive at the number of Erlangs over the busy hour sample period.

total traffic during 1-hour sample period (1 hour/3600 seconds)

$$420 \text{ calls}; \tau = 100 \text{ sec}$$
$$40 \text{ calls}; \tau = 200 \text{ sec}$$
$$20 \text{ calls}; \tau = 300 \text{ sec}$$

$$\text{traffic} = E = \frac{(420 \times 100) + (40 \times 200) + (20 \times 300)}{3600}$$

$$= \frac{56000}{3600} = 15.6 \text{ Erlangs}$$

Table 5-1 summarizes the conditions and definitions of teletraffic. Box 5-1 illustrates a set of relationships between the Erlang and a commonly used unit of telecommunications traffic, the *CCS*, which is 100 call-seconds. A **call-second** represents the use of a facility during a call over a period of 1 second. A 5-minute call, for example, uses the network facilities for 300 call-seconds, or 3 CCS. Since Erlangs are traditionally defined for a busy hour period, and an hour has 3600 seconds, there are 3600 call-seconds—thus 36 CCSs—in an Erlang. Reversing the process, we see that 1 CCS is 1/36 of an Erlang.

Table 5.1. Summary of Teletraffic

Based on *random* nature of "calls"
Different formulations for various modes of operation
Use of average traffic intensity, Erlangs $= \lambda\tau$
λ = call arrival rate, e.g., calls/hr; calls/sec
τ = average holding time, e.g., seconds, hours

$$1\,\frac{\text{CS}}{\text{hr}} = \frac{1}{3600}\,\frac{\text{CS}}{\text{sec}} = \frac{1}{3600}\,\text{Erlang}$$

$$1\,\frac{\text{CCS}}{\text{hr}} = \frac{1}{36}\,\text{Erlang}$$

$$1 \text{ Erlang} = 3600 \text{ CS} = 36 \text{ CCS}$$

Box 5-1

BLOCKING, LINES, AND ERLANGS

With the Erlang as the measure of traffic intensity in a communications network, we can identify the relationship between the offered traffic and the quantity of resources available.

Lost Calls Cleared

The relationship is normally defined through a complex mathematical expression for the probability of blocking as a function of offered load and the number of lines available. The most commonly used formulation is known as Erlang B, or *lost calls cleared.* This means that the derivation of the equation assumes (1) a statistically unrelated population, and (2) any call that is attempted but blocked is not retried until after the measurement period of interest (normally the busy hour). Calls that are lost because of inadequate resources are cleared from the network without being satisfied, at least during the measurement period.

As derived by Erlang, the formula for the lost calls cleared case is given by:

$$P_s = B(s, E) = \frac{E^s/s!}{\sum_{k=0}^{k=s} E^k/k!}$$

where

P_s = system grade of service

$B(s, E)$ = blocking as a function of the number of resources (s) and the number of Erlangs (E)

$s! = s(s-1)(s-2) \cdots (3)(2)(1)$

Clearly this is not an easy expression to evaluate. Even more complex than computing the blocking probability from this expression is reversing the expression so as to determine how many lines are needed in order to satisfy a given level of offered traffic. What is generally done is to tabulate or plot the results of a large number of data points that can be used to estimate the blocking as a function of offered traffic and available resources.

The Formulation of Erlang B. Figure 5-2 shows such a set of curves. The horizontal scale shows the offered traffic, measured in Erlangs. The vertical scale shows the estimated network blocking. Each curve is for a different value of the number of channels available, s, with s ranging from 10 channels up to 70 channels. While this kind of formulation is applicable, with certain modifications, to complex networks, its precise application is really for the situation in Figure 5-1, with the single line replaced by a group of channels. We can use the curves in Figure 5-2 to illustrate several very important characteristics of switched networks and traffic sizing.

Let's focus on the horizontal line at a value of blocking equal to 0.01 (1% blocking). We see that, at a value of 20 Erlangs, 30 lines are required to achieve a level of performance of 1% blocking. The overall efficiency of carrying 20

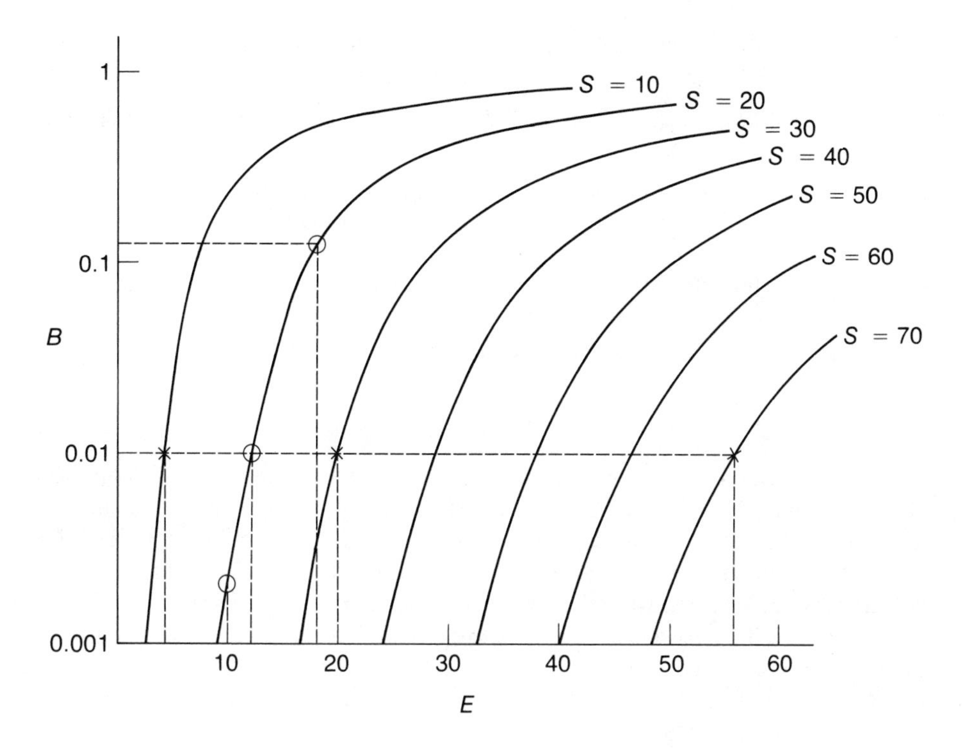

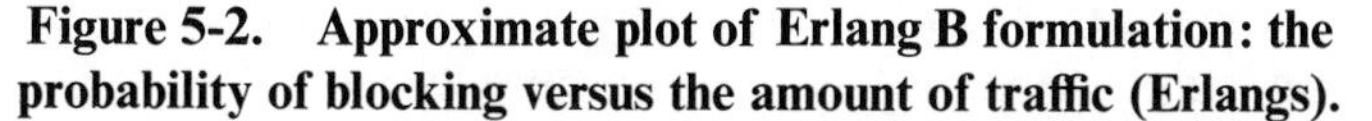

Figure 5-2. Approximate plot of Erlang B formulation: the probability of blocking versus the amount of traffic (Erlangs).

Erlangs of traffic on 30 lines is 20/30 or 67%. If 70 lines are available, about 58 Erlangs of traffic can be carried at a level of performance equal to 1% blocking. This works out to an operating efficiency of about 83%, a considerable improvement compared to the 67% associated with the 20 line situation. At the other extreme, only about 3 Erlangs of traffic can be carried on a 10 line group at a 1% level of blocking. The corresponding efficiency, of course, is 30% in this case. The point is that, as the amount of resources handled as a group increases, the more efficiently they can be utilized.

Another important fact is the steepness of the blocking curves. Consider the curve for $s = 20$. If we use a 20 line group to carry 10 Erlangs of traffic, less than one call in 500 will be blocked. If about 13 Erlangs of traffic are carried, the blocking rises to about one call in 100. If 17 Erlangs are carried, more than one call in ten will be blocked. The point here is that a 70% increase in traffic carried increases the network blocking by a factor of 50 times or more. Relatively small changes in the traffic can have a very large impact on the network grade of service—for example, from a nearly imperceptible blocking rate of one call in 500 to a totally unacceptable rate of one call in ten, or worse. The conclusion is clearly that careful design and consideration must be paid to the overall capacity allocation within a network to insure an acceptable level of performance.

NETWORKS WITH STORAGE AND DELAY INSTEAD OF BLOCKING

When a communications resource, such as a line between two switches, is in use by one pair of users, any attempt by another pair of users to gain control of that line will be blocked. The Erlang B formulation showed us the relationship between the blocking, offered traffic, and number of lines in the system. But what if we can tell the system that, rather than being blocked, we are willing to wait for our turn to use the next available resource. This is common in everyday activities. We wait in line at the supermarket cashier, the post office, the bank, even the gas station. Why not wait our turn in a communications network?

Queueing

Communications networks with waiting—or **queueing**—are not common when transparent techniques are employed. In the first place, circuit switched, voice-oriented networks have generally been sized so blocking occurred infrequently enough that the users were willing to leave the network and try blocked calls at some other time. In addition, since much of the blocking in a circuit network is due to the destination user's being already engaged in communications, the amount of time that we would have to wait for the full path to be free could be very long. Waiting our turn in such a network is simply not practical. However, when we are dealing with data communications, we can make very good use of queueing in the network. In this case the network must be able to store the message or transaction temporarily until its turn for transmission comes.

How Queueing Works. Figure 5-3 illustrates a situation analogous to the one in Figure 5-1. Each user is generating messages for other users, some of which require transmission over the line between the switches. Rather than allowing messages to be blocked whenever two messages arrive at the switch at the same time, the network provides some storage capability at each switch. In order to do this, the

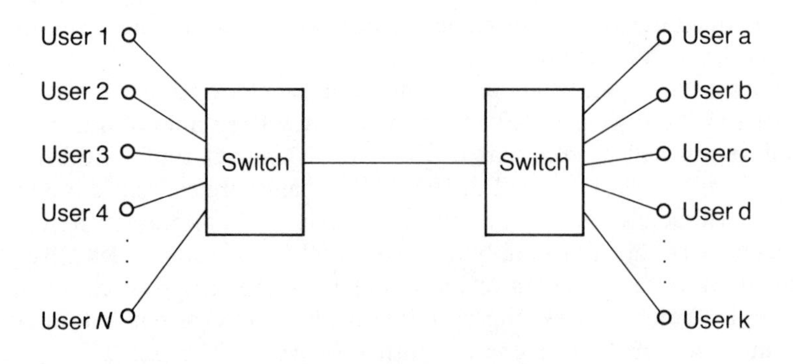

Figure 5-3. A single line connecting two switches with storage and buffering capability.

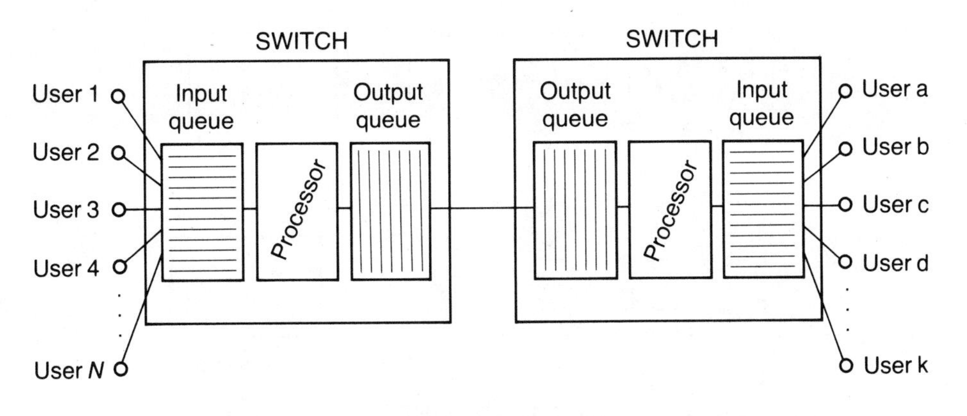

Figure 5-4. Design of switches with storage and buffering capability.

switch has to be functionally designed as shown in Figure 5-4. Messages come into the switch at random times and are stored temporarily in the input queue. The arrival of a new message interrupts the processor, which decides where the message is going, makes a routing decision, and puts the message in the appropriate output queue. If no other message is presently using the line to the other switch, the new message can be transmitted immediately. But if the line is already in use, the message is temporarily stored on the output queue until the line is free.

How will this kind of system perform, especially in comparison to the blocking form of operation described by the Erlang B formulation? Since new messages arriving when the line is already in use are not blocked, but delayed, a formulation is required that relates the offered traffic to the delay encountered. Queueing theory provides such a relationship.

Applying Queueing Theory to a Switch. In order to easily apply fundamental aspects of queueing theory to a switch in a communications network, it is convenient to simplify the switch in Figure 5-4 to the model shown in Figure 5-5. The input queue is shown as a sequential input storage **buffer,** where the messages wait their turn for service by the single element shown as the server. The server in effect combines the processor and output communication line into a single entity with some total capacity—C.

As we learned in Chapter 1, the relative cost of transmission and processing is such that it would normally be cost-effective to insure that system capacity would be limited by the transmission. The processor should always be sufficiently large that it can process the messages rapidly enough to keep the output queue full of undelivered messages. If the output queue were to be emptied by the line while messages were still waiting for the processor to get to them at the input queue, there would be a gross waste of expensive capacity on the output line.

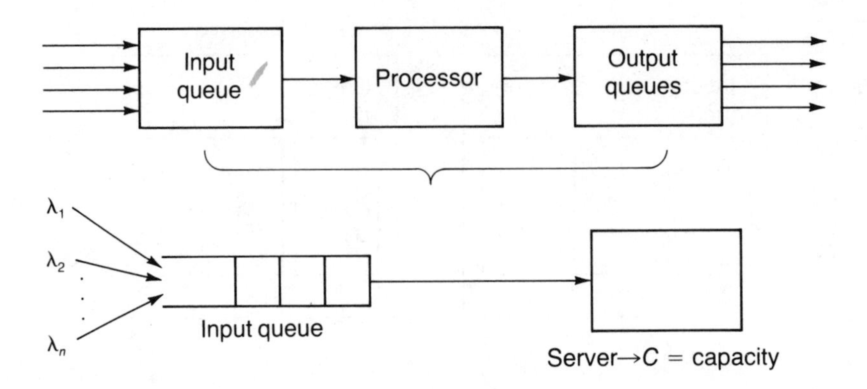

Figure 5-5. Simplified model of a switch with storage and buffering capability.

Some Additional Terminology

In order to understand the relationship between network traffic, capacity, and delay, some additional terms must be introduced. Each of the terms has a partner in the blocking formulation. In the previous formulation we talked about calls with a total average call arrival rate, λ, and average call holding time, τ.

Arrival Rate. For the traffic delayed model of Figure 5-5 each user originates messages at his own average rate, $\lambda_1, \lambda_2, \ldots, \lambda_N$. The total message arrival rate is simply the sum of the individual arrival rates, such that:

$$\lambda = \lambda_1 + \lambda_2 + \cdots + \lambda_N \qquad \textbf{(Eq. 5-1)}$$

The average length of each message, rather than being measured in minutes or seconds, as in the previous case, is measured in terms of bits. The average message length is simply defined as k bits. However, to be consistent with notation that is frequently found in queueing theory, we must alternatively define the term $1/\mu$ as the reciprocal of the message length. Thus:

$$k = 1/\mu = \text{average number of bits per message}$$

The analogy to the number of Erlangs, or offered traffic in the previous case, is similarly the product of the message arrival rate and the average message length. However, assuming that the arrival rate is messages per second, and the length is bits per message, the product simply represents bits per second rather than Erlangs. That is, the offered traffic is:

$$E_a = \lambda k = \frac{\lambda}{\mu} \text{ bits/second} \qquad \textbf{(Eq. 5-2)}$$

as long as λ is given as messages per second and k is given as bits per message.

Capacity and Utilization. The normal way of specifying the capacity of a line used to carry data communications information is in terms of the number of bits per second of information it can carry. As we saw in the blocking case, any single resource could carry at most 1 Erlang of traffic, which represents 100% of capacity. Similarly, even if we allow messages to be temporarily stored, the most that could be carried is 100% of the line capacity, or a total data flow of C bits per second. The ratio of the traffic, E_d, to the capacity, C, gives a measure of the achieved utilization of the resource. The utilization factor, defined as ρ, can then be shown as:

$$\rho = \frac{E_d}{C} = \frac{\lambda k}{C} = \frac{\lambda}{\mu C} \qquad \textbf{(Eq. 5-3)}$$

Service Time. One additional factor we will need is the average length of time it takes to serve a message. If the average message length is k bits, and the server is operating at C bits per second, then the average time of service is simply:

$$\tau = k/C = 1/\mu C = \text{average service time}$$

Relating Traffic Carried to Delay Through the System

The key terms and expressions are summarized in Box 5-2. Our main purpose in introducing these terms is to relate the traffic carried to the delay through the

$\lambda = \lambda_1 + \lambda_2 + \cdots + \lambda_N =$ average number of messages per second

$k =$ average number of bits per message

notation $k = \dfrac{1}{\mu}$

$E_d = \text{traffic} = \lambda k = \dfrac{\lambda}{\mu} =$ average traffic in bits per second

$C =$ capacity in bits per second of system

$\rho = \text{utilization} = \dfrac{E_d}{C} = \dfrac{\lambda}{\mu C}$

$\tau = \text{average service time} = \dfrac{k}{C} = \dfrac{1}{\mu C}$

also $\rho = \lambda\tau$

Box 5-2

system for the delay mode of operation. To do this, all we have to do is determine the average number of messages that are on line in the input queue, waiting for service, when a new message is originated.

Waiting Time. If we represent the average number of messages on the input queue as N, the average time that we have to wait in the input queue is $N\tau$. We can designate the average waiting time as W;

$$W = N\tau = N/\mu C = \text{average waiting time}$$

To estimate the average number of messages waiting at any given time, visualize the operation of the queue in the system as follows. We know by both intuition and experience that the busier a system is, the longer the queue of waiting users will typically be. The queue will remain that length as long as new arrivals come into the system at the same rate that old arrivals are served. For the kinds of systems we are considering there is a well-defined set of probabilities that, within a stated time interval, a certain number of new arrivals will actually enter the waiting queue. That process, known as a *Poisson arrival process*, is mathematically stated as:

$$P(k) = \text{probability of exactly } k \text{ arrivals in } \tau \text{ seconds}$$

$$P(k) = \frac{(\lambda\tau)^k e^{-\lambda\tau}}{k!} \qquad \textbf{(Eq. 5-4)}$$

where

$$k! = (k)(k-1)(k-2)(\quad)\cdots(2)(1)$$

and

$$\lambda = \text{the average arrival rate of new entries into the system}$$

In a system with fixed capacity the number of customers that can be served in a given time depends solely on the time it takes to serve each one, which we previously called the average service time, $\tau = 1/\mu C$. If the arrival rate is high, chances are that more than one new customer will arrive in the length of time, τ, that is takes to serve one customer, and the queue will begin to grow. If the arrival rate is low, then there is a good probability, $P(O) = e^{-\lambda\tau}$, that no new customer will arrive during the service time interval, τ, and the queue will begin to shrink. It should be evident that the probability of the queue's growing or shrinking depends on the size of λ, or the rate that new customers enter the system. If the service time is fixed at a value of τ, but we remember that the system utilization was defined as $\rho = \lambda\tau$, we can see that this probability will depend on the utilization, since the utilization is proportional to λ.

Stable Queue Operation. For a system to operate in a stable condition, the number of customers served over a given time interval has to equal the number of new arrivals. By a *stable condition* we mean one in which the queue has an average length that is neither growing nor shrinking rapidly. The condition of stability is derivable from the definition of the Poisson arrival process and the system service rate, by defining the average arrival and departure rates and letting these averages be equal.* As might be anticipated, the average number of users waiting in the queue is strictly dependent on how busy the system is, and is given by:

$$N = \rho/(1 - \rho) = \text{average number in input queue} \qquad \textbf{(Eq. 5-5)}$$

where

$$\rho = \lambda\tau = \lambda/\mu C$$

The average waiting time is then:

$$W = N\tau = \frac{\rho\tau}{1 - \rho} = \frac{\rho/\mu C}{1 - \rho} \qquad \textbf{(Eq. 5-6)}$$

As a final step, we can estimate the average delay through the system as the sum of the average time a new arrival has to wait for service and the time it takes for that arrival to be served once its turn arrives. In other words,

$$\text{average delay} = T = \text{service time} + \text{waiting time}$$

$$T = \tau + W = \tau + N\tau = \tau(1 + N)$$

$$T = \tau\left(1 + \frac{\rho}{1 - \rho}\right)$$

$$T = \tau\left(\frac{1}{1 - \rho}\right)$$

thus

$$T = \tau/(1 - \rho) \qquad \textbf{(Eq. 5-7)}$$

The average delay is thus estimated by the average service time and the relative loading of the system compared to system capacity.

The Delay-Utilization Tradeoff. This result is shown graphically in Figure 5-6, which clearly illustrates the increase in delay as the system approaches capacity. If the system is very lightly loaded, and ρ is close to zero, the average delay is just the average service time, τ or $1/\mu C$. At one-half capacity, $\rho = 0.5$, the delay has doubled since, on the average, there is one user in the queue ahead of a new arrival. At three-quarters capacity the delay has reached 4τ, and at 90% capacity the average delay has reached 10τ. If we attempt to operate at 99% capacity, the average delay would be 100 times the average service time. The dramatic change

* For an excellent derivation of the average queue occupancy see Mischa Schwartz, *Computer Communication Network Design and Analysis* (Englewood Cliffs, N.J.: Prentice-Hall, 1977), Ch. 6.

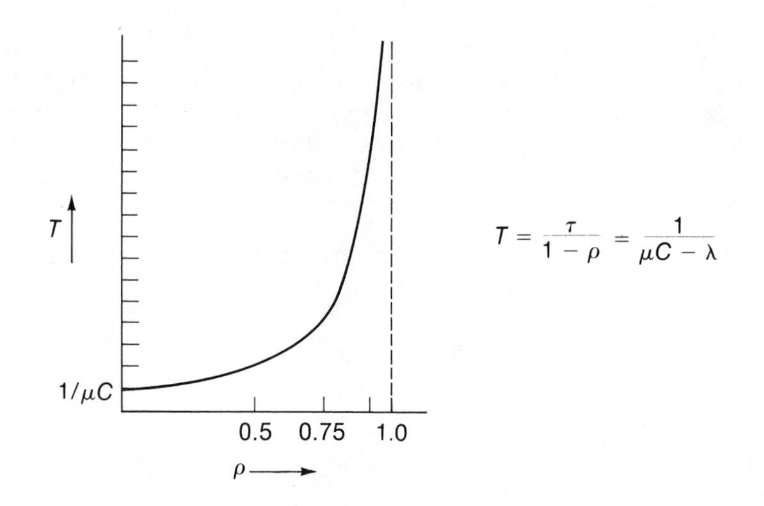

Figure 5-6. Basic queueing theory result: delay plotted as a function of utilization.

of delay with system occupancy is a key factor in the design of efficient systems that embody storage, as packet switched communications networks do.

In Figure 5-6 a commonly used alternative expression for network delay is shown. If we refer back to Box 5-2, we recall that $\tau = 1/\mu C$, and we can see that $\lambda = \rho\mu C$. By substituting these terms into our expression for average delay, we find that the average delay can be expressed as:

$$T = \tau/(1 - \rho) = 1/\mu C(1 - \rho) = 1/(\mu C - \lambda) \qquad \textbf{(Eq. 5-8)}$$

This expression, together with the plot of delay versus system loading shown in Figure 5-6, gives us the basic tools with which to understand the operation of packet switched communications networks. In circuit switched networks our major concern was that, as the system was more heavily utilized, the probability that a new arrival would be blocked and excluded from the system increased. In systems with delay, as the system loading increases, the average delay rises very dramatically, although as long as there is buffering and processor power available, all new users will be allowed into the system (that is, they will not be blocked).

Before we look at specific quantitative tradeoffs between the various switching techniques in the next chapter, let us use some exercises to demonstrate the very important principle that, when it comes to designing a communications system, "bigger is better."

CAPACITY OF NETWORKS—SOME EXAMPLES

To begin, let us refer to the situation depicted in Figure 5-7. There are two communities of users connected to circuit switches with only one line between them.

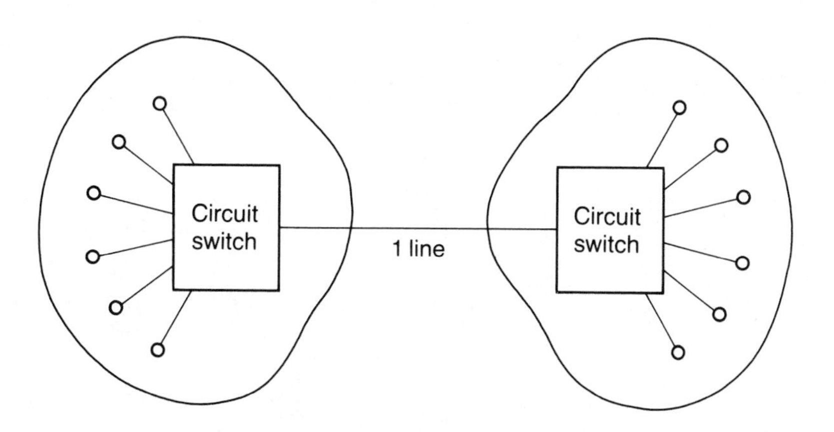

Figure 5-7. Utilization of a single line connecting two circuit switches.

Demand = 0.5 Erlang

After considerable measurement and experimentation we can establish to a high degree of accuracy that the total demand during the busy hour of the day is for exactly 0.5 Erlang of traffic to flow between the two switches. (Of course, additional traffic is flowing entirely within each community.) Assuming that there is a sufficiently large population of users creating the total demand, we can apply the Erlang formula for blocking and determine the grade of service seen by the users during this busy hour. The sample calculation in Box 5-3 shows that the blocking is 33%. In other words, one-third of the call attempts will be blocked by calls already using the single line between the switches.

$$P_s = B(s, E) = \frac{E^s/s!}{\sum_{k=0}^{k=s} E^k/k!}$$

$$E = 0.5$$

$$s = 1$$

$$B(s, E) = \frac{(0.5)^1/1!}{\frac{(0.5)^0}{0!} + \frac{(0.5)^1}{1!}}$$

$$B(s, E) = \frac{0.5}{1 + 0.5} = \frac{0.5}{1.5} = 33\%$$

Box 5-3

Since one line is successfully carrying 0.5 Erlang of traffic, the line utilization is 50% at the calculated level of blocking (33%).

Demand = 2 Erlangs

In Figure 5-8 the same picture is repeated, except there are four times as many users at each switch, generating a total of 2 Erlangs of traffic. As a consequence, we decide to put four lines in service between the two switches. The blocking calculation (Box 5-4) is considerably more complex this time. The result shows a blocking percentage of just about 10%, or a threefold improvement over the previous case. Note, however, that 2 Erlangs of traffic are carried on four lines, resulting in the same 50% achieved line utilization as the previous case.

More Erlangs—Better Performance

The next step in the process could be pictured as in Figure 5-8, except that traffic has now grown by a factor of 4, to a total of 8 Erlangs. As a result we increase the number of lines to 16. At this number of lines calculation of the Erlang formula becomes quite tedious, so we will interpolate a value of blocking from the curves in Figure 5-2 and find that the blocking would be approximately 0.01, or 1.0%—approximately a tenfold improvement over the last case. The group of 16 lines is still operating at a utilization of 8 Erlangs on 16 lines, or 50%. If we were to carry this yet one step further—that is, to 32 Erlangs on 64 lines—we would be way below the scale of Figure 5-2, at a blocking level of 0.01% or less.

Bigger Really Is Better. The conclusion to be drawn from this exercise is clear. Even in the simple case of a single primary route between two switches the larger the amount of traffic that we can aggregate into a single large facility, the more efficiently we can handle that traffic. From another viewpoint, at a fixed level of efficiency (50% in this case) a higher level of performance (as seen by the users) can be achieved in the form of lower blocking probability.

Aggregation in Delay Networks

To look at the analogous situation for a delay network, let us consider the following simple model represented in Box 5-5. We assume an arbitrary delay system, of capacity C_1, which has traffic arriving at a rate of λ_1. Using Equation 5-8, the average delay through the system is given as:

$$T_1 = 1/(\mu C_1 - \lambda_1)$$

where $1/\mu$ is the average message length.

Now the situation is changed slightly (Box 5-6). The total capacity, C_1, is split between two smaller systems, each with one-half the capacity of C_1. Let us thus split the traffic to each of the smaller systems, so that each of the smaller systems carries one-half the total traffic. In this case, $\lambda_2 = \frac{1}{2}\lambda_1$.

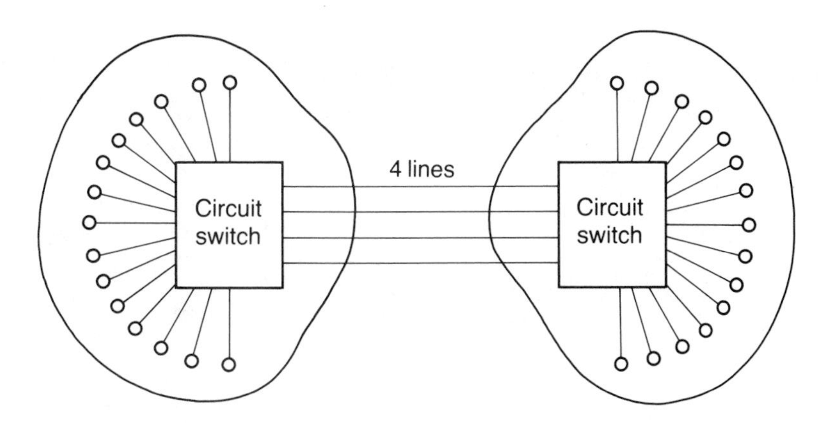

Figure 5-8. Utilization of four lines connecting two circuit switches.

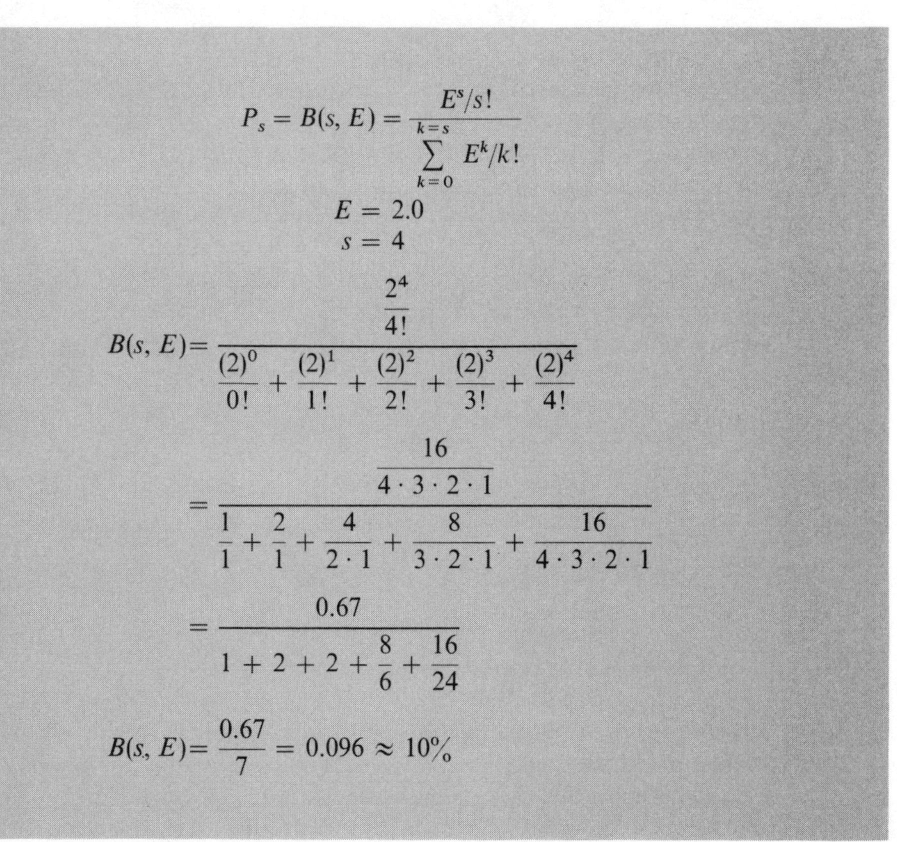

$$P_s = B(s, E) = \frac{E^s/s!}{\sum_{k=0}^{k=s} E^k/k!}$$

$$E = 2.0$$

$$s = 4$$

$$B(s, E) = \frac{\frac{2^4}{4!}}{\frac{(2)^0}{0!} + \frac{(2)^1}{1!} + \frac{(2)^2}{2!} + \frac{(2)^3}{3!} + \frac{(2)^4}{4!}}$$

$$= \frac{\frac{16}{4 \cdot 3 \cdot 2 \cdot 1}}{\frac{1}{1} + \frac{2}{1} + \frac{4}{2 \cdot 1} + \frac{8}{3 \cdot 2 \cdot 1} + \frac{16}{4 \cdot 3 \cdot 2 \cdot 1}}$$

$$= \frac{0.67}{1 + 2 + 2 + \frac{8}{6} + \frac{16}{24}}$$

$$B(s, E) = \frac{0.67}{7} = 0.096 \approx 10\%$$

Box 5-4

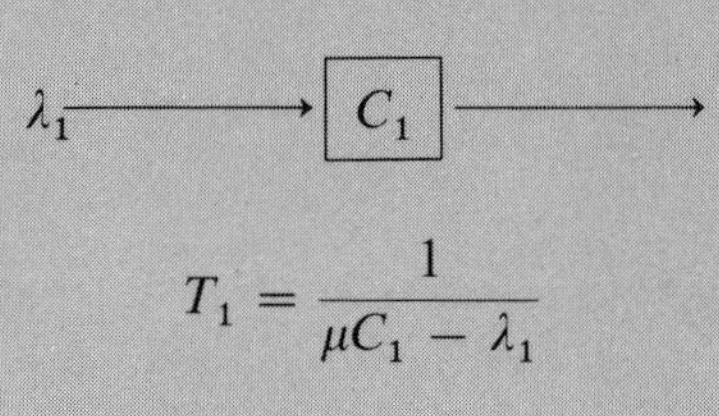

Box 5-5

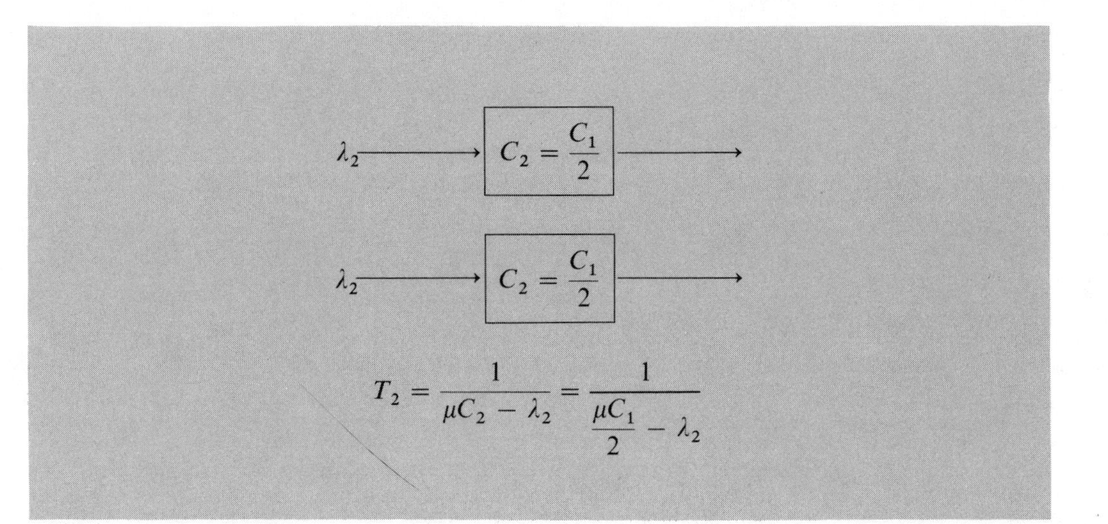

Box 5-6

If we then compute the new delay seen by the traffic arriving at either of the smaller systems, we find:

$$T_2 = \frac{1}{\frac{\mu C_1}{2} - \lambda_2} = \frac{1}{\frac{\mu C_1}{2} - \frac{\lambda_1}{2}} = \frac{2}{\mu C_1 - \lambda_1} = 2T_1$$

In other words, with the capacity and traffic each split in half, the delay through each of the smaller capacity systems is doubled compared to the delay through the original system. The delay seen by new arrivals in a queueing system is thus increased by splitting the total capacity into smaller systems, even if the total capacity remains constant.

LARGE SYSTEM (Box 5-5)

$$T_1 = \frac{1}{\mu C_1 - \lambda_1}$$

SMALL SYSTEMS (Box 5-6)

$$T_2 = \frac{1}{\mu C_2 - \lambda_2} \quad \text{where } C_2 = \tfrac{1}{2} C_1$$

$$T_2 = \frac{1}{\frac{\mu C_1}{2} - \lambda_2}$$

for equal delay: $T_2 = T_1$, thus:

$$\frac{1}{\mu C_1 - \lambda_1} = \frac{1}{\frac{\mu C_1}{2} - \lambda_2}$$

$$\mu C_1 - \lambda_1 = \frac{\mu C_1}{2} - \lambda_2$$

$$\lambda_2 = \lambda_1 - \frac{\mu C_1}{2} = \lambda_1 - \frac{\mu C_1 \rho_1}{2\rho_1}$$

$$\lambda_2 = \lambda_1 - \frac{\lambda_1}{2\rho_1} = \lambda_1\left(1 - \frac{1}{2\rho_1}\right)$$

$$\frac{\lambda_2}{\lambda_1} = \left(1 - \frac{1}{2\rho_1}\right)$$

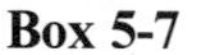

Box 5-7

Alternatively, we can attempt to insure that the delay through the two smaller systems is the same as the delay through the original system by reducing the amount of traffic arriving at the smaller systems. We can do this by writing the expression for the average delay in both of the arrangements (Boxes 5-5 and 5-6) and setting the two delay values equal. This mathematical exercise, which is illustrated in Box 5-7, shows that:

$$\frac{\lambda_2}{\lambda_1} = (1 - 1/2\rho_1) \qquad \textbf{(Eq. 5-9)}$$

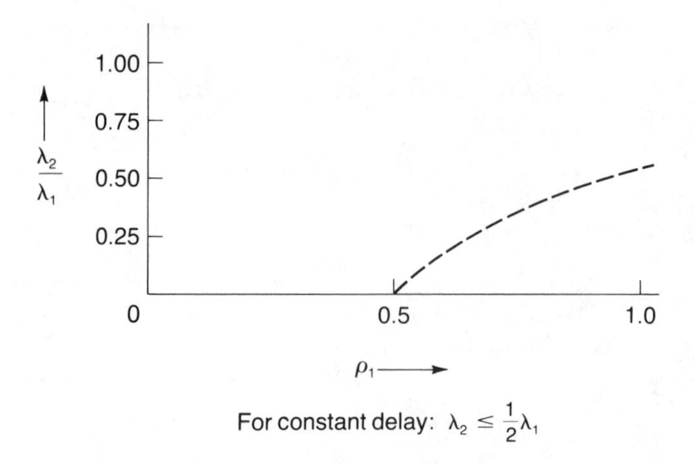

Figure 5-9. Plot of relative allowable traffic for each of two "half-capacity" systems to achieve the same delay as the "full-capacity" system.

This result says that, for the delay in the smaller systems to be equal to the delay in the larger one, the traffic permitted to arrive at the smaller systems has to be less than the amount of traffic arriving at the larger system, by an amount that depends on how heavily loaded the larger system originally was, as given by Equation 5-9. For clarity, Equation 5-9 is graphically portrayed in Figure 5-9. This shows that, if the traffic in the single system is sufficiently high that the original system is operating at an efficiency between 50% and 100%, we can achieve the same delay in the two reduced-sized systems by adjusting the traffic in the smaller systems to a small fraction of the total traffic in the larger system. The traffic through each of the smaller systems would then always be less than one-half the traffic through the larger system. However, if the larger system is operating at a loading of less than 50%, then it is impossible to achieve the same delay in either of the smaller systems, no matter how little traffic is allowed to enter them. In other words, for moderately loaded systems splitting the capacity into smaller systems always causes an increase in delay.

We have all observed this last result in everyday situations, when we have had to decide which line to get into at a bank or a post office. Invariably, we seem to choose the slowest moving line. In recent years many common user facilities have changed their operations slightly. Instead of separate lines in front of each server, there is now a single long line, laid out in serpentine fashion. Each person in line is directed to the first server to become free. Although the line seems much longer when we first join it, it moves much more quickly, and the average delay in getting served is generally reduced.

The source of the improvement in this situation is the one we have just been examining in the delay comparison. When separate lines form in front of each server, we in effect split the total capacity of the system into N separate smaller systems, each with the capacity of that individual server, C_o. When a single line is formed, with each new arrival directed to the first available server, we have in effect created a single, much larger system, with capacity of NC_o. As we can infer from inverting the reasoning in Box 5-7, the result is a substantial reduction in the average delay through the system with the larger, aggregated capacity.

SUMMARY

Whether we are considering blocking systems or queueing systems, the result is comparable. That is, by aggregating the individual demands of many users into combined larger average demands, we benefit not only from economy of scale but also from the higher operating efficiencies that can be achieved in statistically driven systems.

The ideas that have been presented in this chapter are a basic introduction into the complex mathematics of statistical systems. The behavior patterns of real-life systems will differ substantially from these paradigms. What is important is to recognize the fundamental principles and how they apply to the kinds of communications systems we are dealing with. As we move through the book, we will recall the following principles from time to time:

1. The fundamental measure of telephone traffic is the Erlang, which represents 100% utilization of a single facility over a stated measurement period. Erlangs can be computed from the average arrival rate of new calls multiplied by the average duration (holding time) of each arrival.
2. Circuit switched networks are generally characterized in terms of blocking probability, which can be computed by the amount of traffic in Erlangs and the number of facilities (lines) available for service.
3. In systems with temporary storage (or buffering) we can trade blocking for delay through the system. Delay is related to both available capacity and the efficiency with which facilities are utilized. As the utilization increases, so does the average delay.
4. At a specified level of blocking a larger group of lines can be used more efficiently than a smaller group of lines. Therefore, we should endeavor to aggregate communications demand. In other words, bigger is better.

SUGGESTED READINGS

The following four references provide an excellent selection for readers desiring a comprehensive, understandable treatment of basic queueing theory and its applications to telecommunications networks.

The chapters from the Schwartz and Kleinrock texts deal with basic derivations of queueing theory results and their application to examples of computer and data communications networks, including packet switched networks as well as other types of capacity and flow problems. The other two papers deal more extensively with the adaptation of queueing theory to the particular configurations of real-world networks. The paper by Tobagi et al. also presents various measurement techniques and results that can be applied to help validate the models in practice.

KLEINROCK, LEONARD. *Queueing Systems. Volume II, Computer Applications.* New York: John Wiley, 1976, Ch. 5.

KOBAYASHI, H., and KONHEIM, A. G. "Queueing Models for Computer Communications System Analysis." *IEEE Transactions on Communications*, vol. COM-25, no. 1 (January 1977), pp. 2–29.

SCHWARTZ, MISCHA. *Computer-Communication Network Design and Analysis.* Englewood Cliffs, N. J.: Prentice-Hall, 1977, Chs. 6–8.

TOBAGI, F. A., GERLA, M., PEEBLES, R. W., and MANNING, E. G. "Modeling and Measurement Techniques in Packet Communication Networks." *Proceedings of the IEEE*, vol. 66, no. 11 (November 1978), pp. 1423–1447.

6

Quantitative Tradeoffs—Packet, Circuit, and Message Switching

THIS CHAPTER:

will look a bit further at the operational characteristics of packet switching.

will explain the fundamental overhead structure of packet switching.

will compare circuit switching, packet switching, and message switching in terms of delay, loading, and overhead.

will examine a specific example of a large nationwide data network.

Before looking at some of the detailed quantitative comparisons and tradeoffs between packet switching and the commonly available alternatives, let us briefly recapitulate what we have already learned about the three principal forms of switched telecommunications. Box 6-1 summarizes the key operational characteristics of the switched networks. Circuit switching handles calls on the basis of one line per call in progress. There is a call establishment delay associated with the allocation of the needed resources to a new call, and, because of the competition for limited resources, blocking within the switches or communications lines can occur.

For message switching, each switch in the network stores the network messages in their entirety, making the service very reliable. Competition occurs between messages, but rather than traffic being blocked when all resources are occupied, traffic is delayed (stored) until capacity is available to deliver currently stored traffic. Since long messages and short messages tend to become mixed together on the lines and in the switches, the variance in the average delay is very large, even if the average delay is quite small.

Packet switching, which is, in large measure, a special case of message switching wherein the maximum message length is severely restricted, is also based on

OPERATIONAL CHARACTERISTICS OF SWITCHING TECHNIQUES

Circuit Switching	Message Switching	Packet Switching
Blocking	Storage	Queueing
One line per call	Delay variance	Logical multiplexing
Establishment delay		Packet transit delays

Box 6-1

the principles of queueing and blocked messages delayed. However, since individual packets have a short lifetime in the system, both average delay and delay variance can be kept quite small. Consequently, packet switching is ideally suited for transactional and interactive communications. The lines between the users and the switches can be used to perform logical multiplexing; in fact, the packet switched network appears to most users as a sophisticated time-division statistical **multiplexor.** As a result of the multiplexing function communications associated with a number of users can be combined on a single line. However, since the technology is based on queueing processes, packets are subject to delays, which must be controlled to prevent errors and faults.

In this chapter we will compare the switching techniques in terms of delay, processor loading, and system overhead. We will then combine these aspects in order to arrive at an overall cost and efficiency comparison of complete network designs. We will introduce some additional details related to the normal operation of packet switching, as well as some of the major possible fault conditions. We will talk fairly generally about these issues, leaving detailed discussion of network protocols until Part Two.

PACKET SWITCHING—OPERATIONAL DETAIL AND POSSIBLE FAULTS

Figure 6-1 illustrates a portion of a hypothetical packet switched network. User A is a subscriber attached to switch 1, and user B is a subscriber attached to switch 3. As an example, we shall trace the flow of a three-packet message from user A to user B, focusing on switches 1, 2, 3, and 4. It is important to remember, however, that many other packets, flowing between other users, are simultaneously moving throughout the network.

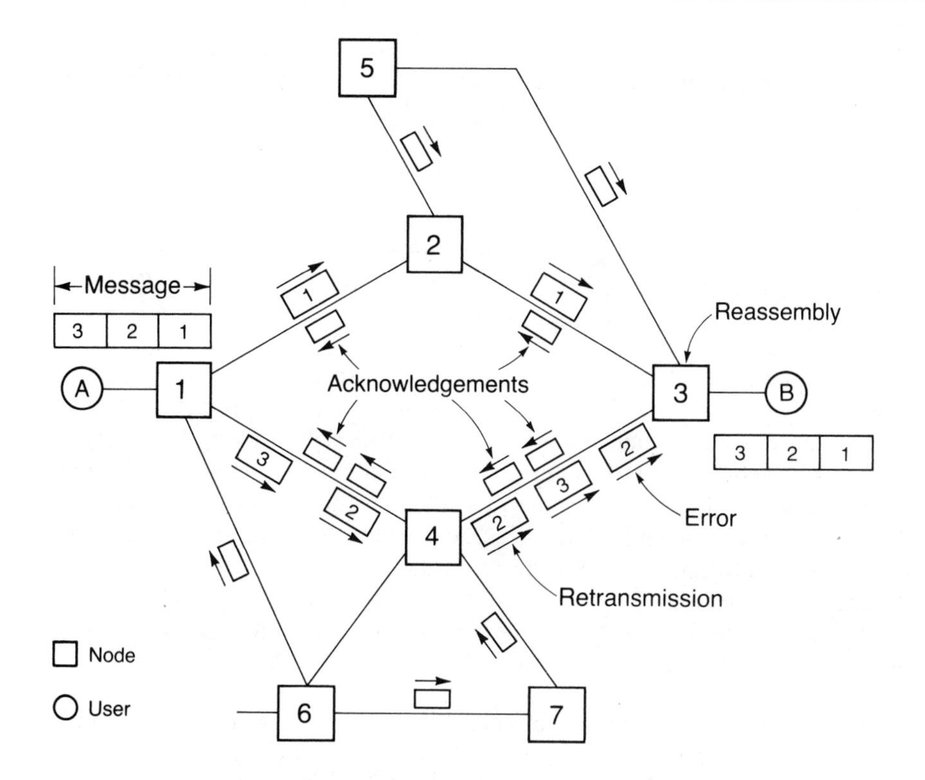

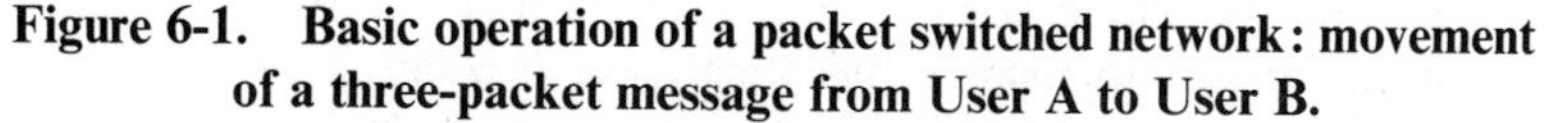
Figure 6-1. Basic operation of a packet switched network: movement of a three-packet message from User A to User B.

Basic Transmission Operation

The flow of the message is initiated by the transmission of packet 1 between user A and switch 1. When it fully receives the first packet, switch 1, following a set of routing rules, transmits packet 1 toward its destination by sending it via switch 2. In the meantime, packet 2 is moving from user A into switch 1. During this time the conditions in the network change (for instance, a large amount of traffic from switch 5 arrives at switch 2), so the second packet of the message from A to B is routed via switch 4. The third packet of the message, arriving at switch 1 soon after the second packet, is similarly routed via switch 4.

After being received correctly by switch 4, the second packet is transmitted to the destination switch, switch 3. But during that transmission an error occurs. When switch 3 receives packet 2, the error-detection mechanism is able to detect the error and requests a retransmission of packet 2. However, while this is occurring, packet 3 has been transmitted immediately behind the first (and errored) copy of packet 2. As a result, the second (correct) copy of packet 2 is received at switch 3 after packet 3. If we look at the network from the perspective of switch 3, first packet 1 is received, then packet 3, and finally packet 2. If switch 3 delivered the packets to the destination—that is, to user B—in the same order they arrived at switch 3, user B would receive the packets in a different order than they entered the network.

Network-Introduced Errors

Packet sequencing is but one of a number of possible network protocol–introduced problems that can occur in the packetizing process. The other two most serious problems are the undetected loss of a packet, and the duplication of a packet that has been successfully transmitted.

Packet Sequencing. The problem of packet sequencing is a direct result of the hold-and-forward mode of operation, along with the need to protect each transmission from network-introduced errors. Differential delays along the many paths through the network introduce the possibility that packets will be received out of proper sequence. For users to be protected from this form of error, the packets have to be reassembled into the same basic message structure they had upon initial transmission into the network. The process of packet reassembly is done at the destination switch—in this case, switch 3—using packet sequence information (such as a serial number) that must be carried through the network along with the user-introduced information.

Acknowledgements. A number of acknowledgements are shown in Figure 6-1, flowing on the various links in the network in the reverse direction from the information packets. These acknowledgement packets are the key to the error-detection mechanism that insures the integrity and accuracy of the transmitted data.

Any information packet that is properly received is immediately acknowledged back to the sender with one of these short acknowledgement packets. In this way the sending switch knows that the information packet has been received properly by the next switch along the path toward the destination. If an acknowledgement is not received within a certain time (known as the **timeout period**), the sending switch presumes that the packet was received in error and retransmits the packet. This presumption is necessary because a transmitted packet could be so badly garbled that the receiver could not even make enough sense from it to intelligently ask for a retransmission. If a packet is received with only a minor error, a negative acknowledgement, asking for a retransmission, avoids having to wait for the timeout period to elapse.

Lost Packets. Figure 6-2 shows the same network with a slightly different sequence of events. User A enters the first packet into the network via switch 1. Switch 1 routes packet 1 via switch 2, which receives packet 1 correctly and immediately acknowledges it. However, before packet 1 is transmitted from switch 2 toward switch 3, something goes wrong, and switch 2 fails. Having received an acknowledgement for packet 1, switch 1 is no longer concerned about it. However, on the

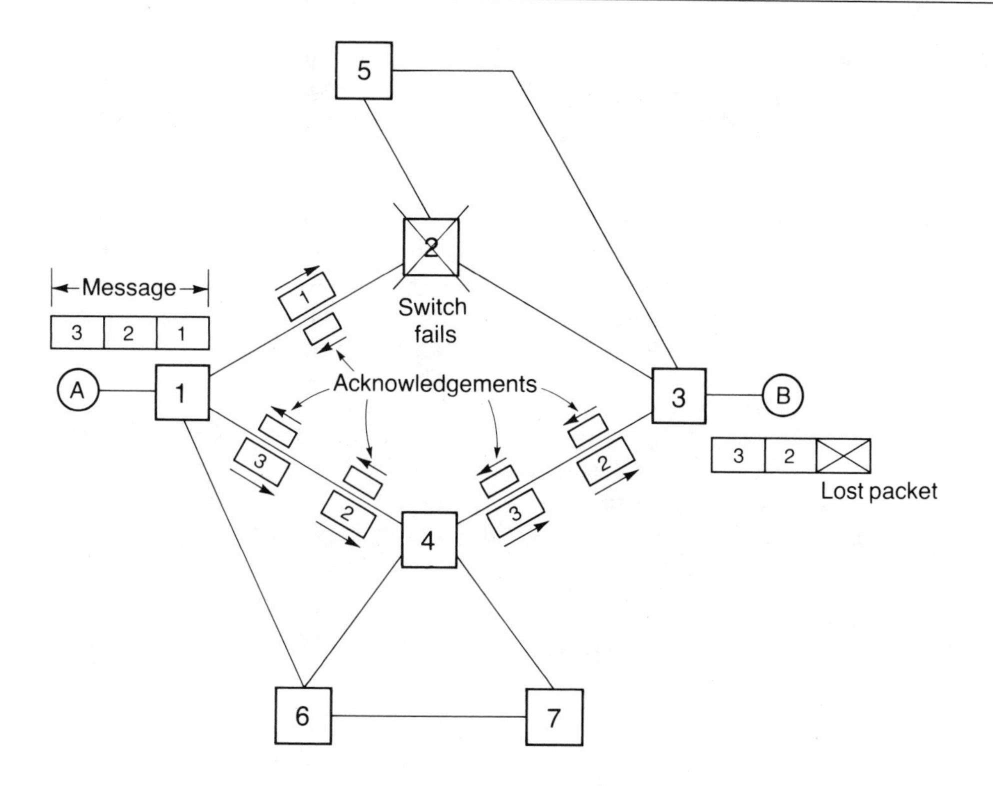

Figure 6-2. Basic operation of a packet switched network with a switch failure: loss of a packet.

failure of switch 2 the network routing plan changes the packet routing to avoid the failed switch. Thus packet 2 and packet 3 are routed via switch 4, and soon thereafter they are received properly at the destination switch, switch 3.

But what about packet 1? It was lost when the switch failed. Having acknowledged packet 1, switch 1 was no longer responsible for this packet. Yet switch 2 failed before it had a chance to relay packet 1 on through the network. From the point of view of user B, packet 1 is irretrievably lost; he receives only packets 2 and 3. The basic packet switching protocol has thus introduced the possibility of lost packets.

There are many ways for the network to protect against this problem. For example, a switch could be restricted from sending an acknowledgement until it has actually relayed the packet on. Or ultimate responsibility for the packet could rest with the originating switch. Or the sending user could be required to fill in missing (lost) packets at the request of the destination switch. We shall explore the advantages and disadvantages of these and other possibilities further in Part Two. The key point now is that the operation of the packet network requires the layering of multiple protective checks and protocols.

Duplication of Packets. In Figure 6-3 another set of conditions is introduced. Here again, packet 1 is routed via switch 2. Packet 1 is received correctly by switch 2 and is immediately acknowledged. However, just as the acknowledgement for packet 1 leaves switch 2, the line from switch 2 back to switch 1 fails, in the process destroying the acknowledgement for packet 1. Not receiving an acknowledgement for packet 1 within the timeout period causes switch 1 to retransmit packet 1. Having detected that the line between switch 1 and switch 2 has failed, the network routing plan naturally sends the retransmission of packet 1, together with the transmission of packets 2 and 3, via switch 4.

In the meantime the first transmission of packet 1 was really received properly by switch 2, even though the acknowledgement of that first copy was never received by switch 1. Having no way to know about this, switch 2 relays packet 1 to the destination switch, switch 3. Soon thereafter comes the second copy of packet 1, together with packets 2 and 3. User B is therefore likely to see a duplicate packet 1 together with the rest of the message in its entirety.

Here again, there are ways to avoid the problem of duplicate packets. Duplication may not even be a big problem to user B, unless the duplicate packet

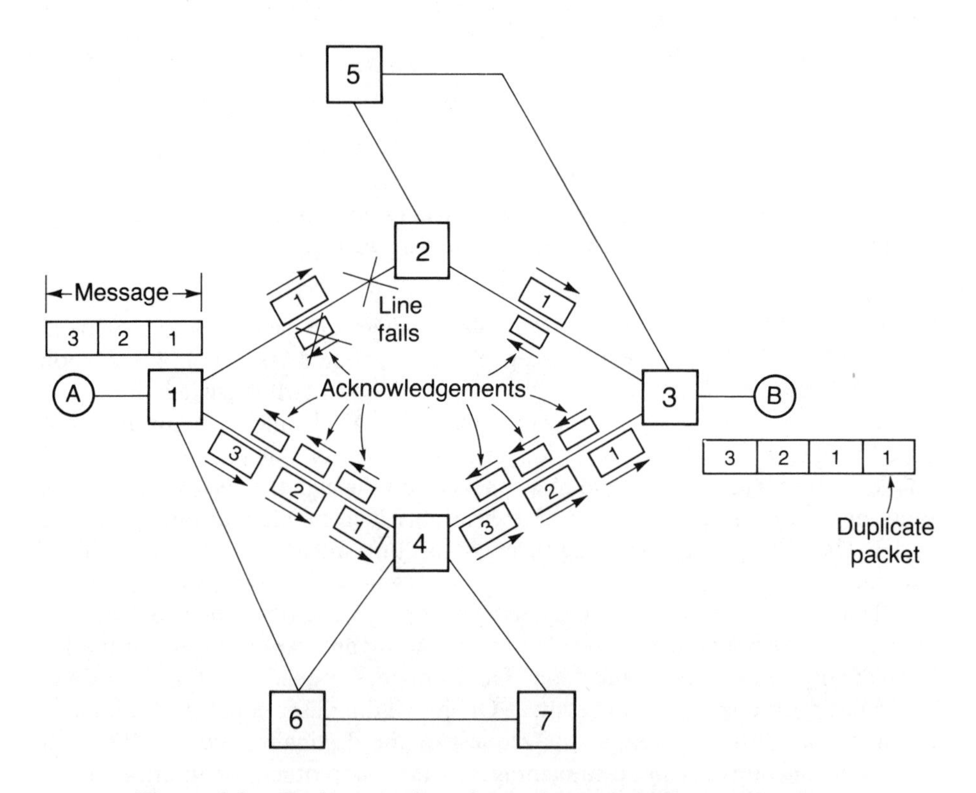

Figure 6-3. Basic operation of a packet switched network with a line failure: duplication of a packet.

happens to be something like a second $500 withdrawal from a checking account or a second $200 charge added to a charge account. A simple protocol protection might be to acknowledge successful acknowledgements. But this may mean acknowledging the acknowledgements of acknowledgements, ad infinitum. Again, the key point is that we must be prepared for the protocol's inherent ability to cause the creation of duplicate packets.

In order to protect against the several network-introduced errors we have examined, the network has to contain a certain amount of redundant **overhead** information. Overhead is the information that has to be transmitted through a network in order for the switches to properly handle the transmission and delivery of a call or message. Overhead takes different forms, depending on the type of network and the switching techniques employed. This information requires a percentage of the overall network capacity. The capacity is, in effect, dedicated to the operation of the network and thus is not available for the transmission of user (revenue-producing) information.

PACKET SWITCHING—OVERHEAD STRUCTURE FUNDAMENTALS

Overhead information exists in two basic forms: (1) appended to each user packet, and (2) in the form of self-contained acknowledgement packets or control packets flowing among the switches.

Protocols in Packets and Segments

Figure 6-4 is a general representation of the overhead structure associated with user packets. User transactions or messages can be of arbitrary length, ranging from just a few up to many millions of bits. Because of the way packet

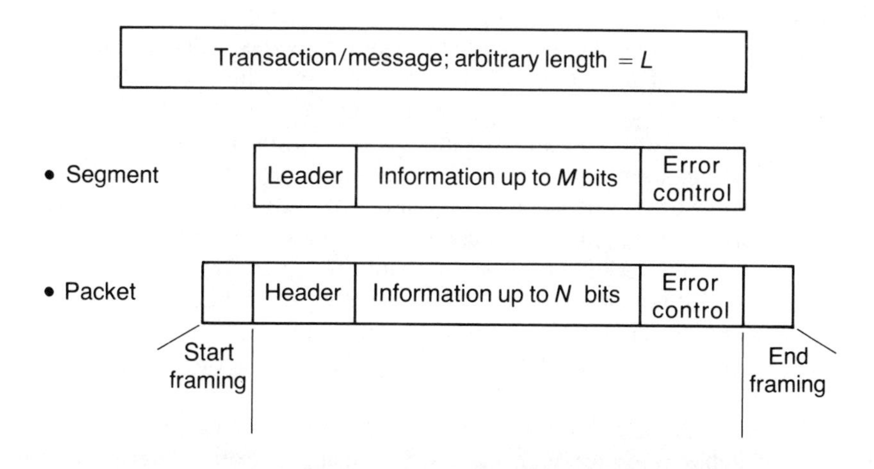

Figure 6-4. Generalized packet switching overhead structure.

networks operate, it is generally not possible—nor would it be advantageous—to transmit the entire message, of length L bits, at one time. The user-network interface protocol restricts the user to message or transaction **segments,** which have a maximum length specification of up to M bits. Within the network the lines and switches exchange packets, each with a maximum length specification of up to N bits. Depending on the design of the network protocols, the value of M and N may be equal, or M may be significantly larger than N. (If M is larger than N, it will usually be an integer multiple of N.) Of course, since the user messages or transactions are arbitrarily long, clearly L is larger than either M or N. Expressing this situation mathematically, we can say:

$$L > M \geq N$$

If the network protocol requires that M and N be equal, we have what is called a single-packet-per-segment protocol. That is, each segment introduced into the network creates just a single packet. All messages or transactions longer than a single packet have to be divided by the user into single packet–length segments. When M is permitted to be larger than N, we have a multiple-packet-per-segment protocol. User segments of length M that arrive at the network are then divided by the switches into packets of information length N. The advantage of a multiple-packet-per-segment protocol is that, for user messages of length M bits or less, the packetizing function is transparent. That is, if the user always restricts his messages to less than M bits, he does not have to divide the messages into smaller subunits.

It is also possible to conceive of protocols where the user-to-network segment length is arbitrarily large—where, in effect, M and L are permitted to be equal. In such a case, the packet network is actually capable of emulating a full-period, point-to-point circuit, with continuous transmission across the network a possibility.

Types of Protocols. There are examples of each of these types of protocols in actual operational networks, and popular terminology has evolved to categorize the differing modes of operation. Single-packet-per-segment protocols have generally been termed **datagram** networks, where each individual packet is a discrete entity. Multiple-packet-per-segment protocols, or packet networks that operate on arbitrarily long user inputs without user segmentation, are termed **virtual circuit** packet networks. Virtual circuit implies that, although the internal operation of the network is based on packets, to the end user the network is indistinguishable from a **full-period,** end-to-end circuit. The packetized operation must be essentially invisible to the user, with data coming out of the network in exactly the same sequence it went in. Though true virtual circuit operation is quite difficult to achieve in practice, the general concept is useful in packet network operation. We will discuss it in more detail in Chapter 7.

In any case, the user interfaces with the network switches, using segments, while the switches interface with each other, using packets. The information contained in the segments and packets may or may not be the same, depending on the protocol in use.

Protocol Information. If we look again at Figure 6-4, we see that the user's information segment consists of a segment leader, information field, and error-control block. The **leader** contains the destination address to which the segment is to be delivered, together with control information required by the network. Examples of this control information are the segment sequence number, **logical channel number** used to separate user information that has been multiplexed together, designation of the first or last segment of a transaction, and a wide range of protocol information related to the user-to-user control of the circuit.

The packets, which flow among the switches, contain framing patterns to designate the beginning and end of each packet, a packet header, information field, and an error-control block. The packet **header** contains all the same information that the segment leader contained, plus other information that the switches need to control the movement of the packets through the network. Examples of additional network information in the packet headers are source address, packet sequence number, and control blocks (to prevent looping, loss, or duplication of packets).

The error-control blocks use principles of mathematical coding to allow the switches to recognize if any single bit or group of bits in the packet are in error due to electrical interference. The error-control process is part of the detailed operational protocols.

The amount of information contained in the segment leaders and packet headers is significant. Typically, anywhere from 64 to 256 bits of total overhead information is required with each packet. If we consider packet networks with the value of N equal to 1000 bits, we can see that some packet overheads may be as much as 25% of the total data transmitted. This 25% of the network capacity is not available for revenue-producing movement of user information through the network, so the percentage of overhead is of concern in comparisons of the efficiency of switching techniques.

An Example of Overhead Structure. To clarify overall overhead structure, consider the example in Figure 6-5. The user's message is 70,000 bits in total length ($L = 70{,}000$). The network segment length is shown as 8192 bits ($M = 8192$), and the packet length is 1024 bits ($N = 1024$). On the basis of this structure the message is divided into nine segments. Each of the first eight segments is filled with 8192 bits, and the remaining 4464 bits are in the ninth segment. Each of the first eight segments is divided by the network switches into eight packets, each with 1024 bits. The ninth segment is divided into five packets, the first four containing 1024 bits each. The fifth, and final, packet of the message contains the remaining 368 bits of data.*

---

* If these numbers seem strange choices for an example, they are based on numbers that are Modulo 2. Since computers deal most conveniently with binary numbers composed of ones and zeros, and since computers are commonly built with structures that are multiples of 16 or 32 bits at a time, network structures try to conform to units of information based on the number series of 2^N, where $N = 1, 2, 3, \ldots$ etc. For example, $2^4 = 16$; $2^8 = 256$; $2^{10} = 1024$; etc.

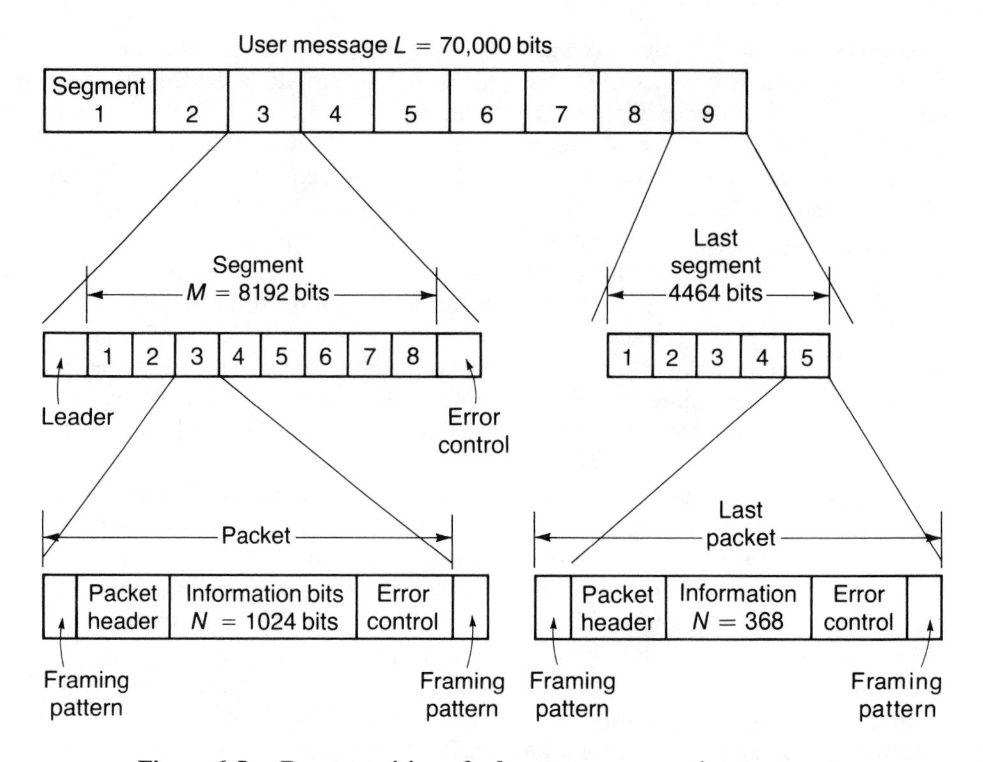

Figure 6-5. Decomposition of a long user message into segments and packets.

We are now in a position to apply quantitative analysis to the different switching techniques.

DELAY COMPARISON—IDEAL CASE

The first quantitative comparisons we shall look at are concerned with the delay of transmitting a long message through a switched communications system. We shall start by looking at an ideal case. By ideal, we mean that we will not be concerned with the overhead portions of the communications; we will deal with these separately later in the chapter. For circuit switching, neglecting overhead means that we do not consider the time it takes the switches to establish the path through the network as part of the message transmission delay. We measure the delay from the time the switches complete the call establishment phase. For message and packet switching neglecting overheads means that the bits associated with the packet headers and error control are not counted as part of the capacity utilizing data flows.

The General Configuration

Figure 6-6 presents the generalized network configuration and the key parameters we will consider in the delay comparison. The general case consists of a message that is transmitted between two users in the network. The message enters through the originating switch, passes through tandem switches, and reaches the destination user via the destination switch.

The total number of switches in the user-to-user path will be represented by N, and in the expressions we derive we will not place any limitation on the value of N. If we let N be equal to 1, the originating and terminating subscriber are homed on the same switch. If N is equal to 2, the message passes from the originating switch to the terminating switch without passing through any tandem switches. Values of N of 3 or more imply an increasingly large number of tandem switches.

In Chapter 5 we found that the delay through a single-server queueing system was given by Equation 5-8 as:

$$T = \tau/(1 - \rho)$$

where

$$\tau = K/C = \frac{\text{average message length}}{\text{server capacity}}$$

and

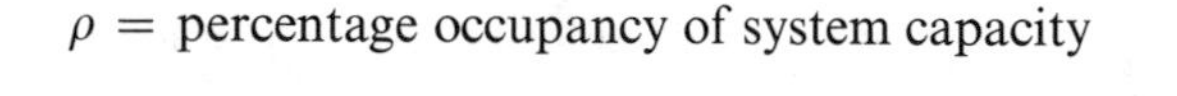

$$\rho = \text{percentage occupancy of system capacity}$$

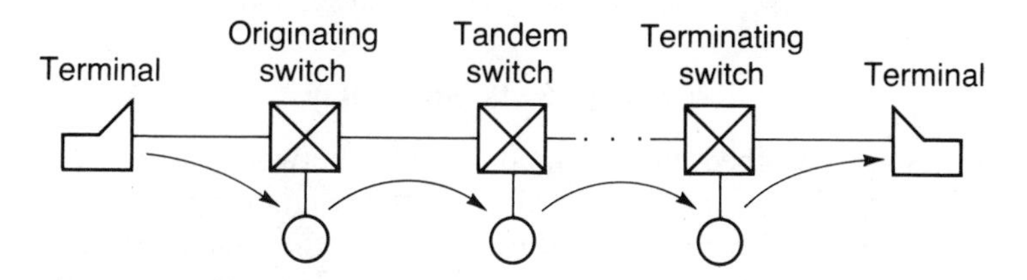

D_{cs} = Circuit switch delay (seconds)

D_{ps} = Packet switching delay (seconds)

D_{sf} = Message switching (store-and-forward) delay (seconds)

K = Message length (bits)

P = Packet length (bits)

C = Line transmission rate (bits/second)

N = Number of switches in path

Figure 6-6. Switching configuration and key parameters for delay comparison of circuit, packet, and message switching.

The situation shown in Figure 6-6, for both message and packet switching, is, in effect, a sequence of single-server queueing systems in each of the switches. We can simplify the analysis further by assuming that all the systems are very lightly loaded, which would be represented by values of ρ close to zero. In that case the delay in each element of the network would be given by the service time alone. Thus, in general, the delay will be approximated by:

$$T = \tau = K/C$$

since

$$\rho \approx 0$$

For the general case that we are considering we are going to determine the total delay through the system from the time the first bit leaves the sending terminal until the last bit is received by the receiving terminal. We are going to consider a long message of total length K; for the packet switching case we will break the message into a number of packets, each of length P.

Circuit Switching Delay

In a circuit switched network the delay we are considering is measured after the switches establish the end-to-end path between the user terminals. The delay from the beginning of the first bit transmitted to the end of the last bit received then can be determined only by the transmission rate and the total message length as:

$$D_{cs} = K/C$$

For example, if the message is 1,000,000 bits, and the lines have capacity of 50,000 bits/second:

$$D_{cs} = 1{,}000{,}000/50{,}000 = 20 \text{ seconds}$$

Message Switching Delay

In a message switched, or store-and-forward, network the entire message is transmitted from the originating terminal to the originating switch and stored there; then relayed from the originating switch to the first tandem switch and stored there; then relayed to the next switch; and so on, until the message finally reaches the terminating switch. At that point it is finally stored and relayed to the destination terminal.

Each time the message is relayed, it incurs a delay equal to:

$$D = K/C$$

If there are a total of N switches between the two user terminals, the message is actually relayed or transmitted a total of $N + 1$ times. The total delay for the message switched case is thus:

$$D_{sf} = \frac{K}{C}(N + 1)$$

For example, if the message is 1,000,000 bits long, the lines have capacity of 50,000 bits/second, and there are four switches in the system, the total store-and-forward delay is:

$$D_{sf} = \frac{1{,}000{,}000}{50{,}000}(4 + 1) = 100 \text{ secs}$$

Remember that this approximation assumes negligible processing and storage time in each store-and-forward switch, and light loading of the network. Therefore, there is essentially no queueing delay at each switch.

Packet Switching Delay

In the case of packet switching the message is divided into packets, each P bits in length. The first packet is transmitted from the originating terminal to the originating switch, which then relays the first packet to the next switch, which relays it to the subsequent switch, and so on, until the first packet reaches the terminating switch. However, while the first packet is being relayed by the originating switch, the second packet is being transmitted from the originating terminal into the originating switch. As soon as the originating switch has completed relaying the first packet, it should have completely received the second packet. It can begin relaying the second packet while it is receiving the third packet, and so on.

First let us look at the delay seen by the first packet. The first packet is P bits long and is transmitted at C bits per second. Each relay then takes P/C seconds, and, like the store-and-forward case, there are $N + 1$ relays between the originating and terminating users. The delay seen by the first packet is thus:

$$D_1 = \frac{P}{C}(N + 1)$$

After the first packet arrives at the destination terminal, there are $K - P$ bits of the message yet to be delivered. However, if the network has been operating properly, and there has been no queueing delay at the intermediary switches, the second packet should be at the terminating switch ready for final transmission at the same time that the first packet has been received by the destination terminal. Furthermore, this should be the case with each subsequent packet, so that the remaining $K - P$ bits should flow out of the network with no further delay or interruption. The delay for the remaining $K - P$ bits after the first packet is received is thus:

$$D_2 = (K - P)/C$$

The total packet switching delay is thus:

$$D_{ps} = D_1 + D_2 = \frac{P(N + 1) + (K - P)}{C}$$

For example, if the message is 1,000,000 bits long, the lines have capacity of 50,000 bits/second, there are a total of four switches in the path, and the packet size is 2000 bits, the delay for the packet switched network is:

$$D_{ps} = \frac{2000(4 + 1) + (1{,}000{,}000 - 2000)}{50{,}000}$$

$$= \frac{10{,}000 + 998{,}000}{50{,}000} = \frac{1{,}008{,}000}{50{,}000}$$

$$D_{ps} = 20.16 \text{ seconds}$$

From the examples we can see that the delay (when we neglect call set-up time and overhead) is approximately equal for circuit and packet switching, and that the delay of the message switched system is considerably larger.

The Delay Comparison Summarized

In Box 6-2 the three delay expressions we have just derived are summarized. In Box 6-3 we perform some mathematical manipulations on these expressions in order to compare them more easily.

From the results shown in Box 6-3 we see that the packet switching delay is approximately equal to the delay in the circuit switched network. In addition, the packet switching delay is approximately $N + 1$ times smaller than the delay through the equivalent message switched network. Since N must always have a value of at least 1 to be a switched network, packet switching is always at least

SWITCHING DELAY—NEGLECTING OVERHEAD:

Circuit Switching

$$D_{cs} = \frac{K}{C}$$

Message Switching

$$D_{sf} = \frac{K}{C}(N + 1)$$

Packet Switching

$$D_{ps} = \frac{P}{C}(N + 1) + \frac{K - P}{C}$$

$$D_{ps} = \frac{P(N + 1) + (K - P)}{C}$$

Box 6-2

Start with packet switching delay:

$$D_{ps} = \frac{P(N+1) + (K-P)}{C} = \frac{PN + P + K - P}{C}$$

$$= \frac{K + PN}{C} = \frac{K}{C}\left(1 + \frac{PN}{K}\right)$$

But:

$$D_{sf} = \frac{K}{C}(N+1) \quad \text{or} \quad \frac{K}{C} = \frac{D_{sf}}{N+1}$$

and

$$D_{cs} = \frac{K}{C}$$

thus:

$$D_{ps} = \frac{D_{sf}}{N+1}\left(1 + \frac{PN}{K}\right)$$

and:

$$D_{ps} = D_{cs}\left(1 + \frac{PN}{K}\right)$$

In a packet network where the packet length is much smaller than the maximum allowable message length, and where N is relatively small, then:

$$\frac{PN}{K} \ll 1$$

thus:

$$D_{ps} = \frac{D_{sf}}{N+1}\left(1 + \frac{PN}{K}\right) \approx D_{ps} \approx \frac{D_{sf}}{N+1}$$

$$D_{ps} = D_{cs}\left(1 + \frac{PN}{K}\right) \approx D_{ps} \approx D_{cs}$$

Box 6-3

twice as fast as message switching unless the message is shorter than one packet. In the latter case the delays would be approximately equal.

Furthermore, neglecting call set-up time, the delays through the packet network are approximately the same as through a circuit switched network. Therefore, the time it takes the circuit switches to select and establish a physical path becomes a key characteristic that separates circuit and packet network delays.

In Figure 6-7 the values of delay for each of the switching techniques are plotted as a function of message length. The parameters used are similar to the

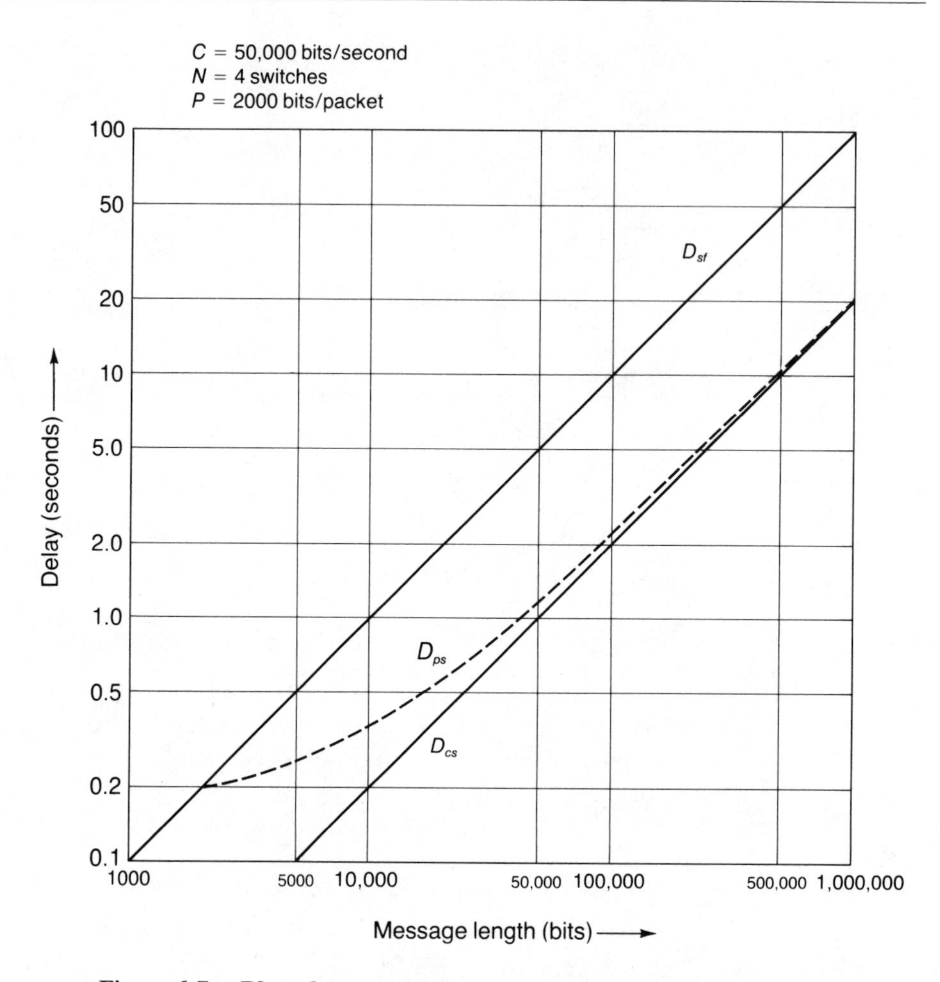

Figure 6-7. Plot of message delay versus message length for various switching techniques.

examples—that is, 50,000-bit/second lines, four switches in the path, and 2000-bit packets. For messages shorter than one packet in length, the message and packet switching delays are equal. As the message length becomes longer than one packet, the packet switching delay becomes increasingly closer to the circuit switching delay, as our approximation in Box 6-3 shows.

DATA TRAFFIC IMPACT ON PROCESSOR LOADING

Next we will look at the impact of computer and data terminal communications on a conventional circuit switched network. By conventional circuit switched network we mean one that is basically configured to handle voice communications, but which can operate sufficiently quickly that individual calls are originated for each new data transaction. The call set-up delay is negligible compared to the actual message transmission delay.

The processor loading impact computations recognize that the switches do most of their work during the process of establishing a call. Once the circuit has been set up between the users, the circuit switch has little to do for that call except detect when the call has been completed so that the disconnect process can take place.

Defining the Situation

Figure 6-8 is a simplified picture of the circuit switched network environment and the key parameter definitions we shall be using. We can recall from the discussion of traffic and blocking in Chapter 5 that the total traffic that can be carried is related to the number of Erlangs of available capacity. We can estimate the total traffic carried by multiplying the number of calls completed by their average length. This traffic is what produces revenue for the network owner. On the other hand, the number of calls to be originated is the main factor in the work performed by the switches. We shall attempt to derive a relationship that shows the impact of introducing a small amount of data traffic into a circuit switched network designed primarily for voice traffic.

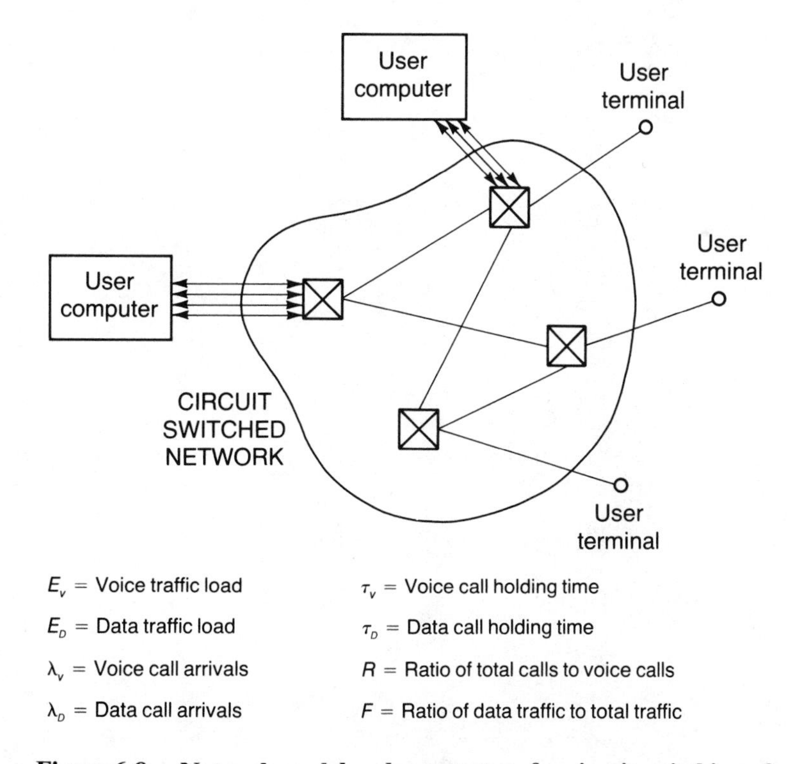

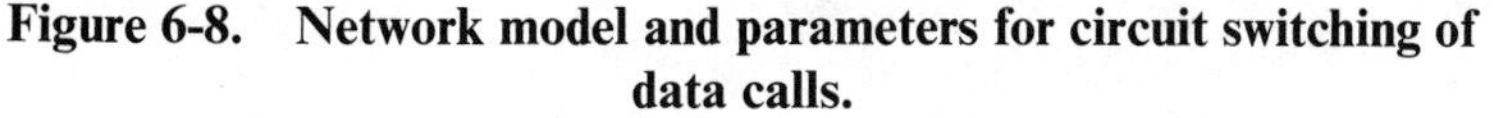

Figure 6-8. Network model and parameters for circuit switching of data calls.

Deriving and Evaluating the Relationships

The derivation of such a relationship is shown in Box 6-4. The call processing ratio, R, is the ratio of the total voice and data call arrivals divided by the voice arrivals alone. The data traffic ratio, F, represents the percentage of total revenue-producing traffic that results from data communications transactions. Given these basic definitions, we can perform algebraic manipulations, resulting in the relationship:

$$R = 1 + (\tau_v/\tau_D)(F/(1 - F))$$

In Box 6-5 this rather difficult to visualize relationship is evaluated for a range of values. In order to evaluate this expression, we have to look at plausible values for the voice and data holding times, τ_v and τ_D. For simplicity we will assume a

E_v = voice traffic = $\lambda_v \tau_v$

E_D = data traffic = $\lambda_D \tau_D$

$$R = \text{call processing ratio} = \frac{\lambda_v + \lambda_D}{\lambda_v} = 1 + \frac{\lambda_D}{\lambda_v}$$

$$F = \text{data traffic ratio} = \frac{E_D}{E_D + E_v} = \frac{\lambda_D \tau_D}{\lambda_D \tau_D + \lambda_v \tau_v}$$

$$F = \frac{1}{1 + \dfrac{\lambda_v \tau_v}{\lambda_D \tau_D}}$$

or

$$\frac{\lambda_v \tau_v}{\lambda_D \tau_D} = \frac{1}{F} - 1 = \left(\frac{1 - F}{F}\right)$$

$$\frac{\lambda_v}{\lambda_D} = \frac{\tau_D}{\tau_v}\left(\frac{1 - F}{F}\right)$$

or

$$\frac{\lambda_D}{\lambda_v} = \frac{\tau_v F}{\tau_D (1 - F)}$$

thus:

$$R = 1 + \frac{\tau_v F}{\tau_D (1 - F)} \approx 1 + \left(\frac{\tau_v}{\tau_D}\right)\left[\frac{F}{1 - F}\right]$$

Box 6-4

$$R = 1 + \frac{\tau_v}{\tau_D}\left[\frac{F}{1 - F}\right]$$

$$\tau_v \approx 300 \text{ seconds}$$

$$\tau_D > \frac{20}{9600} \approx 0.002 \text{ sec}$$

$$\tau_D < \frac{10{,}000}{300} \approx 30 \text{ secs}$$

$$10 \leq \frac{\tau_v}{\tau_D} \leq 150{,}000$$

$\frac{\tau_v}{\tau_D} = 10$

F	R
0.1	2.11
0.2	3.5
0.5	11.0
0.9	91.0

$\frac{\tau_v}{\tau_D} = 150{,}000$

F	R
0.1	1.67×10^4
0.2	3.75×10^4
0.5	15.0×10^4
0.9	135×10^4

Box 6-5

value of 300 seconds for the average voice call holding time; thus $\tau_v = 300$. Estimating τ_D is difficult because of the wide variety of possible data applications. However, we can define a range of possible values for data call holding time. At the low end the shortest logical length might be associated with a message of nominally 20 bits transmitted at a 9600-bit/second data rate. At the other extreme we may have data messages of 10,000 bits (a full CRT screen of data) at a 300-bit/second data rate. The logical values of τ_D thus range from $20/9600 = 0.002$ sec to $10{,}000/300 = 30$ secs. Using the value of 300 seconds for τ_v, the resultant values for the ratio of τ_v/τ_D is from 10 to 150,000.

The two tables in Box 6-5 show the relationship of call processing ratio and traffic ratio at either end of this range of parameters. For example, at a value of $\tau_v/\tau_D = 10$, if only 10% ($F = 0.1$) of the traffic in the circuit switched network is data traffic, the call processing load (R) on the switches is 2.11 times greater than for voice calls only. If 50% of the traffic is data, the call processing load is 11 times greater. The ratio of $\tau_v/\tau_D = 150{,}000$ produces extreme values, which are totally unachievable. For example, if only 10% of the traffic were data, the call processing load would increase by more than 16,000 times.

In Figure 6-9 the relationship of F and R is plotted for the two different values of τ_v/τ_D. A precise value for this ratio is difficult to predict for an average network, but actual values will certainly lie between the two curves plotted in Figure 6-9.

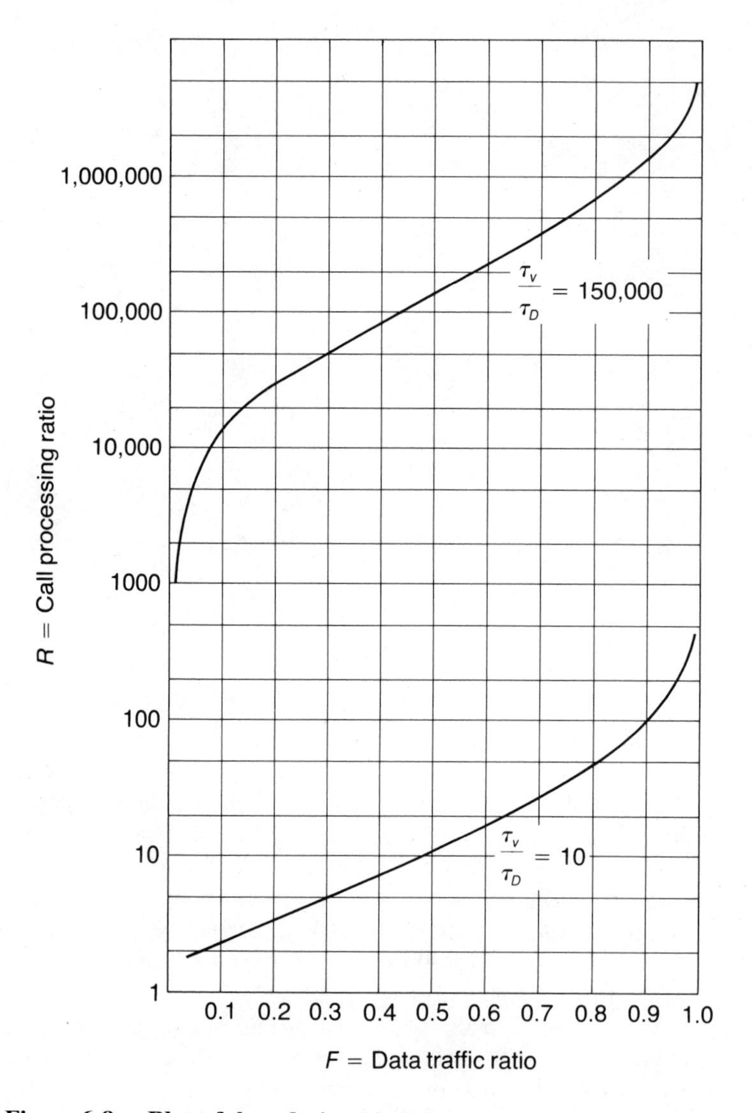

Figure 6-9. Plot of the relationship between call processing loading and data traffic percentage.

The Results Summarized

These results highlight the fact that even moderate data communications traffic loads applied to a conventional circuit switched network will cause a very large increase in the amount of processing the network switches must perform. This is because of the much shorter average length of the data communications calls, even at the upper extreme of a full CRT screen of data sent over a fairly low-speed line. The picture improves when we deal with very long messages, commonly associated with the data exchanges between computers. However, across a large, nationwide network the quantity of data calls coming from individual user terminals will likely far exceed the number of large computer messages.

Even if the data circuits can be established in a short enough time that the efficient flow of data communications is not interrupted, the load on the circuit switches would have to be accommodated in the fundamental design of the switch processors. Switches presently used in voice-based circuit switched networks are not capable of either the call connection speed or the processor loads necessary for efficient handling of data communications calls. As a result, data calls are generally handled in a highly inefficient way: A data circuit through a circuit network is held over an extended period of time. During the total connection only a small part of the total transmission capability is actually used. The price of this inefficiency is borne by both the user and the supplier of the communications services.

OVERHEAD IN CIRCUIT SWITCHING AND PACKET SWITCHING

In analyzing delay and processor loading, we did not consider the overhead associated with the various switching techniques. We shall now compare the relative capacity utilization attributable to the overhead portions of each switching technique.

Components of Overhead

For circuit switching the overhead consists of the call destination (e.g., the telephone number), which has to be relayed from switch to switch. Each switch, in turn, selects a route and transmission channel for the connection. Upon termination of the call, similar signals have to be transmitted to instruct the switches to perform the disconnect function and transmit any billing or cost information to the originating toll office. The key factor in circuit switched networks is that the overhead information is independent of the length of the message or the holding time of the call.

In packet switched networks every packet transmitted contains a fixed amount of overhead, which is appended to the information portion of the packet. The overhead consists of the packet header, the error control, framing patterns at the beginning and end of each packet, and any other control information. In addition, other forms of packet switching overhead are transmitted apart from the user's information—control messages, routing table changes, and acknowledgements of other packets.

The total amount of overhead associated with a particular message or transaction is proportional to the length of the message. If a message or transaction is long, it has to be broken into a large number of packets, each packet containing a fixed amount of overhead. As the message length increases, so does the number of packets that are needed to transmit the message, and therefore the total amount of overhead. The overhead in a packet network, especially if most of the packets are filled to the maximum number of allowable bits, can be estimated as a percentage of the total useful information capacity of the network. For example, if the maximum packet size is 2000 bits, and the packet header, error control, flags, and acknowledgements average 250 bits per packet transmitted, we can say that the overhead is 250/2250 or 11.1%. In other words, 11.1% of the useful capacity of the network is consumed in transmitting the network overhead information.

A Quantitative Comparison

We can now make the quantitative comparison of overhead in circuit switching and packet switching illustrated in Figure 6-10. The overhead is plotted as a function of message length (in bits). The horizontal lines, indicating constant overhead regardless of message length, represent various circuit switching cases. The lines tending upward to the right represent the packet switching cases, where

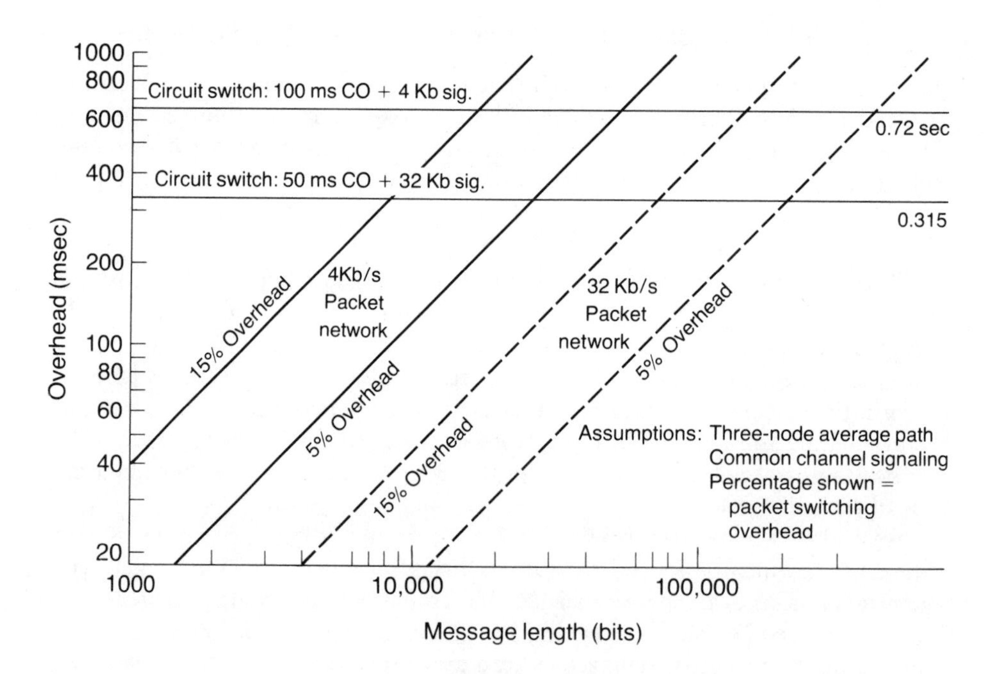

Figure 6-10. Plot of transmission overhead versus message length for circuit and packet switching.

the total amount of overhead increases linearly with increasing message length. In order to make this comparison on a common set of axes, it is necessary to convert the overhead in the packet switched network (which is measured in bits per second of network capacity) to a measure compatible with circuit switching (where the overhead is measured in seconds of time during which capacity is unavailable while the circuit is being established).

The compatible measure chosen for Figure 6-10 is time attributable to the overhead functions. For the circuit switched network this is computed as the time it takes for the originating user to send the destination address (telephone number) to the originating switch. Added to this is the time it takes for the switch to process the call information, select a route, find an available channel, establish the circuit, and relay the call information to the next switch along the path to the destination. This process is repeated at each switch along the path, until the destination is reached.

Present techniques generally require 5 to 15 seconds to complete the activities needed to establish a connection. If overhead times of this order were used for comparison, packet switching would be more overhead efficient for all but the very longest data messages. In order to make a realistic comparison, we will hypothesize a new generation of circuit switching, using very fast switch processors and high-speed interoffice digital switching. Figure 6-10 uses two specific sets of parameters, in each case assuming that three switches are used in the path between the subscribers.

In one case shown the switches are estimated as being able to process the call and perform the switching function in 100 milliseconds (0.1 second). These switches are teamed with a 4000-bit/second digital signaling channel. In the second case the processing time of the switches (also known as the switch **cross-office time,** CO) is assumed as 50 milliseconds (0.05 second). These higher speed switches are assumed to be interconnected using 32,000-bit/second digital signaling channels. The overhead time is computed using an 80-bit signaling message to represent the number of the destination; it is measured from the initiation of signaling to the first switch until the destination switch has completed the circuit and begun ringing.

This time is overhead because, as soon as signaling has begun, resources of the network become allocated to this call, even though additional time must elapse before actual transmission of data can begin. This is idle time of the network facilities—time that cannot be used for revenue-producing service. In addition, at the end of a call a similar time is lost while the switches process the disconnect order and signal each other terminating the call. Consequently, the overhead times shown for the circuit switched cases in Figure 6-10 are twice the call connect time. The computed values are approximately 720 milliseconds for the 100 ms switches and approximately 315 milliseconds for the 50 ms switches with the higher speed signaling.

The overhead computation for the packet switched networks is based on dividing a fixed percentage of the message length by the channel rate to estimate the equivalent amount of time lost to actual user data transmission. For example, look at the line representing a packet network operating at 15% overhead, using 32,000-bit/second lines. For a 100,000 bit message, there would be 15,000 bits of overhead. 15,000 bits at 32,000 bits/second would take approximately 500 milliseconds (0.5 sec) of time on the network resources. This 15% overhead, 32 Kb/sec line thus passes through 500 milliseconds on the vertical axis. The other plots for the packet switched cases are drawn similarly.

The implication of this comparison is that, over a wide range of message lengths and assuming a new generation of circuit networks, packet switching utilizes less total overhead than circuit switching. Even under the conditions most favorable to circuit switching, with very high-speed switches and signaling, and a rather low-speed packet network, messages less than about 8000 bits in length would have less total overhead using packet switching than circuit switching. Using more realistic figures puts the crossover message length on the order of 100,000 bits. Given present actual parameters, the crossover point falls at a 1 million-bit or more message length.

A NATIONWIDE DATA NETWORK

Let us now look at a hypothetical case—a large, common user network for data communications, in an operational environment where about one-half the traffic is short and bursty and the other half is long data messages. For such a mix, obviously neither packet switching nor data switching is ideal.

The Model

The user model consists of 2500 computers and 25,000 user terminals geographically distributed throughout the contiguous United States. This community of 27,500 users generates a total of 1,600,000 transactions during the busiest hour of the day, accounting for a total data flow of 3.27×10^{10} bits of data. For simplicity we can group the users into three classes: low-speed terminals (averaging 450 bits/second), high-speed terminals (averaging 3600 b/s), and computers (averaging 8000 b/s).

The network size is 70 switching centers in a nonhierarchical network, with no imposed minimum connectivity. A distributed network with a relatively large number of switching **nodes** is necessary to provide a cost-effective and reliable network with an acceptable level of redundancy for the projected user community.

Circuit Switching and Packet Switching Compared*

The ability of each network structure to efficiently use the available network resources is reflected by the number of lines and quantity of transmission facilities (channel-miles). Table 6-1 presents a direct comparison of several network designs, derived from analytical data, for the condition of 8000-bits/second transmission

* For further details see Rosner and Springer, 1976.

Table 6-1. Comparison of Network Designs

	CIRCUIT SWITCHED NETWORKS						PACKET SWITCHED NETWORK
NETWORK	CASE A PERFECT, BUFFERED		CASE B PERFECT, UNBUFFERED		CASE C DIAL AND HOLD		
Blocking probability	P.10	P.01	P.10	P.01	P.10	P.01	.5 sec delay
Channels	2600	3600	9400	12500	9900	13000	2800
Channel-miles	$.95 \times 10^6$	1.2×10^6	4.2×10^6	4.7×10^6	4.4×10^6	4.9×10^6	$.8 \times 10^6$

lines, a 70-node network, 1% and 10% blocking probabilities (P), and an achieved delay of one-half second for the packet switched network.

Note that only in the case of the perfect, buffered circuit switched network does the overall network efficiency of circuit switching approach that of packet switching. Even at this level the circuit switched network requires about 20% more channel-miles than the packet switched network. If figures for a perfect circuit switched network operating at 1% blocking were used, a differential of nearly 50% would exist between the "perfect" circuit switched network and the packet network.

Figure 6-11 plots the connection overhead time (in milliseconds) versus the data transaction length in bits for both packet and circuit switched networks. The specific parameters used are 8000-bit/second lines for both networks and an average path through the network for all user pairs of approximately three nodes. The packet structure is 1144 bits per packet, of which 144 bits are packet overhead. Further, 300 bits of signaling information are needed to set up and later disconnect a circuit switched call.

Since the amount of overhead needed to establish a circuit switched call is independent of the transaction length, the circuit switch overhead appears as horizontal lines in Figure 6-11. Overhead curves are shown for "perfect" circuit switches, as well as circuit switches with 10-, 50-, and 100-millisecond cross-office times. The overhead of the packet switched network is constant until the first packet is filled, and then is a step function of the transaction length as the transaction is composed of more and more packets.

The intersection of the packet switched network curve with the circuit switched network curves defines the crossover between the preferred handling techniques for data transactions, based on overhead considerations. Even in the case of "perfect" circuit switches, packet switching would be the preferred technique for transactions shorter than 6500 bits. The crossover point moves up to 25,000 bits per transaction for a more realistic 100-millisecond cross-office time circuit switch.

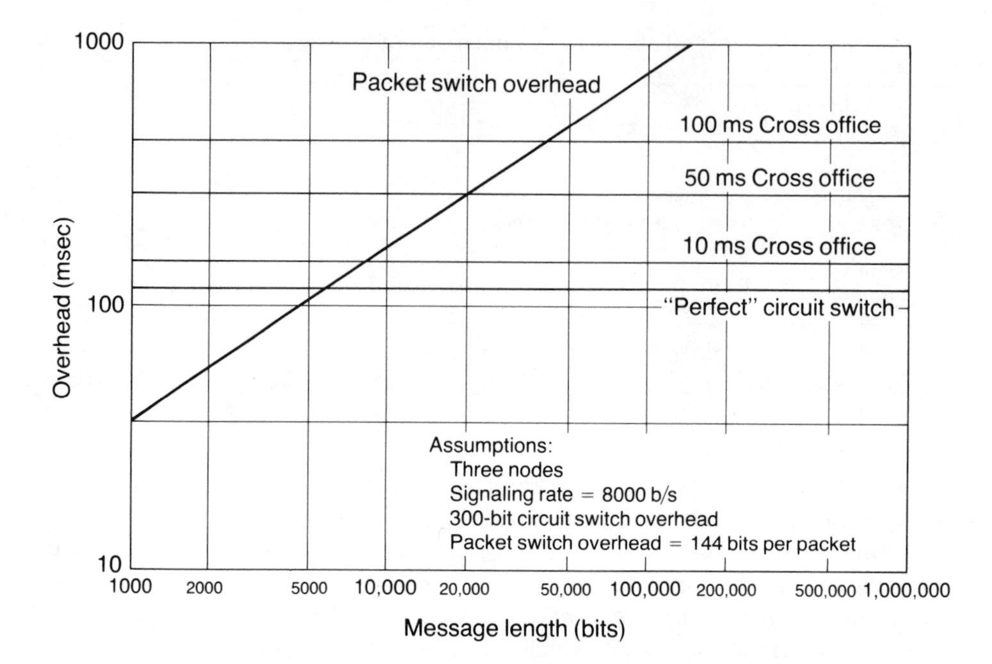

Figure 6-11. Plot of network overhead versus message length for packet and circuit switching.

Potential Economic Impact of Network Differences

The absolute measure of the difference between the two communications networks is the potential cost difference that would result from greater transmission efficiency. While detailed cost analysis is too complex an undertaking for us to attempt here, and specific figures change too quickly to be meaningful, we can nevertheless make some judgments from relative costs. Currently, the annual cost of a circuit switched network is approximately $1.3 million more than the equivalent packet switched network, even under the most ideal assumptions. This amounts to a saving of about 5% of the total annual cost for the "ideal" circuit switched network. However, since "ideal" conditions don't actually exist, a more realistic cost differential estimate is anywhere from $10 to $100 million, or 50 to 400 percent. The conclusion seems clearly to be that a common user data network can be more efficiently implemented using packet switching rather than circuit switching, even over a broad range of assumptions and traffic loads.

SUMMARY

In this chapter we have been able to establish some quantitative comparisons of the performance of packet switching, circuit switching, and message switching. In order to do this, we looked in some detail at the operation of packet switched networks.

The normal operation of packet switched networks can easily lead to the loss, duplication, or missequencing of packets. Consequently, an overhead structure must provide the switches with the information to insure the proper flow of the user information through the network.

Quantitative analysis of the switching techniques enabled us to make comparisons in the areas of message delivery delay, processor loading, and protocol and overhead. First of all, delay for circuit switched networks and packet switched networks could be about equivalent, but in any case was much less than for a message switched network. Secondly, data calls applied to a voice-based circuit switched network develop extremely large processor loads on the switches, even for modest amounts of data traffic. Finally, the crossover between packet and circuit switching with regard to network overhead occurs in the range of 5000 to 100,000 bits per message, when parameters associated with a next generation circuit switched network are assumed.

We have seen that packet switching is the most efficient technique for a common user data network. In the next part of the book we will look more closely at the details of packet network design and applications.

SUGGESTED READINGS

CLOSS, FELIX. "Message Delays and Trunk Utilization in Line-Switched and Message-Switched Data Networks." In *Computer Communications*, ed. P.E. Green and R. W. Lucky. New York: IEEE Press, 1974, pp. 602–608.

Closs compares circuit and message networks for data communications and concludes that "signalling may become the dominant factor in network performance when short messages . . . and high call rates determine customer traffic." He further finds that, for messages longer than a few thousand bits, circuit switching is preferable to message switching.

COVIELLO, G. J., and ROSNER, R. D. "Cost Considerations for a Large Data Network." *Proceedings of the ICCC '74*, Stockholm, August 1974, pp. 289–294.

This paper looks at some of the cost issues in data network design, including that used in determining the economic comparisons between circuit and packet switching in this chapter. The paper was reprinted in the IEEE Computer Society publication Computer Networks: A Tutorial, *edited by Marshall Abrams, Robert Blanc, and Ira Cotton. Long Beach, Calif.: IEEE, 1975.*

DAVIES, D. W., and BARBER, D. L. A. *Communication Networks for Computers.* London: John Wiley, 1973, chs. 9 and 10.

The relative advantages and disadvantages of the various switching techniques for data communications are discussed, and the structure and general operation of packet networks described. These chapters are of interest because much of the description is set against the operational background of European telecommunications.

ROSNER, ROY D. "A Digital Data Network Concept for the Defense Communications System." *Proceedings of the National Telecommunications Conference*, NTC '73. Atlanta, November 1973. Pp. 22C-1–22C-6.

This paper illustrates the considerations in a worldwide data communications network. An example of a basic network structure using packet switching is given. Overhead comparisons and packet switching/circuit switching comparisons that support the discussions in this chapter were presented publicly here for the first time.

ROSNER, ROY D., and SPRINGER, BEN. "Circuit and Packet Switching—A Cost and Performance Tradeoff Study." *Computer Networks Journal*, vol. 1, no. 1 (January 1976), pp. 7–26.

This paper describes the computer-based network model that was the basis of some of the results in this chapter. It also contains several useful references to source material on specific design algorithms.

PART TWO

Operational Protocols—Packet Switching in Networks

In Part One we examined the principles of packet switching and compared it to a range of other techniques for telecommunications networking, with emphasis on high-speed data communications users.

In this part we will deal exclusively with packet switching, looking in some detail at the operation of a packet network. Inasmuch as the switching functions in a packet network result from data manipulations inside a high-speed digital computer, the implementation of packet switched networks depends on rules and procedures for moving the data from one user, through the switches, to another user. These rules and procedures are collectively termed *protocols*.

Communications protocols are not unique to packet switched networks. Our use of the everyday common user telephone network is very constrained by standard protocols, which we can recognize in the ordinary procedure of making a phone call. Initiation of a call requires lifting the receiver, listening for the dial tone, dialing a set of address digits, listening for a ringing tone to signal proper operation of the network connection process, and the destination addressee answering the phone with a proper salutatory phrase, such as "Hello." In fact, the protocol even extends beyond the mechanics of the connection process. Established "protocol" requires the initiator of the call to then give some indication of his identity and purpose of the call. To see the power of these implied protocols, try calling a friend sometime and, after he answers, saying to him: "What do you want?" This disruption of normal telephone "protocol" is likely to leave your friend speechless, at least temporarily.

Protocols play an even more critical role in packet switched networks, however, since most of the interaction is between machines at either end of the network. Thus protocols must define the full range of network conditions that may occur during normal, as well as predictable abnormal, operations of the network and its users. So let us proceed to learn more about how packet switched networks operate, both internally and across the user-to-network interface.

7

Protocol Structures in Packet Switched Networks

THIS CHAPTER:

will look at various types of network protocols.

will define the several layers of standardized protocol structures.

will study the protocol operation of the original ARPANET and the evolution of the virtual and logical circuit form of packet switched network operation.

will examine the datagram and see how it differs from the virtual circuit.

In order to communicate through an electronic medium, both the people and the machines involved must operate according to a set of well-defined rules and procedures known as protocols. Such protocols are particularly critical in a data-oriented network, where much of the network usage is between "machines" at either end of the network. A large measure of mutual agreement between the network supplier and network users is required to insure that the operations will be mutually satisfactory over a wide range of conditions.

TYPES OF NETWORK PROTOCOLS

Since many different functions are performed under the definition of protocols, we are concerned with a number of different protocol types. Figure 7-1 illustrates three distinct functional levels of system operation. For the purposes of this figure we are portraying the communication between two intelligent computers, each capable of supporting several user programs simultaneously. In the terminology of automated data processing (ADP) the programs in a computer running under a real-time operating system are referred to as *user processes*, and the computer is referred to as the *ADP* **host.**

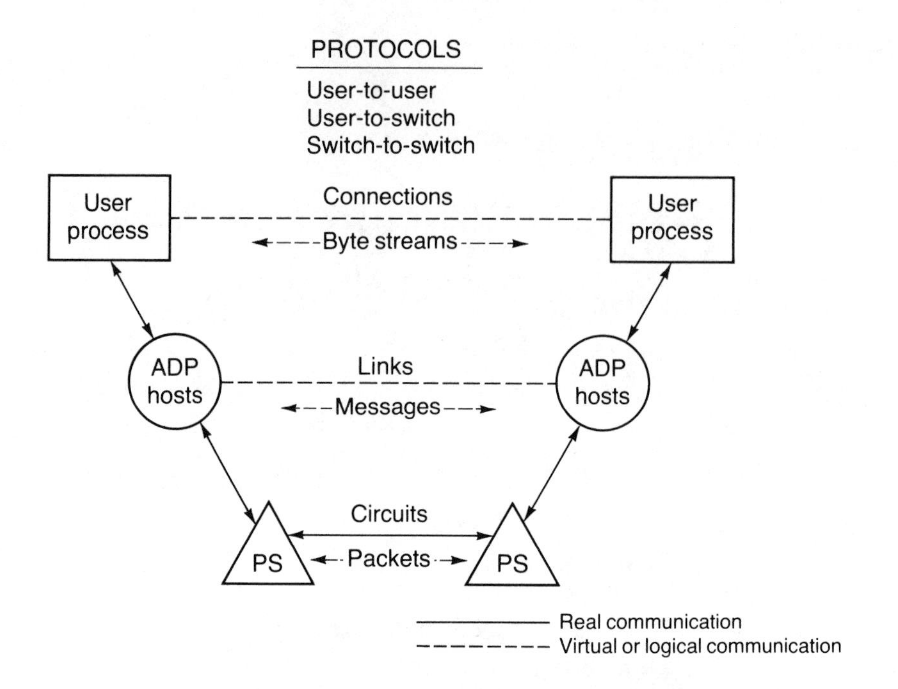

Figure 7-1. User and network protocol hierarchy.

From a protocol standpoint, the basic communications requirement illustrated here is for the user processes in each computer to exchange information with each other. The information exchange may be to update a distributed data base, request a seat reservation on another airline, or transfer funds between two accounts at a distant bank. In any case, the user processes are designed to exchange data with like processes, using a data flow at the byte- or bit-stream level, over what appear to be physical connections between the processes.

Physical connections between the processes do not actually exist, however, and unless the processes happen to coexist within the same computer, they cannot exist. The user processes are like tenants who depend on the ADP hosts to effect the needed exchange of data. It is generally impractical, if not impossible, for the hosts to be directly connected to each other, although their operating systems are functionally capable of exchanging messages over host-to-host links. In the case of a packet switched network the packet switches provide the physical circuits over which the packets can carry the host-to-host messages.

The real communication path is thus from the originating user process to its host computer to the originating packet switch, across the network to the destination packet switch, to the destination host, to the destination user process. In general, the data exchange is bidirectional, so that a path in the reverse direction is required as well. Despite this physical reality, there is logical or virtual communication at the process-to-process level and the host-to-host level. In order to simplify the software associated with this model, the appropriate protocols are kept as unique to a single level as possible. Thus we can identify the three major types of protocols as user-to-user, user-to-switch, and switch-to-switch. We will look at each in some detail below.

User-to-User Protocols

The user-to-user protocols involve a large range of functions from authentication of the data exchange to the remote management of complex user files or processing algorithms. From a network operational viewpoint these protocols are considered as ADP functions, within the host software or host operating system.

These protocols are of such complexity that standards are evolving that define at least four different levels of user-to-user protocol. We will discuss this multilevel structure in more detail later in the chapter.

User-to-Switch Protocols

The user-to-switch protocols define the relationship between the host computers and the packet switches with which they interface. These user-to-switch protocols can be adapted to deal with many different terminal devices besides computers.

In general, these protocols are concerned with controlling the physical communications circuit between the user and the network, assuring data transparency, detecting and recovering from failures and errors, and controlling information flow.

User-to-switch protocols for packet networks fall into two broad categories—the packetized virtual circuit and the packetized datagram mode. The virtual circuit mode assures the sequencing of user data. It attempts to streamline the flow of long messages that routinely exceed the capacity of a single packet. The datagram mode tries to utilize the high-speed, real-time capabilities of the packet network to speed data across the network one packet at a time. It is not concerned with initiation of the data connections or the sequence of the data. This mode is particularly suited to the exchange of information that will fit within a single packet.

User-to-switch protocols have evolved to a standard definition involving three distinct levels—the physical level, the data-link level, and the network control level.

Switch-to-Switch Protocols

The switch-to-switch protocols deal with the internal operation of the packet network and the way the network controls the flow of information from user input port to user output port. The various ways user information can be controlled and protected within a packet network define the switch-to-switch protocols.

The switch-to-switch protocols are to some extent related to the user-to-switch protocols. However, there are many design decisions and tradeoffs to be made in the definition of the internal protocols of a packet network.

As a result of this layered approach to structuring the networking protocols, a seven-level standardized definition of protocol functions has evolved. In the following section we will further define the functions of the various levels in the protocol hierarchy.

THE INTERNATIONAL STANDARDS ORGANIZATION (ISO) PROTOCOL HIERARCHY

In order to facilitate the exchange of data communications, international standards for data network activity evolved quickly between 1975 and 1980. One such standard is the protocol hierarchy adopted by the International Standards Organization (**ISO**) to facilitate the intercommunication of data-processing devices (Figure 7-2).

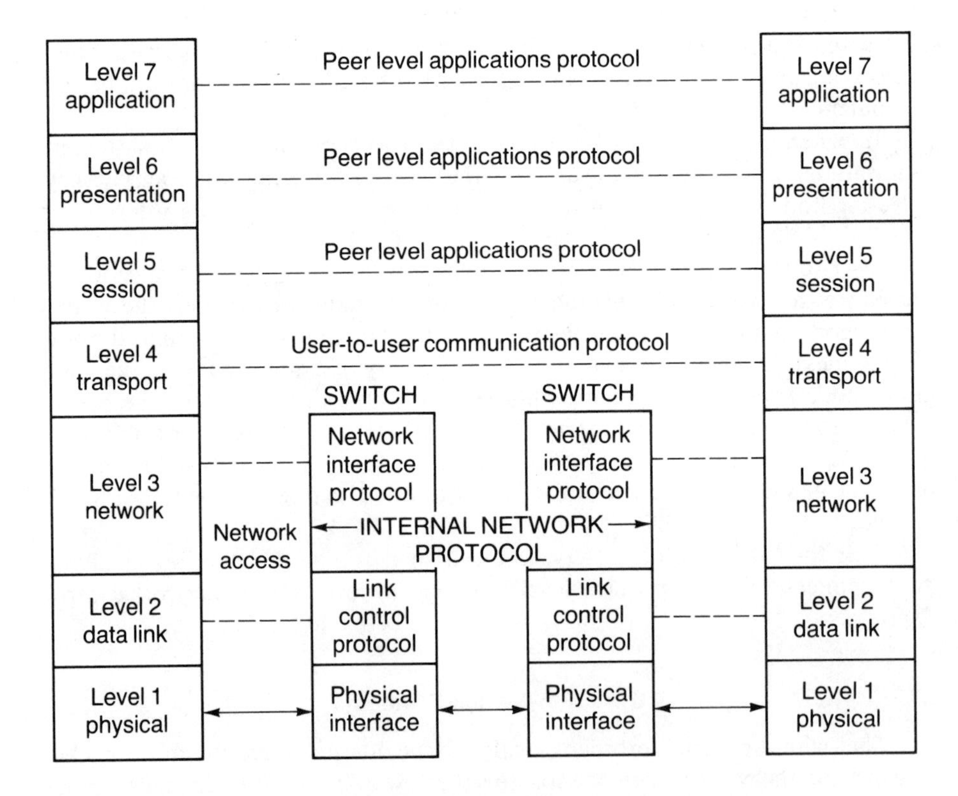

Figure 7-2. International Standards Organization (ISO) protocol hierarchy.

The lowest level of the ISO protocol hierarchy is the physical level, where previously defined standards were applied to define the physical interface. By physical interface to the network we refer to the pin connections, electrical voltage levels, and signal formats. Level 2, known as the data-link level, controls the data link between the user and the network. This level defines data format, error control and recovery procedures, data transparency, and implementation of certain command sequences. For nonswitched networks, or the interface of simple terminals with computers through point-to-point services, generally only levels 1 and 2 are required. Networks designed by a single manufacturer around a single product line, generally do so with a combination of level 1 and level 2 protocols.

Level 3, the network level, defines most of the protocol-driven functions of the packet network interface, or the internal network. It is at this level that the flow-control procedures are employed and where switched services are initiated through a data call establishment procedure.

Level 4, known as the transport level, assures the end-to-end flow of complete messages. If the network requires that messages be broken down into segments or packets at the interface, the transport level assures that the message segmentation takes place and that the message is properly delivered.

Level 5, the session control level, controls the interaction of user software, which is exchanging data at each end of the network. Session control includes such things as network log-on, user authentication, and the allocation of ADP resources within user equipment. Level 6, the presentation level, controls display formats, data code conversion, and information going to and from peripheral storage devices. Level 7, the user process or user application level, deals directly with the software application programs that interact through the network.

Although at levels 5, 6, and 7 the protocol is defined from a functional viewpoint, implementation of standard software that can operate at these levels has been slow. The software at all of these levels (often referred to as peer-level software) tends to be both equipment and application dependent. However, the the layered approach to protocol development achieves a degree of isolation and modularity between the various layers, so that changes in one level can be made without changes in any other level.

Let us now focus on examples of communications protocols, with primary emphasis on the network and transport levels and the interaction between them.

THE ARPANET APPROACH TO PROTOCOLS

The ARPANET is an operational digital network within the U.S. Department of Defense. It serves a wide range of research and development institutions within this country and at selected sites in Europe. The ARPANET is generally credited with making packet switching a practical reality, and it acts as a prototype for many commercial ventures into packet switching. Though originally conceived as a resource-sharing network among a large number of computing centers funded and sponsored by the U.S. Defense Advanced Research Projects

Agency (DARPA), the network itself became a popular object of advanced research in data communications techniques and distributed data processing.

Configuration of the Network

The ARPANET is configured with network switches—known as **IMPs** (for interface message processors) and **TIPs** (for terminal interface message processors)—interconnected with 50,000-bit/second lines. IMPs are designed to interface only with computers, while the TIPS are capable of interfacing with a combination of computers and individual user terminals. Each node is programmed to receive and forward messages to the neighboring nodes in the network, with end-to-end delays of about one-quarter second. Host computers communicate with each other via network messages up to 8159 bits in length.

The first 96 bits of a network message, known as the message leader, specify the message destination and handling information. The leader information uniquely specifies a connection between the source and destination hosts. A message identification number is used to uniquely control and manage the flow of a message through the network until message delivery is confirmed back to the originator. In addition to the user data each packet contains a packet header, which includes the address information that is in the message leader and network control information needed to protect the flow of the packet through the network.

Message-Handling Procedure

Messages received by the network are divided into packets of up to 1008 data bits each. When all the packets associated with a particular message arrive at the destination switch, they are reassembled to form the original message, which is passed to the destination host. The destination switch then returns a positive acknowledgement (**ACK**) to the source host.

Rather than simply confirming delivery, the message acknowledgement assumes that a subsequent message will follow immediately on the same connection. Therefore, the delivery acknowledgement is known as a **RFNM** (request for next message). If a message cannot be delivered owing to a line failure, a node failure, missing packets, or the like, an incomplete message will be returned, thus initiating a retransmission. If neither a RFNM nor an incomplete message arrives back at the source during a suitable timeout interval, the message will be reinitiated.

Figure 7-3 illustrates the message-handling procedure. This figure functionally represents two host computers exchanging messages through an originating packet switch, a tandem packet switch, and a destination packet switch. Message segments from the originating host, or source computer, arrive at the originating packet switch either as a single high-speed local transfer between the host and the switch, or via a block-by-block transfer over a data-link controlled-access line between the host and the switch.

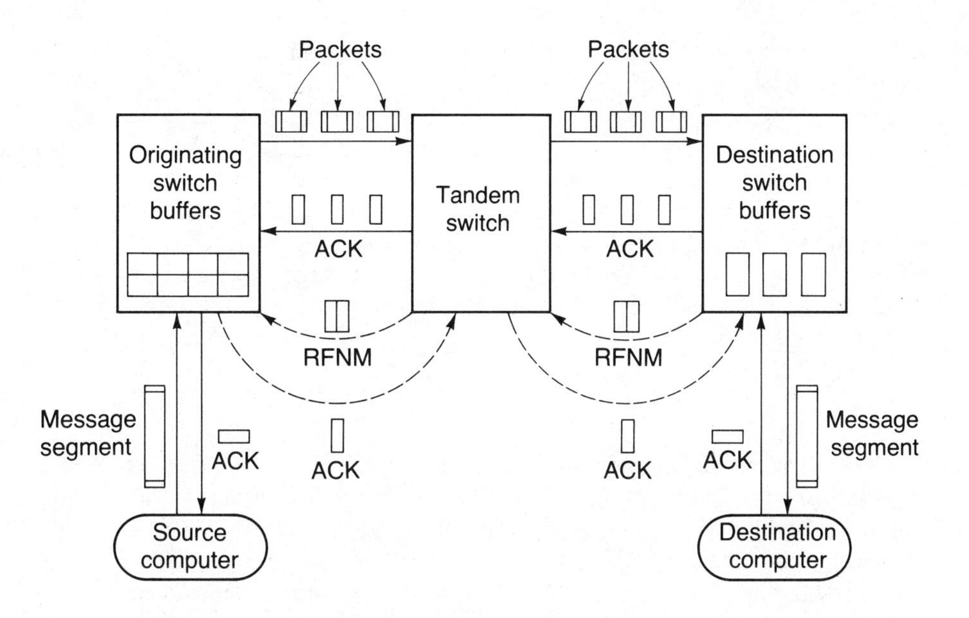

Figure 7-3. Operational model of an ARPANET-like network.

Commonly, blocks of data are transmitted from the host to the switch under control of the data-link level (level 2) of the protocol hierarchy. The information field of these blocks may or may not correspond to the information content of the packets. In any case, the complete message segment, containing the segment leader and the data bits, is temporarily stored in the originating packet switch. This message arrival is treated as a request for service. The switch therefore initiates several actions at the network and transport levels of protocol, as well as at the switch-to-switch protocol levels.

The new message flow generates the entry of information into a "pending leader table" (PLT), which the originating switch uses to create the packet headers for the individual packets of the message from the user-supplied packet leader information. At the same time, the originating switch requests the destination switch to allocate a set of buffers for the reassembly of the individual packets into the complete message segment for delivery to the destination host. The allocation request (**REQALL**) transits the network rapidly as a short control/information packet. If the required buffers are available at the destination switch, the destination switch returns an allocation (**ALL**) control message to the originating switch, which can then commence transmission of the individual packets composing the message.

Despite the fact that the packets are transmitted in sequence by the originating switch, the network operation—employing error correction, dynamic routing, and other features that affect the end-to-end delay—may cause the packets to arrive at the destination switch in a different sequence. When the message is fully reassembled in the destination switch buffers, it can then be relayed to the destination host. The destination switch then transmits an acknowledgement in the form of the RFNM, together with an allocation for additional message segments. If for some reason no buffers are available after delivery of the current message segment, it is possible to return the RFNM without an allocation.

While all of this is occurring on an end-to-end basis, each packet transmitted within the network is acknowledged on a link, switch-to-switch basis, to insure proper delivery and error-free transmission. The control packets, such as the REQALL and RFNMs, flow through the network in much the same way the user data packets do.

VIRTUAL CIRCUIT

The basic mode of operation in the ARPANET, in conjunction with the required host-to-switch protocol functions, is known as *virtual circuit.* Virtual circuit can best be defined by the properties it must have. These properties include sequenced data transfer, data transparency, a full-duplex path, in-band and out-of-band signaling, flow control, error control, interface independence, and a switchable form of operation. Let us look at each of these individually.

Sequenced Data Transfer

All data bits delivered to the destination host must be in the same order they were delivered to the network by the source host. This property implies the need for the message reassembly process.

Data Transparency

Data bits in the user data fields must be accepted in any sequence of ones and zeros. No sequence may be prohibited, despite the fact that special bit groups are needed to "flag" the beginning and end of packets. This property implies the need for special handling of the data stream to protect against inadvertent flag sequences.

Full-Duplex Path

Data has to be able to flow in both directions between the end users simultaneously. Thus the initiation of a connection and buffering for a message in one direction requires a similar process in the opposite direction.

In-band/Out-of-Band Signaling

Signals have to move between the users and the switches in order to control the flow of information, to inform the user of status information, to respond to network or user inquiries, etc. This signaling can take place as part of the normal user data stream (in-band) or outside the normal user data transmissions (out-of-band).

Flow Control

The network must be capable of reducing the allowed input rate of information. This is important to prevent **congestion** to the point where normal operation may become impossible.

Error Control

All network transmission must be error protected, so that the probability of an undetected network-introduced error will be negligible.

Interface Independence

Operation of the network must be independent of the physical and electrical properties of the user interface. It must be consistent with the logical data structures.

Switchability

Network operation in the virtual circuit mode allows information to be exchanged among various user pairs by modifying the address field of the user segments.

Figure 7-4 depicts three user computers interchanging information through an ARPANET-like, virtual circuit–based packet network. For this example we assume that computer A has three transactions, two destined to computer B and one destined to computer C. Each of these transactions is identified within the operating system of computer A with unique transaction numbers.

Transaction 1, intended for computer B, consists of approximately 22,000 data bits. Transaction 2, destined for computer C, consists of about 12,000 data bits. Finally, transaction 3, consisting of about 7000 data bits, is headed for computer B. Notice that, while both transactions 1 and 3 are headed for the same physical destination, they are likely to be associated with different user processes (at the application level, level 7), and therefore are handled as completely independent transactions.

Network Control Program

Resident within each computer tied to the packet network is a network control program (**NCP**), which operates at the network and transport levels of the protocol hierarchy. The network control program controls the flow of the data transactions across the network, as well as the flow of message segments

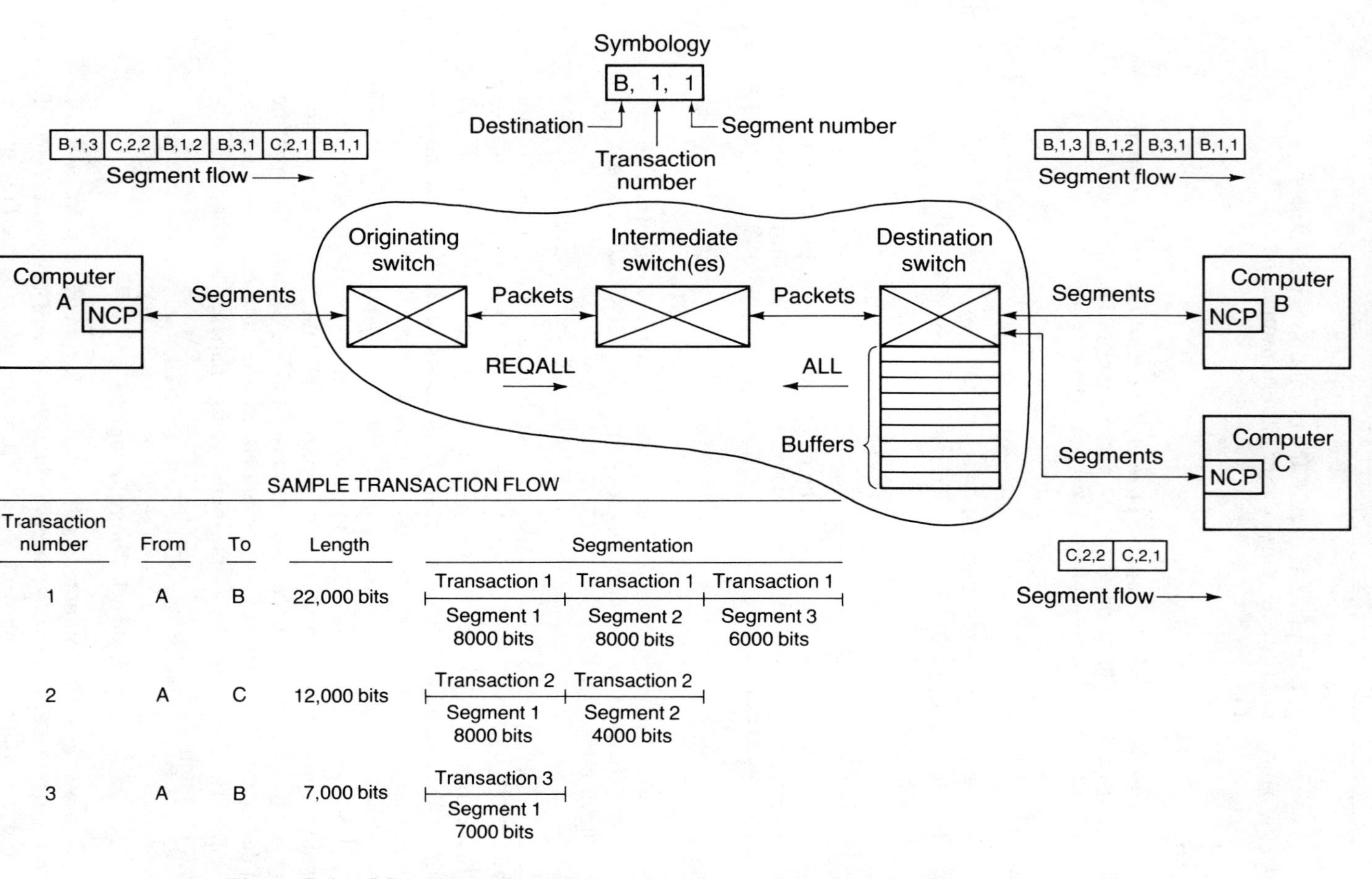

Transaction number	From	To	Length	Segmentation		
1	A	B	22,000 bits	Transaction 1 Segment 1 8000 bits	Transaction 1 Segment 2 8000 bits	Transaction 1 Segment 3 6000 bits
2	A	C	12,000 bits	Transaction 2 Segment 1 8000 bits	Transaction 2 Segment 2 4000 bits	
3	A	B	7,000 bits	Transaction 3 Segment 1 7000 bits		

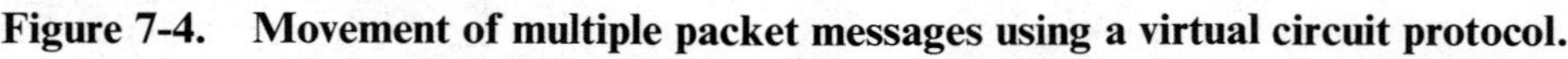

Figure 7-4. Movement of multiple packet messages using a virtual circuit protocol.

between the computer and the packet switch. That is, the NCP divides the transactions into the network segments, according to the network level host-to-switch protocol, and manages the composition of full transactions from the various message segments.

The network control program also manages the interleaving of segments of different messages over a single access line between the computer and the network. This process, known as logical multiplexing, is facilitated by a set of logical channel identification numbers. Together with the destination and source address, these numbers identify a unique, logical (virtual) connection through the packet network for the duration of the transaction flow.

The Flow of Data

Following the flow of data shown in Figure 7-4 will demonstrate these principles. We will assume that the three transactions shown are presented to the NCP at approximately the same time. Transaction 1 is initiated first, by computer A sending segment 1 of transaction 1 to the source switch. The source switch then attempts to open a connection by sending the request for allocation (REQALL) to the destination switch. Assuming that destination buffers are available, an allocate (ALL) confirmation will be returned. While this connection process is going on, computer A initiates transaction 2 by sending segment 1 of transaction 2 to the source switch. A similar connection process has to occur between the source switch and the switch serving destination computer C.

Note that, in addition to assigning buffers at the destination switch, the connection processing at the destination switch will check to make sure that the destination computer is currently able to receive traffic. It will also verify that the line between the network and the destination is operating properly. We would not want to establish connections in the network if, after arrival at the destination switch, the traffic cannot be immediately delivered to the addressee. While the connection process for transaction 2 ensues, computer A initiates transaction 3 by sending the one and only segment of transaction 3 to the source switch.

Assuming that connections are successfully established for each of the three transactions, the NCP would have considerable versatility in deciding how best to utilize the channel between computer A and the network. If, for example, the line from computer A and the network (source switch) is 9600 bits/second, it will take nearly 1 second for an 8000-bit segment to be transmitted to the network. By the time the first segments of transaction 2 and transaction 3 are transmitted, we would expect to have received confirmation of receipt of the first segment of transaction 1 at computer A, in the form of a RFNM for transaction 1, segment 1. Therefore, computer A then sends segment 2 of transaction 1, followed by segment 2 of transaction 2. At this point, the only remaining undelivered segment is segment 3 of transaction 1, which can be sent as soon as the RFNM for segment 2, transaction 1 is received.

At the output side of the network we see the operation of the logical multiplexing on the lines between the destination switches and the destination computers. The line to computer C sees only transaction 2, which happens to be the only message destined for computer C at this particular time. The line to computer B, however, sees the interleaved, logically multiplexed stream of segments associated with transactions 1 and 3. As a result, we find the one and only segment of transaction 3 inserted between the first and second segments of transaction 1. It is the responsibility of the NCP at the destination end to separate the multiplexed information of the multiple messages arriving at the destination into the separate user-level transactions, and to deliver them to the correct user process.

DATAGRAMS

It is possible to envision a number of applications where the user's data will generally, if not always, fit into single packets—for example, point-of-sale terminals, credit card transactions or credit verification, remote sensor systems, many types of inquiry-response systems such as reservations, and brokerage quotation systems. In addition, passing voice communications through packet networks (which we shall discuss more fully in Part Six) depends on packetizing data that has no value after a brief time, so there is little problem if an occasional packet gets lost. Such demands, requiring packets that can be sized to individually carry most of the user's information, can be met by a datagram type of network.

DATAGRAM	VIRTUAL CIRCUIT
Self-contained	Long term
Fully identified	Initially set up and formally terminated
Highly (but not absolutely) reliable	Absolutely reliable
Unsequenced	Sequenced
Uncontrolled	Highly controlled

Figure 7-5. Comparison of datagram and virtual circuit modes of network operation.

The Distinction between Virtual Circuits and Datagrams

Before we elaborate the differences between virtual circuits and datagrams, it is important to emphasize that our definition is based on the user's perception of the network and the protocol processes that take place at the user interface to the network. There is a widespread misunderstanding in the literature that these modes require a specific set of internal protocols. This is not the case. What is required is that the internal operation of the network be sufficiently sophisticated to guarantee to the user the interface characteristics prescribed by each mode of operation.

Figure 7-5 summarizes the datagram and virtual circuit characteristics. Essentially, they are functional opposites. As we have examined the virtual circuit in detail, let us now look at the datagram mode of operation.

Self-contained

The information contained in a datagram is complete and useful in and of itself. It does not depend on the contents of preceding or following datagrams to have utility to the end users.

Fully Identified

The beginning and end of the datagram are readily identified by the destination, and are recognizable as a complete entity. Any needed control, numbering, and routing information are fully identified within each datagram.

Highly (but Not Absolutely) Reliable

Datagram delivery has a very high probability of success. However, there is a chance that one will become lost, with the destination having no knowledge that it has ever been sent. It is also possible that a duplicate datagram may arrive at the destination.

Unsequenced

Sequentially transmitted datagrams may arrive at the destination in a different order. Since the datagrams are considered to be self-contained, the network makes no attempt to check or preserve entry sequence.

Uncontrolled

The network will attempt to advise the datagram originator of failures to deliver a datagram or of network conditions that may have resulted in the loss of a datagram. However, after the datagram leaves the source node, the flow is uncontrolled (except for network-induced error checking). It may not be possible for the network to keep the source user informed of the progress of the datagram.

While the datagram and virtual circuit operational concepts imply substantially different physical and protocol implementations of the internal network structure, they do not define the exact operation of the switches. In designing a network to handle packet-oriented data, we first have to decide what kind of user data will dominate the traffic. Particularly, will most of the user transactions fit into a single packet, or will they be significantly longer? Single-packet transactions are ideal for datagrams; long messages are best handled in virtual circuit.

To approach this dichotomy, most packet networks use network protocols that can be described as single-packet-per-segment protocols. Unlike the ARPANET, where 8000-bit user segments are subdivided by the network into eight packets, the single-packet-per-segment protocols require the user hosts to provide the data in segments that correspond to single packets (at most). Transactions longer than a single packet are handled by a cooperative effort of the user and the network.

Analytical comparison of various network protocols, too detailed to include here, reveal that the single-packet-per-segment protocol enjoys operational advantages in terms of network simplicity and efficient use of network buffering.

TWO CLASSES OF TERMINALS

Up to this point we have talked about data exchanges between intelligent host computers. A very large percentage of all data traffic transits a network between user terminals and host computers. Figure 7-6 depicts this case and functionally represents two general classes of terminals.

Low-speed, asynchronous character-by-character terminals operate in typical speed ranges of 75 to 600 bits/second. This class of terminal is a nonintelligent device and thus cannot respond to the protocol features of a packet switch interface. High-speed, synchronous block-by-block terminals operate in typical speed ranges of 1200 to 9600 bits/second. This class of terminal can range from nonintelligent—which can only respond to a very limited set of level 2, link-control commands—to highly intelligent, processor-controlled terminals—which can support all packet switched network protocol features with the possible exception of multiple simultaneous logical connections.

In ARPANET-like practical networks a "pseudo-host" function in included within the packet switches. The pseudo-host performs a packet assembly/disassembly function for character or bit streams emanating from various types of user terminals. It is implemented as transparently as possible from the user's viewpoint. It takes groups of characters and bits, buffers them, forms packet headers, and sends complete packets (though not necessarily full packets) through the network.

Upon receipt of packetized information destined for user terminals, the pseudo-host functions in reverse, taking the user data from the packet, buffering it at the destination switch, and metering it out of the network according to the line rate and line protocol associated with that particular terminal type. In addi-

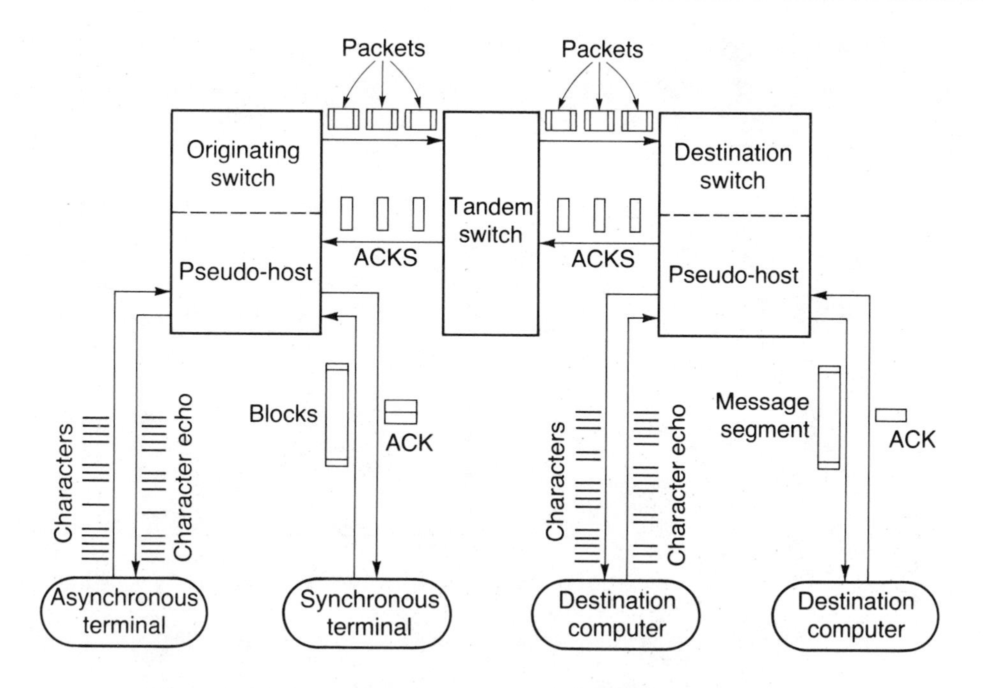

Figure 7-6. Operational model of an ARPANET-like protocol serving synchronous and asynchronous terminals.

tion, the pseudo-host interfaces are often designed to perform local functions that would have to be performed by the destination host computer in the absence of a packet network. Such functions include local echo of user-entered characters, control of multidrop terminal chains or polling groups, and conversions between different signaling codes.

Results

Some experimentally derived results can help establish how well the network we have been studying can perform in actual operation. Figure 7-7 graphically depicts ARPANET measurements taken in May 1975.* This figure shows the achieved host-to-host throughput as a function of the number of nodes (switches) between the two host computers.

Since the interswitch data lines operate at a rate of 50,000 bits/second, the reduction in network throughput below this figure is largely attributable to the effects of the network, user-to-switch, and switch-to-switch protocols. It can be seen from Figure 7-7 that throughputs of about 37,000 bits/second can be achieved

* Figure 7-7 is based on data from Kleinrock and Opderbeck, 1977.

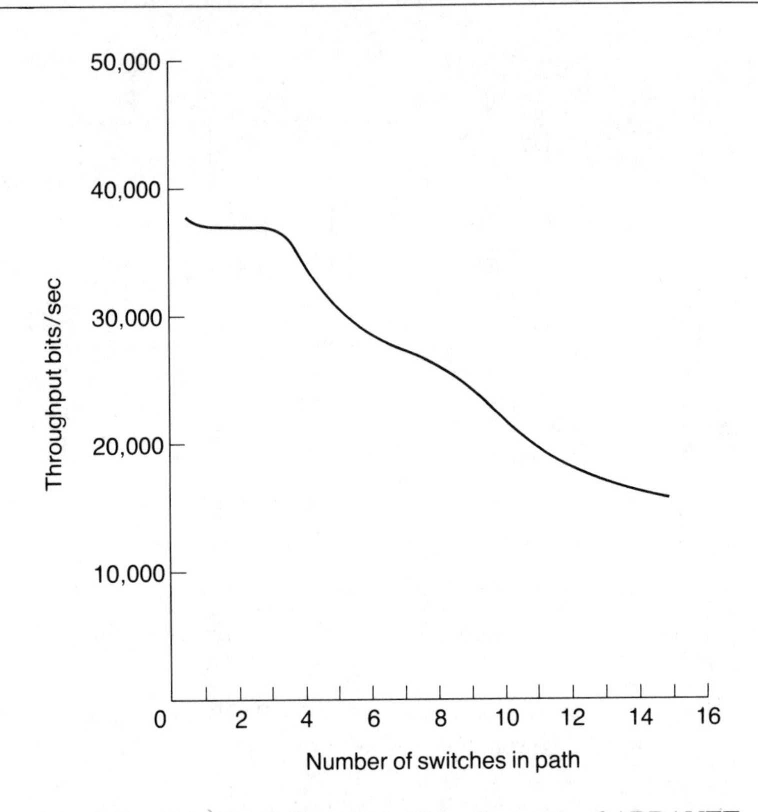

Figure 7-7. Experimental (measured) performance of ARPANET: computer-to-computer throughput versus number of switches in the end-to-end path.

up to about three or four nodes. The throughput falls off rather dramatically as the number of nodes increases. This effect is due primarily to the additional delay as the REQALL and RFNM have to proceed through more and more nodes in order to initiate the flow of segments.

SUMMARY

This chapter has developed the concept of layered network protocols. The International Standards Organization (ISO) has defined a seven-level protocol hierarchy for open data communications systems. The ISO protocol deals with the interconnection of users with each other and between users and the network. Basically, it divides the user-to-network function into four discrete protocol levels: the physical, link, network, and transport levels.

We then examined two modes of network connectivity that take place at the user-network interface. The operation of the protocols inside the ARPANET is an example of virtual circuit connectivity with respect to the handling of user segments. The virtual circuit has the properties of sequenced data transfer, data transparency, a full-duplex path, in-band and out-of-band signaling, flow control, error control, interface independence, and a switchable form of operation. Virtual circuits are best for handling long messages. By contrast, datagrams are best for single-packet transactions. Datagrams are self-contained, fully identified, highly reliable, unsequenced, and uncontrolled.

We concluded this chapter by showing experimental results that relate the achieved network throughput to the operation of the network protocols. As we proceed with our consideration of protocols, we will emphasize the delicate balance between network throughput, protocol complexity, and protection of the user from potential network anomalies such as lost, duplicated, or misordered packets.

SUGGESTED READINGS

CARR, S., CERF, V., and CROCKER, S. "Host-to-Host Protocol in the ARPA Computer Network." *Proceedings of the Spring Joint Computer Conference*, Atlantic City, N.J., American Federation of Information Processing Societies (**AFIPS**), May 1970, pp. 589–597.

Written and published in the earliest days of the ARPA network, this paper provides an excellent description of the initial host-to-host protocols used in the network.

CROCKER, S., HEAFNER, J., METCALFE, R., and POSTEL, J. "Function-oriented Protocols for the ARPA Computer Network." *Proceedings of the Spring Joint Computer Conference*, Atlantic City, N.J., American Federation of Information Processing Societies (AFIPS), May 1972, pp. 271–280.

This paper reviews the development of protocols constructed to promote the substantive use of the ARPANET at the user/functional level. The layering and isolation of the various protocols in the ARPANET, as well as the unavoidable interactions among those layers, are described.

KLEINROCK, LEONARD. "Principles and Lessons in Packet Communications." *Proceedings of the IEEE*, vol. 66, no. 11 (November 1978), pp. 1320–1329.

This paper summarizes a decade's experience with the operation of the ARPANET and the wide range of improvements that were implemented in response to these experiences. Conditions that can result in network deadlocks (the inability to deliver traffic), throughput degradations, and failures of the routing algorithms are emphasized. A number of design principles, some of which we have discussed, are summarized.

KLEINROCK, LEONARD, and OPDERBECK, H. "Throughput in the ARPANET-Protocols and Measurement." *IEEE Transactions on Communications*, vol. COM-25, no. 1 (January 1977), pp. 95–104.

This paper describes the internal protocols of the ARPANET, with particular emphasis on the relationship between the internal protocol operation and the network throughput. The protocol implementations of handling multipacket and multisegment messages are discussed. Protocol changes in the ARPANET that were necessitated by problems discovered in earlier network protocols are described.

POUZIN, LOUIS, and ZIMMERMANN, HUBERT. "A Tutorial on Protocols." *Proceedings of the IEEE*, vol. 66, no. 11 (November 1978), pp. 1346–1370.

This excellent tutorial article includes 85 references on the development and design of networking protocols. Emphasis is placed on the user-to-user environment and the relationship between protocols, distributed information systems architecture, and programming languages.

ROSNER, ROY D., BITTEL, RAYMOND H., and BROWN, DONALD E. "A High Throughput Packet-Switched Network Technique Without Message Reassembly." *IEEE Transactions on Communications*, vol. COM-23, no. 8 (August 1975), pp. 819–828.

This paper describes in detail the single-packet-per-segment network protocol introduced in this chapter. It also provides much more mathematical detail on the analytical comparison of the network protocols, and the derivation of the relationships between the protocol operation, the network delays, the network buffering, and overall performance of the packet switched network.

8

The User-Network Interface—The X.25 Standard

THIS CHAPTER:

will look at packet switched networks from the perspective of a user who is fully compliant with the functions defined by the standards.

will explain the features and facilities provided to the user of a network that complies with the international CCITT X.25 standard.

will summarize the actions that users must take in order to make efficient use of an X.25-compliant packet network.

In the previous chapter in this part we looked at protocols associated with the internal operations of packet switched networks. We now turn to an external perspective of packet switched networks.

The **X.25** standard of the Consultative Committee for International Telephone and Telegraph (**CCITT**) has been a strong influence on the development of packet networks around the world and on the design of user equipment to operate with those networks. With strong leadership by participants from Great Britain, France, Canada, and the United States the X.25 standard for packet switching was developed, ratified, and implemented in a very short time compared to most international standards. This cooperation was motivated by the very high cost of international communications facilities and the exceptionally large efficiency improvements that could be achieved for individual data communications users through shared, switched, common user networks.

Most of the work on X.25 took place between 1974 and 1976, with initial ratification in 1976. However, the 1976 version addressed only major principles of packet network interfaces, leaving room for considerable differences in specific implementations. Between 1976 and 1980 additional work significantly refined the

X.25 standard, better defined many operational details, eliminated several logical defects, incorporated guidelines for datagram service, and significantly enhanced the utility of the standard. Our discussion here will concentrate on the overall structure of packet switching defined by the standard and concentrate on the "core" elements that have proven themselves to be complete and correct. Standards meetings continue to take place, and refinements to the standard will be made from time to time. At the time of this writing, the latest version of X.25 was approved at CCITT meetings in November 1980. Copies of this version can be obtained on request from the Manager of the National Communications System, Office of Technology and Standards (NCS-TS), Washington, D.C. 20305.

The X.25 standard is strictly a user-network interface standard. It does not require a specific mode of operation among the network switching centers, nor does it define a particular switch-to-switch protocol. Rather, it implements the concept of a packetized virtual circuit form of operation, although the latest version also includes provision for datagram service. As a consequence, certain protocol operations are implied in order for the network to satisfy the requirements of the standard.

THE PACKET SWITCHED NETWORK USER'S PERSPECTIVE

Defining the Interface Point

As we turn our attention to the user's view of an X.25-compliant packet switched network, we must first define the specific interface point we are dealing with. Figure 8-1 illustrates a typical network interconnection. The user operates from a piece of equipment referred to as the data terminal equipment (**DTE**). The network supplier provides the data circuit terminating equipment (**DCE**), which provides not only the interface control but also the means of digital transmission between the user premises and the network switches. In most cases the

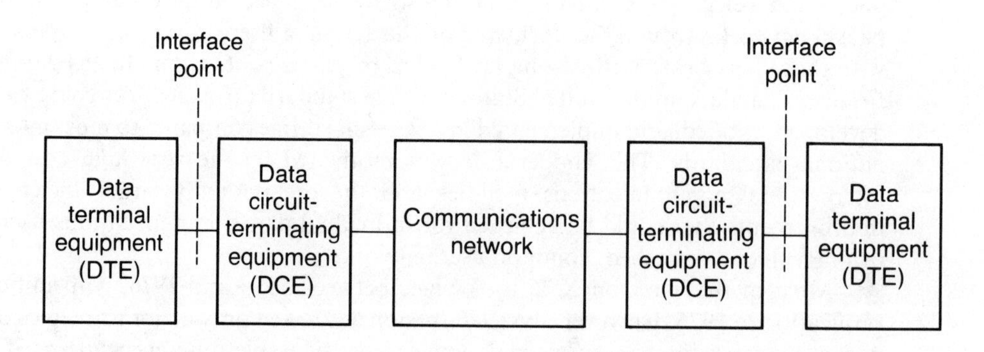

Figure 8-1. The elements of a typical network interconnection.

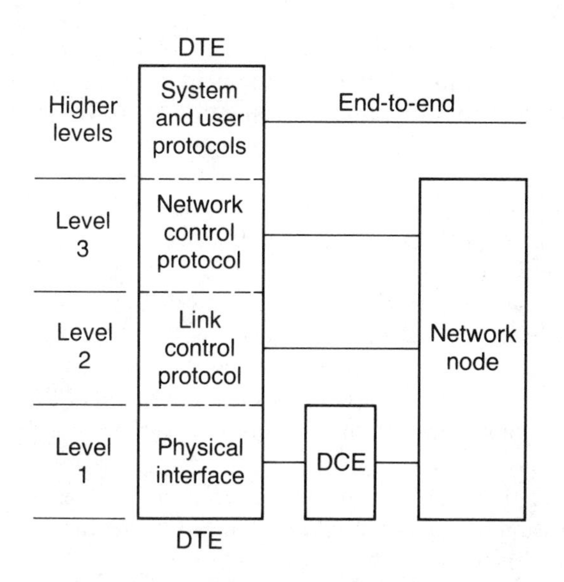

Figure 8-2. The lower levels of the network interface protocols.

DCE is principally a data modem to allow the digital user signals to be transmitted over analog access lines to the digital packet switches.

Figure 8-2 illustrates the network protocol hierarchy. Levels 1 to 3 correspond to the lower three levels of the ISO protocol hierarchy discussed in Chapter 7. Level 1 defines the terminal equipment physical and electrical interface. Level 2 defines the link control procedures to insure an error-free exchange of serial bits between the user equipment and the switching node. Level 3 provides the high-efficiency communications control to direct the flow of information through the packet network. Above level 3 are the numerous higher levels, which reside in the user's applications software, including the operating system, functional, and applications protocols. It is at level 3 that the X.25 standard makes its unique contribution in defining the service capabilities and features the network provides to the user.

Using Existing Standards

A key feature of the X.25 standard for packet switching is that it capitalizes on existing standards for public network interface wherever applicable. For example, at the level 1 electrical interface, the packet network standard calls for the use of the CCITT X.26 (for electrically unbalanced connections) or CCITT X.27 (for electrically balanced connections) interface standards. These are identical to standards within the United States defined by the Electronic Industries Association (EIA) as standards RS-423 and RS-422, respectively.

While we need not examine the specific details of the level 1 interfaces at this point, it is useful to know that the equipment manufacturers can work with agreed-on national and international standards in order to insure the interoperability of the data users' equipment with the public networks. (For interested readers, the Folts (1979) paper in the Suggested Readings for this chapter provides circuit and connection details of all the CCITT interfaces.)

At level 2 the CCITT X.25 again capitalizes on existing standards for link control procedures. It presumes that the user will interface the network using a relatively high-speed, synchronous form of data-link control. Over the years **link** level standards have evolved internationally as the ISO High Level Data Link Control (**HDLC**), and, in the United States, as the American National Standards Institute (**ANSI**) Advanced Data-Communications Control Procedure (**ADCCP**). These standards control data terminal equipment operating over point-to-point full-period circuits.

The heart of the data-link control procedures involve the use of bit synchronous transmission of distinct blocks of data. Many features of these procedures are superfluous for packet network access, since HDLC and ADCCP include addressing and control features used for polling and multidrop chains, and various shared arrangements of point-to-point circuits. The main features of the link level control that are relevant for packet network access are the block format and the error control.

The ADCCP-defined frame format is shown in Figure 8-3. Notice the similarity between the ADCCP frame structure shown here and the basic structure of the packet that was illustrated in Figure 6-4 (Chapter 6). Use of the ADCCP or HDLC link control at the packet switch-user interface allows the hardware to transmit any sequence of ones and zeros to achieve bit transparency. The beginning and end of a packet (or ADCCP frame) are delineated by the flag sequence, consisting of 01111110. Since this sequence could possibly occur as part of the user information, the ADCCP-compliant transmitter has to scan the outgoing bit stream. If five consecutive ones are detected, and the data set logic does not signal an end of the frame, the transmitter inserts a zero. At the receiver, if five consecutive ones are followed by a zero, the zero is deleted. If they are not, then the sequence is recognized as the true flag sequence indicating the end of the frame or packet.

Flag (F)	Address (A)	Control (C)	User data (information) (I)	Frame check sequence (FCS)	Flag (F)
01111110	8 bits	8 bits		16 bits	01111110

Figure 8-3. Packet network frame structure based on the Advanced Data Communications Control Procedure (ADCCP).

Error-Control Mechanism

The other major "standard" feature of the level 2 link protocol retained from the ADCCP for the X.25 interface is the error-control mechanism. Errors in the transmitted bit stream are detected and corrected by a process called a cyclic redundancy check (**CRC**) for **error detection** and block retransmission. The principle of operation of this mechanism is conceptually quite simple, although the mathematical basis and derivation of the error probabilities are quite sophisticated. We can see how it operates by reference to Figure 8-4 as well as by recalling how we carry out long division.

The CRC Procedure. The CRC concept treats the binary string of ones and zeros contained in the frame (packet) address, control, and information fields as a single long binary number, and uses that number as the dividend in a long-division problem. The dividend is divided, in every case, by the binary divisor represented by the binary sequence 10001000000100001. This digital operation results in a binary quotient plus a remainder (which, because of the size of the divisor, may contain up to 16 bits). As in decimal arithmetic, the divisor may go into the dividend an integral number of times, so that the remainder could be zero (in binary terms 16 zeros), but the zero remainder is a valid result.

This division process takes place in the transmitter at the sending end of the link, and the 16-bit remainder is appended to the frame (packet) in the frame check sequence, or CRC field. When the packet or frame is received at the distant end, the last 16 bits before the final flag are removed, and the same division, using the same divisor, is repeated at the receiver. The remainder computed at the receiver is compared to the one received with the data over the link. If they are the same, it is presumed that all the data was correct. If they differ in any bit, it is presumed that an error was made in transmission, and the receiver asks the transmitter to repeat the errored frame (packet).

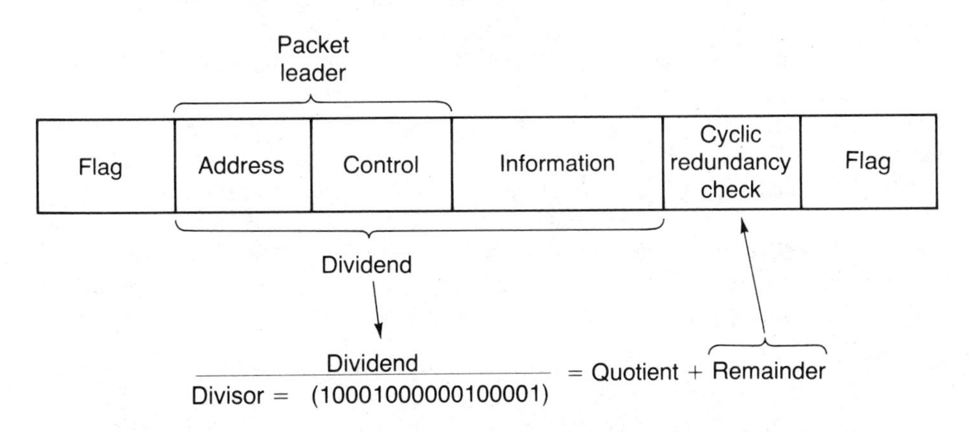

Figure 8-4. Operation of the cyclic redundancy code (CRC) error-checking procedure.

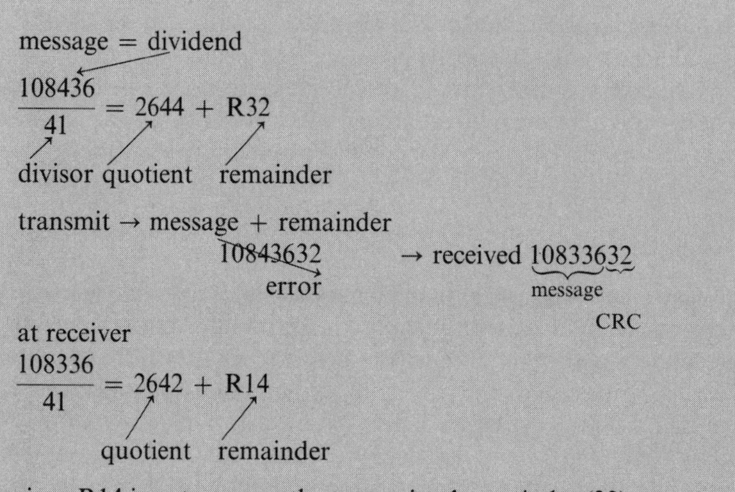

Box 8-1

An example, using decimal numbers rather than binary numbers, is given in Box 8-1. In this example the transmitted remainder was 32, but the remainder computed at the receiver was 14, indicating an error. Remember, the quotient itself is not transmitted since this would effectively double the number of bits transmitted. Only the remainder is appended to the information bits. The remainder is the only part of the process that is checked, although, obviously, an error in the transmitted data also changes the computed quotient.

Using the divisor shown, and packet lengths of up to several thousand bits, quite good error-control probabilities are achieved. This process will detect all single-, 2-, or 3-bit errors, all error bursts (consecutive errors) of 16 bits or fewer, and all error bursts with an odd number of errors. There is a chance that certain error combinations will cause the remainder to come out the same as the initial bit sequence, in effect fooling the error-control mechanism. However, the probability of this occurring with the 16-bit CRC is one out of 10^{10} packets. (This means that, if a user transmits 10,000 packets per day, an errored packet will slip through undetected, on the average, once every 3000 years!)

The link control functions at level 2, particularly the error control and bit transparency, are readily implemented in hardware or in microprocessor-controlled interface equipment. However, the real advantages of packet switching lie in a flexible set of programs implemented in software in the users' computers. The remaining functions and features that occur at the user-network interface reside at the third level of the protocol hierarchy. These functions are implemented in software in the network front end or network operating system in a computer using a packet network.

THE X.25 PROTOCOL—LEVEL 3 FEATURES AND FACILITIES

The main contribution of the X.25 is the set of switching functions that fall under the control of the user computer. Table 8-1 summarizes the functions at the user-network interface and the protocol levels that control and implement those functions. X.25-compliant networks have to insure a high probability of successful transmissions, which generally implies a means of alternate routing around network failures and a method for controlling network congestion. The network side of the level 3 protocol interface must also insure the proper sequencing of delivered packets and the accounting for packets actually delivered. Finally, the network is required to implement the fundamental switching functions, permitting multiple connections between different combinations of network ports.

Operation

Operation of the X.25 network is the fundamental virtual circuit we examined in Chapter 7. However, two possible modes are used—permanent virtual circuits (**PVC**) and switched virtual circuits (**SVC**). The only difference between these modes is in the interface function of call set-up and call clearing.

A permanent virtual circuit functionally replaces a permanent point-to-point circuit and guarantees connection, on demand, between a fixed pair of network endpoints. Since all data flows between the same two endpoints, there is no need for the user to tell the network what the destination of each new message or segment is when initiating a new network flow of data. For a user of this type, the interface is somewhat more efficient since the connection, essentially guaranteed, does not

Table 8-1 Functions at the User-Network Interface

Level 1—Synchronization
Level 2—Error detection
2—Correction by retransmission
2—Transparency
Level 3—Sequencing
3—Flow control
3—Multiplexing
3—Call set-up and clearing
3—Provision for network interworking
3—Logical in-band signaling
3—Logical out-of-band signaling

suffer a call initiation delay. The switched virtual circuit uses the maximum flexibility of the packet network, which is, in effect, its main distinguishing feature.

The features provided by the network can best be defined by reference to the structure of the call-request packet that must be sent from the user to the switch in order to initiate a new call request. The general structure of the call-request packet is shown in Figure 8-5.

The Call-Request Packet. If we tap the line between the user and the switch and watch the initiation of a new call, we will see the beginning of the call-request packet marked by the 01111110 flag sequence. Following the flag sequence, each data field has either a predefined or a specified length, and a set of agreed-on codings. We will show the functions of the various fields but will not try to define every possible coding feature. If we were actually to implement the X.25 interface in a user computer, the network supplier would provide a precise coding manual for that particular network. The basic structure, of course, would have to follow the format shown in Figure 8-5, as described below.

Flag (8 bits)

The sequence 01111110 denotes the beginning and end of a packet. Two successive packets need only a single flag between them to mark the end of one packet and the beginning of the next.

Link Address (8 bits)

Part of level 2 link control. The address is limited to the devices at each end of the connecting link between the DTE and the DCE. The network address is contained in the packet header.

Link Control (8 bits)

This is used for local counting and control, primarily for error control and correction between the DTE and the DCE.

Format Identifier (4 bits)

The format ID defines the nature of the packet that follows—such as a new call request, a data packet, or a previously established call—and defines the sequence numbering range (8 or 128).

Logical Channel Identifier (12 bits)

This designates the logical channel number for this particular data call. It can take any value from zero to 4095, in principle permitting a single user to simultaneously control up to 4096 individual data flows over a single line. The user selects the highest available number for new outgoing calls, while the network selects the lowest possible number for completing incoming data calls.

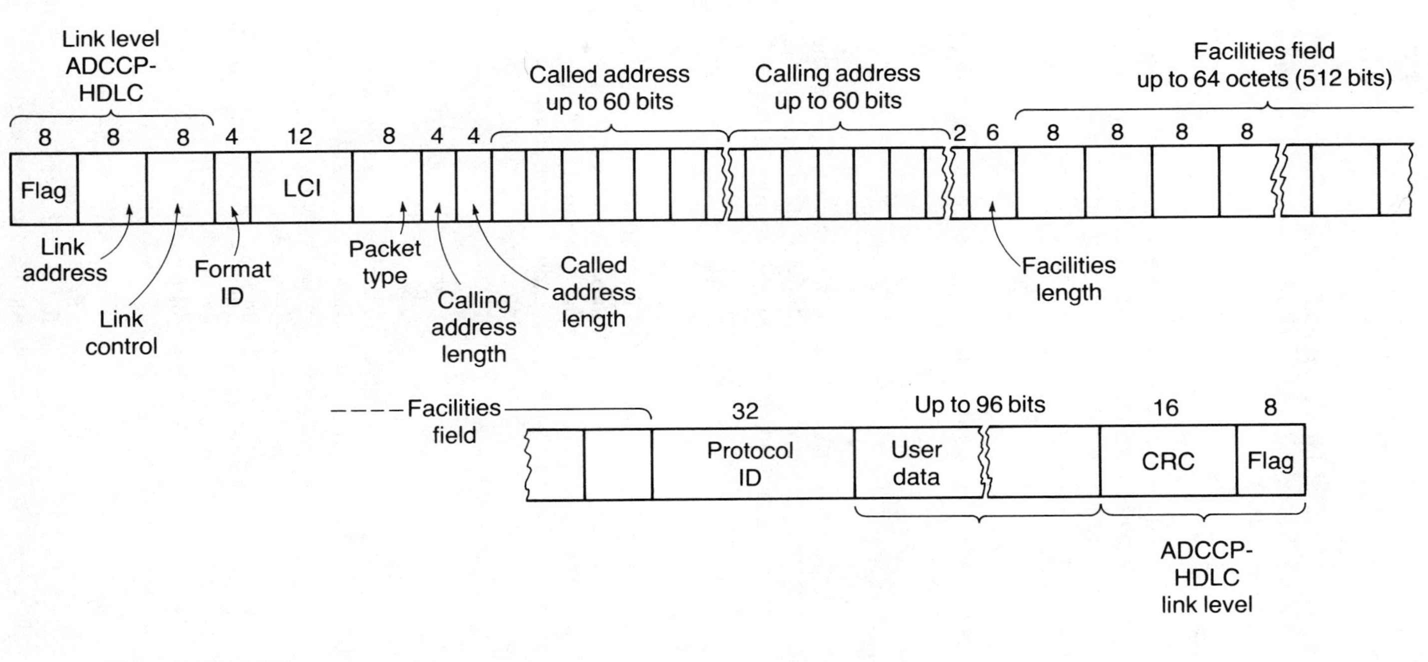

Flag = 01111110
Format ID—4 bits; designates packet characteristics such as connection type, numbering scheme, etc.
Logical channel no.—12 bits; 0-4095
Packet type—8 bits; call request, call clear, data, flow control, signaling, etc.

Figure 8-5. The call-request frame and packet structure as defined by the CCITT X.25 standard.

Packet Type (8 bits)

The packet type further defines the function and content of this packet, such as call clear, call reset, identify various actions under flow control or data restrictions, and the like.

Calling Address Length (4 bits)

This defines the length, in digits, of the address of the **calling party,** contained in a subsequent field. Like the called address length, this is a binary representation of a number from zero to 15 that defines the size of the user address field.

Called Address Length (4 bits)

This defines the length, in digits, of the address of the **called party,** contained in a subsequent field.

Called Address (up to 60 bits)

This field, whose length is determined by the previous field, is the address (in effect, the phone number) of the destination party. Each 4-bit group of the bits in this field represents one digit of the address, using binary coded decimal representation.

Calling Address (up to 60 bits)

This field, whose length is determined by the calling address length field, is the network address of the calling party. It is coded in the same way as the called address and provides, in effect, the return address for information sent back to the calling party. It also acts to verify, to the destination, where the call is coming from in the network. The calling address field is followed by two zeros.

Facilities Length (6 bits)

This field is a binary representation of a number from zero to 63. It tells the length, in 8-bit octets, of the facilities field that follows.

Facilities Field (up to 512 bits)

This field, whose length is determined by the previous field, provides for a large number of optional network facilities, such as reverse charging, a closed user group, one-way (transmit or receive) connections, and so on.

Protocol Identifier (32 bits)

This field can be used by the subscriber for certain user-level protocol features, such as the user identification and log-on procedure for initiating a new connection.

User Data (up to 96 bits)

The user can transmit up to 96 bits of data with the call request packet. This data may be, for example, the user's password to be used as part of the call set-up procedure within the destination computer.

Cyclic Redundancy Check (16 bits)

This is the error check applied to all the bits following the flag, up to the end of the user data field.

Flag (8 bits)

The sequence 01111110 denoting the end of the packet.

Data Transfer. Once the call has been established, an abbreviated version of this packet structure is used in the data transfer function. The acceptance of the call by the destination user is indicated back to the originator, with appropriate codes in the format and type identifier fields. The network, in effect, establishes a call in the reverse direction to provide a two-way, duplex, bidirectional path, and the flow of the user data can now ensue. The structure of the user data packets follows the format shown in Figure 8-6.

The overhead associated with the data packets is considerably less than that required for a call-initiation or call-request packet, thereby permitting efficient use of the circuit between the user and switch. The necessary information for routing the packets through the network is now stored in the switches, referenced to the source subscriber and the current logical channel identifier. Thus when a

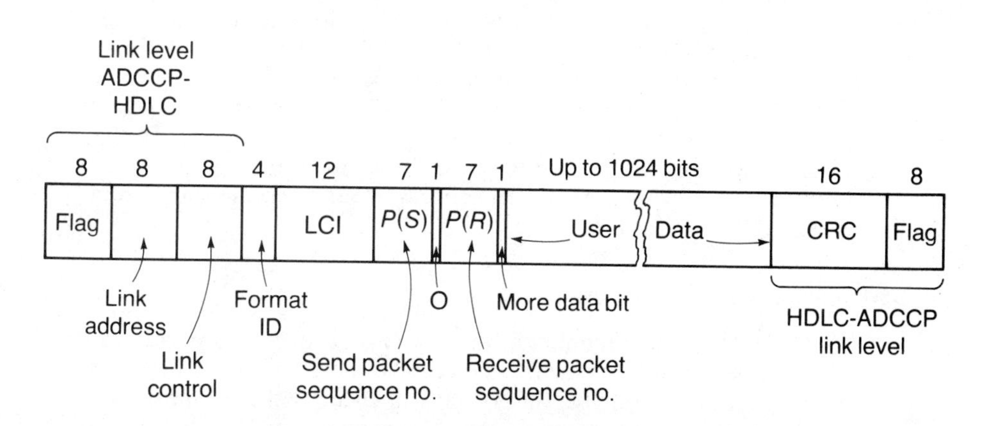

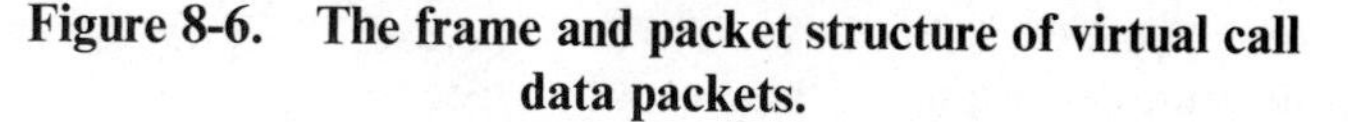

Figure 8-6. The frame and packet structure of virtual call data packets.

user data packet arrives at the source switch from a particular user, it follows the relatively simple structure of Figure 8-6, which is described below.

Flag (8 bits)

The sequence 01111110 denotes the beginning of the packet.

Link Address (8 bits)

Part of level 2 link control. The address is limited to the devices at each end of the connecting link between the DTE and the DCE. The network address is contained in the packet header.

Link Control (8 bits)

This is used for local counting and control, primarily for error control and correction between the DTE and the DCE.

Format Identifier (4 bits)

The format ID, which defines the nature of the packet, in this case would be coded to indicate a data packet in a previously set-up call.

Logical Channel Identifier (12 bits)

This refers to the logical channel assigned to this particular call during the call set-up process.

Send Packet Sequence Number (3 or 7 bits)

This is the sequential number assigned to each successive packet on this logical channel. The counter increments from zero to 7 or from zero to 127 depending on the coding in the format identifier field. The size of this field depends on the network speed of operation and the maximum number of packets that a given user is permitted to have outstanding in the network at any given time. In most cases, the extended field should be used to prevent any ambiguity in the sequence number by limiting the count to the range of zero to 7. The bit following the send sequence number serves no function but should be set to zero.

Receive Packet Sequence Number (7 bits)

This is the packet number of the last packet successfully received on this connection. It is the main method of acknowledging data. The network switches tell the users the sequence number of the last packet successfully received and acknowledged by the other user on the connection. Similarly, it allows the user to signal the switch the number of the last packet it has received.

More Data Bit (1 bit)

This bit is set to a one if the user is prepared to receive more data. It is set to a zero if this is an acknowledgement only, and no further data is desired at this time.

User Data Field (up to 1024 bits)

This is the transparent data field for up to 1024 bits or 128 data characters. The length need not be known in advance, as the packet end is marked by the flag, and the 16 bits before the flag are presumed to be the CRC.

Cyclic Redundancy Check (16 bits)

This is the error check applied to all of the bits following the flag up to the end of the user data field.

Flag (8 bits)

The sequence 01111110 denotes the end of the packet.

A number of special purpose packets can be constructed from these two basic packet formats, using a combination of the format identifier field, packet type identifier, and address fields. Such packets can handle a wide variety of anomalous network conditions, such as network- or user-initiated restarts or interrupts, clear connection requests, and signaling inquiries and responses. Of course, normal responses, connection confirmations, and packet acknowledgements are also handled as subsets to the basic packet formats.

It is important to emphasize once again that this standard addresses the interface between the user and the network. The structure of the packets internal to the network are up to the network designer and are not specified by the X.25 standard. For example, the packet format shown in Figure 8-6 does not contain the destination address of the data packets. The switches can correlate the address of each packet with the proper destination by use of the physical port and the logical channel identifier. When the switch builds the packet for transmission over the network, it is likely to insert the packet address, as well as other network control, into the interswitch packets. At the destination switch, the packets (segments) delivered to the destination user are transformed back to the X.25 defined interface formats.

USER ACTIONS UNDER THE X.25 PROTOCOL

The many features and facilities provided under the X.25 protocol specification accommodate a wide range of user capabilities. The X.25 is an intelligent interface, that is, designed for the interface of programmable computers to an intelligent packet network. The features can be readily extended to microprocessor-controlled user terminals since the X.25 features for a single logical channel can be readily programmed in software or firmware for single user terminals operating in a point-to-point mode through the packet network.

Additional standards, designated X.28 and X.29 have been developed to provide for the network emulation of point-to-point connections for nonintelligent terminals operating through a packet network. These interface features are imposed on the network switching centers, which have to provide the packet assembly-disassembly (**PAD**) functions to accommodate the various commercially available nonintelligent terminals.

We will continue our discussion of the user-network interface in the context of the intelligent user and look now at the actions users must take in response to network commands and instructions in order to operate efficiently through the X.25-compliant network.

Of first priority is the software implementation of the network control program that builds the packets and generates the needed control commands and responses. A standard of this type makes it cost-effective for computer manufacturers to develop standard communications X.25-compliant interface packages for their equipment. In fact, a significant number of computer manufacturers today can supply X.25-compliant networking packages. The specific details for network interfaces are generally supplied by the network operator in cooperation with the specific implementation details of each computer mainframe.

Besides the basic function of division of the messages into packets, the user interface has several other important functions. Current implementations of the X.25-compliant network operation are virtual circuit, although the 1980 version of X.25 now includes standards for the datagram mode of operation. Packets are controlled within the network to insure that they are delivered in sequence, but the user must place a sequential number on each packet on a particular logical channel. As we saw in Figure 8-6, this is achieved by the send packet sequence number, $P(S)$. The packet sequence number is independent for each direction on a particular logical channel, and must increment for each new packet, as long as that channel is in use between the initial connection request and connection termination.

Flow Control

The send packet sequence number is used in conjunction with the receive packet sequence number, $P(R)$, to control the flow of data into the packet network. The network flow control prevents the network from accepting traffic more rapidly than it can deliver it and thereby creating congestion. The flow is controlled so that all packets accepted into the network will be available for delivery at the destination within a fraction of a second. (Notice that we can't say they *are* delivered in any specific time. Actual delivery time is dependent on other traffic going to that destination and the speed of the line out to the destination user.) The X.25 flow control is implemented by what is called the **window** method.

The window flow control matches the data entrance rate with the delivery rate. It also slows the entry rate of new packets in case problems in the network, such as a failed line or failed switch, cause the network to become temporarily congested. As we saw in our examination of queueing theory in Chapter 5, we want to keep the network loading well below capacity to preserve the delay characteristics for the traffic that is already in the network.

The Window. The window defines the maximum number of undelivered packets a given user may have outstanding on a particular logical channel at any given time. When the network switch transmits any information to a user on a particular logical channel, the value of the receive packet sequence number, $[P(R) - 1]$, is the sequence number of the last packet that was successfully delivered by the network. The incremental updating of $P(R)$ by the network acts as the positive acknowledgement of transmitted packets, up to and including that packet number.

The window flow-control mechanism requires that the highest sequence number transmitted by the user be less than the sum of the current $P(R)$ plus W, the maximum transmit window size. For example, let us say that W is equal to four, meaning that the user has at most four packets in process through the network at any given time. If the last packet acknowledged by the network was $P(R) = 16$, then $P(R) + W = 16 + 4 = 20$. This means the user is now permitted to transmit packets number 17, 18, and 19, but may not transmit packet 20 (or higher) until the value of $P(R)$ is again incremented. Any attempt to transmit a packet that is outside the currently allowed window is considered a procedure error, and it will either be rejected or cause a reset of the logical channel.

Window size is a feature that is agreed on by the network supplier and the network user at the time network service is initiated. The larger the window, the greater traffic loading a given user can place on a single logical channel. Moreover, as the window size increases, more network resources in the form of channel capacity and switch buffering must be allocated to that user. Therefore, the window size has to be a balance between the user's required maximum throughput and the cost of providing the service.

Operation of the Flow-Control Mechanism. The operation of the flow-control mechanism is based on the fact that, if the network must limit the flow of new packets into the network, it can slow down the rate at which the $P(R)$ value is incremented or temporarily reduce the value of W by transmitting an appropriate control message to the user. The operation of the window mechanism is illustrated in Figure 8-7, which shows the receipt of packet 16 and a window of 4. Notice that it is not necessary for the value of $P(R)$ to be incremented one unit at a time. For example, we see $P(R)$ going from 17 to 20 in one jump. Such a phenomenon

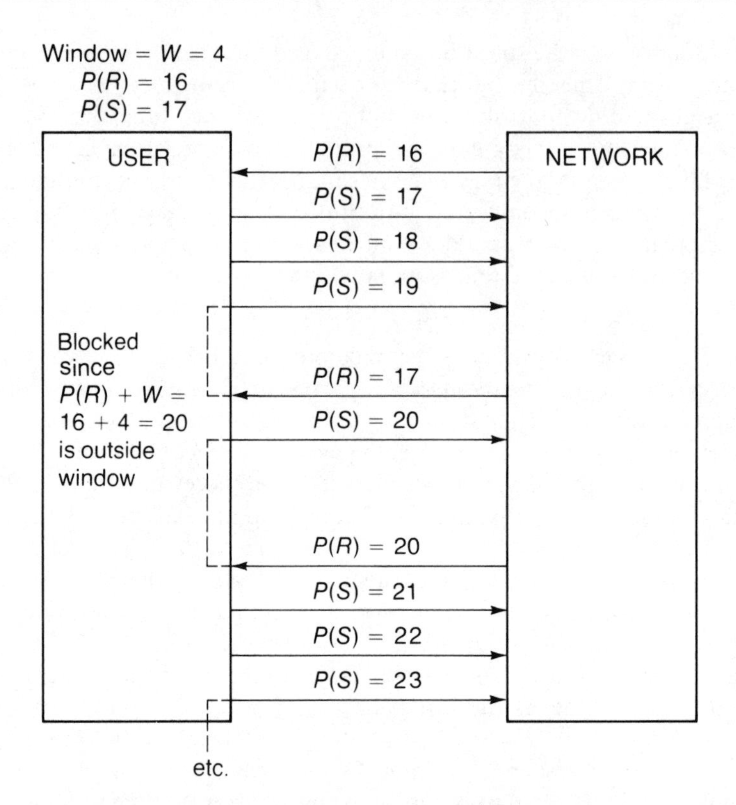

Figure 8-7. Operation of the user-to-network flow-control window.

could occur, for instance, if congestion temporarily slows packet delivery, and three packets are received and acknowledged in a very short period, so that all three can be handled with a single incremental change of $P(R)$.

Out-of-Band Signaling

Since data is subject to flow control independent mechanisms are provided to permit the transfer of control information outside the normal flow of data. In effect, this provides a capability for out-of-band signaling that cannot be constrained by flow control. The relevant mechanisms include interrupt packets, reset requests, restart requests, and status inquiries into the network. Only one such packet can be outstanding in the network at any time, and it must be confirmed before any other such packets can be transmitted.

Signaling packets can make inquiry on direct actions by the network, or they can be used to send a single 8-bit coded message across the network to the distant user. Interrupt packets, that is, packets containing up to 8 bits of user data, will be transmitted by the network ahead of all other packets awaiting transmission.

Furthermore, the network will attempt to transmit interrupt packets to the distant user even when that user indicates that it is not accepting data packets. The signaling packets contain no sequence numbers but are associated with a specific user pair connection with the logical channel indicator.

Optional Facilities

Through the facilities field the X.25 protocol provides for a variety of optional user actions. The optional facilities, which may be implemented by the network supplier in a number of different ways, include reverse charging, one-way only channels, priority classes of service, closed user groups, and retransmission of packets to multiple destinations. These facilities and network actions give the users considerable flexibility in establishing an apparently private network within a public switched network. The capability for reverse charging, for instance, can make network services very attractive to a large customer base, much as the 800 series of inward **WATS** lines serves the voice network–based business community.

SUMMARY

This chapter has explored the interface of the "standard" data user with an X.25-compliant public packet switched network. We have learned about the features and facilities that the X.25 interface makes available to the users. We looked at the structure of the two fundamental packets that flow between the users and the networks—the call set-up packet and the data-flow packet. From this structure, and the operation and meaning of the various fields, we can summarize the key packet functions.

Logical multiplexing over a single-access line is achieved by use of the logical channel indicator field. Packet sequencing is achieved by requiring the user to assign a send sequence number to each data packet. Flow control is maintained by use of an allowable transmit window, linked to the current send packet sequence number and the last acknowledged packet. In-band signaling is provided by a large number of possible format, packet type, and facilities identification fields. Out-of-band signaling is provided by interrupt and special control and command packets. There are also a large number of reset and progress commands and signals to keep the users of the network informed on network problems. Finally, we have gained some familiarity with the many current optional features that make packetized public networks using the X.25 protocol very responsive to many different classes of users.

In conclusion, it should be reiterated that the X.25 protocol, despite all of its features and capabilities, is solely a network interface protocol. It does not provide for any of the user-to-user level protocols. However, the layering of the protocols makes solution of these user-level protocol issues essentially independent of interfacing the network. The result is an efficient, effective data transport mechanism.

SUGGESTED READINGS

FOLTS, HAROLD C. "Interface Standards for Public Data Networks." *NCS Technical Information Bulletin 79-2*. March 1979. Available from National Communications System, NCS-TS, Washington, D.C. 20305.

Folts reviews all of the CCITT interface standards applicable to public data networks. These include standards applicable to nonpacket networks, such as public networks that provide point-to-point leased private-line services. The paper shows the broad applicability of such standards, the physical/electrical arrangements of standard data terminal equipment, the link level and electrical level interface standards, and the subordinate standards on which the X.25 protocol layers are based.

"Revised CCITT Recommendation X.25—1980." *NCS Technical Information Bulletin 80-5*. August 1980. Available from National Communications System, NCS-TS, Washington, D.C. 20305.

The X.25 standard is a living capability. Revisions, extensions, and modifications are always being suggested by user and supplier experience and new service requirements. This technical information bulletin presents the text of the X.25 standard, incorporating the latest revisions as of the publication date.

"Standard Network Access Protocol Specification for Datapac." Ottawa: The Computer Communications Group, Trans-Canada Telephone System, 1976.

This document was prepared to explain a specific implementation of the X.25 protocol to potential users of Datapac, the Canadian nationwide packet switching network. It illustrates some of the versatility of this protocol.

9

Control and Monitoring in a Network—Some Considerations

THIS CHAPTER:

will look at the important role of packet network control and management in different user environments.

will examine several approaches to control implementation and evaluate their advantages and disadvantages.

will consider network control centers as examples of control implementation.

THE ROLE OF CONTROL

Development of a control and management structure for a communications network is rather imprecise because of the many unpredictable and statistical conditions that can affect its operation. In this chapter we will look at many of the considerations that a network manager must include in the establishment of a control structure for a packet switched network. To begin, we will distinguish between the two basic network environments.

Two Network Environments

Table 9-1 summarizes the key characteristics of two network environments—the common user network and the single user network. The public, common user network must deal with a large number of functionally disjoint users who have different characteristics and statistics. Because of the public access to the network, control and protection must be able to cope with both intentional and accidental threats to the accuracy and integrity of the data carried by the network. The network also has to be adaptive to a variety of user needs, changing demands, and service requirements. This type of user environment is typified by the worldwide **AUTODIN II** network of the U.S. Department of Defense, or any of the commercial packet switched networks, such as Telenet in the United States.

Table 9-1. Characteristics of the User Environment

Common User Network	*Single User Network*
Functionally disjoint	Functionally homogeneous (not necessarily operationally)
Vulnerable to various threats	Benign
Adaptive to user needs	Legislative to user needs
Typified by AUTODIN II and Telenet	Typified by ARPANET and internal corporate net

A single user network is designed to accommodate many functionally homogeneous users, although they may not be operationally homogeneous (that is, they may have a wide variety of equipments and user statistics). As the only users with access are part of the same organizational entity, the environment can be assumed to be relatively benign. Such a network can be legislative to user needs, in that it can be restricted to a limited set of user characteristics, equipment manufacturers, and interface protocols. A single user network is typified by the ARPANET or by a network serving the needs of a single corporation, organization, or agency. Compared to the common user network, the controls can be simpler, less protective, and concerned with achieving maximum throughput.

Control Functions

While the implementation of control varies with the user environment, the role of control in respect to network operation is essentially uniform. Table 9-2 summarizes the various functions that must be performed in a packet switched network. Let us take a closer look at these functions.

Assure Network Integrity

The most fundamental control operation is to insure that users can communicate with each other. This involves maintaining user and switch connectivity and restoring service if any network components fail. The control system has to insure user and network reliability and availability, monitor adverse trends in service quality, and take corrective action to restore service when necessary. When service is required to a location presently unserved, the control system must be able to manage the network extension, with appropriate routing plan and numbering plan changes, and additions to the directory and reference listings.

Maximize Traffic Throughput

Since the effectiveness and revenue production of a network will be related to the amount of data traffic carried, one objective of the

Table 9-2. Role of Control in a Packet Switched Network

ASSURE NETWORK INTEGRITY
Connectivity
Reliability/availability
Restoration
Extension
MAXIMIZE TRAFFIC THROUGHPUT
Routing
Flow control
MONITOR ABUSES AND MALFUNCTIONS
Privacy
Misrouting
Disruption and spoofing
Data accuracy
PROVIDE BASIS FOR BILLING OR COST RECOVERY
Traffic quantity
Delay and precedence data
PROVIDE DATA FOR PLANNING AND ENGINEERING
Traffic growth
User sensitivities
Traffic patterns

control system is to make sure that the network is passing as much traffic as possible. This involves traffic routing and flow control to use the parts of the network efficiently and prevent excessive delays.

Monitor Abuses and Malfunctions

The control system of a packet network has to watch for both intentional and inadvertent abuses or malfunctions that may occur as a result of switch problems or improper actions by users. Many user errors, if not detected, could cause network problems such as interruption of data connections in progress, loss of data packets, or temporary refusal to accept new traffic.

The control system thus has to monitor protocol errors, insure the correct identification of traffic sources and delivery points to protect the privacy of user information, detect and correct misroutes of traffic, and insure the accuracy of all data delivered by the network. Finally, especially in public networks, the control system must be capable of detecting disruptive users, spoofing, or users attempting to extract information to which they are not entitled.

Provide Basis for Billing or Cost Recovery

A key control issue in both public and private networks is the ability to allocate the cost of the network to the actual users. Billing and cost recovery should make use of traffic quantity information, as well as network performance factors such as achieved delay, precedence service, or the distance over which the information is carried.

Provide Data for Planning and Engineering

A final control function is to collect information on traffic patterns, traffic growth, and desired and needed ranges of user performance. This information is required in order to plan for network growth, to organize network facilities efficiently, and to provide for efficient allocation of transmission capacity among the switching centers.

These control functions can be implemented in many ways. Some are generally integrated as part of the normal network level operation and interswitch protocols. The rest are best included in specialized facilities that can collect and process the required information and, when necessary, alert control and management personnel of problems. The next section will discuss various possible implementations of the control functions in a packet switched network.

Implementation of Control

Figure 9-1 presents a decision tree approach to control implementation alternatives. In terms of a packet network as a whole, the control facilities may be either centralized or decentralized, and either active or passive. Decentralized control is also referred to as distributed control.

Centralized versus Decentralized Control. By **centralized control** we mean that, at any given time, the controlling element is the responsibility of a single element within the network. In order to assure reliability, a number of different elements probably can assume the role as the network controller. At any particular moment, however, only one such element is actively carrying out control duties, with the others available as backup.

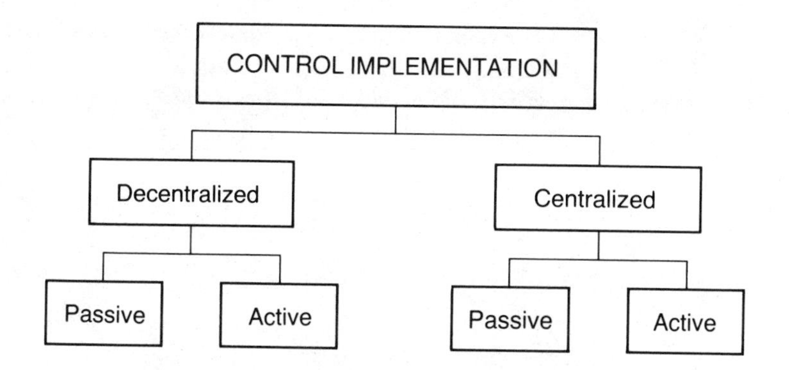

Figure 9-1. Decision choices in packet network control implementation.

Decentralized control means that the control over network activities is divided among many elements of the network. For example, part of the functions of each switch of a packet switched network could be to maintain control over the network's own operation and its relationship with its directly connected neighbors. Decentralized, or distributed, control tends to be very **robust,** since its operation does not depend on receiving direction from any single outside control entity. At the same time, however, it tends to be less than optimum since each control element makes decisions based on very limited visibility of the entire network.

Active versus Passive Control. The terms *active* and *passive* control are nearly the same as *dynamic* and *static*. By **active control** we mean a dynamic implementation where the control elements are continually evaluating network performance and are taking real-time actions to insure that the network is operating in a (near) optimum fashion. Further, actions taken by any network element, or user, must, in some sense, be actively checked and "validated" by the control elements. Active control allows the limited resources of the network the maximum utility to meet the total demands of the users. On the other hand, the entire network operation may be seriously impaired if the control capability becomes defective.

With **passive control** the control elements are effectively outside the network, looking in and observing the network's operation. The network is designed to be self-sufficient, and the information associated with the control elements has only a non-real-time, or long-term, impact on network operation. In general, the advantages and disadvantages of passive control are essentially the reverse of those of active control. Continued network operation does not depend on the presence of control elements. However, network operation is likely to be far from optimum since the operating parameters can be changed only over a long, non-real-time period.

Table 9-3. Evaluation of Control Approaches

	Centralized Passive	*Centralized Active*	*Decentralized Passive*	*Decentralized Active*
Network integrity	2	1	4	3
Traffic throughput	4	2	3	1
Abuse monitoring	1	3	2	4
Cost and billing	1	3	2	4
Engineering data	2	1	4	3
Cost	3	4	1	2
Adverse impacts	3	4	1	2
Total	16	18	17	19

(Best = 1)

Evaluating Control Approaches

Table 9-3 ranks the various approaches in terms of the five major control functions, the cost of control, and the adverse impacts on the network if the control elements fail to operate. A "1" indicates the approach that is best for that function, and a "4" the least desirable approach.

Network Integrity. Network integrity, the assurance that connected users can communicate with each other, can best be managed and controlled by a centralized, active control structure. A centralized control maintains maximum visibility of conditions throughout the network, and an active control allows the network to rapidly adapt to changes in facilities, conditions, or user demands. Inventory and management of all network assets is most effectively maintained by the centralized control facilities, as are the assignment of resources and the cataloging of users and services. The least desirable control approach for the maintenance of network integrity is a decentralized, passive control structure. Passive control cannot overcome network problems rapidly enough to insure network integrity for all users. Similarly, a decentralized structure requires more time for control directives to be propagated among the network elements, further delaying the actions required to maintain integrity.

Traffic Throughput. Maximum traffic throughput of the network can best be achieved by a decentralized, active control. To achieve maximum network throughput, each switch must be able to find capacity in the network that is available to deliver new user traffic. Active control is thus required, and decentralized control is desirable since loading conditions can change so rapidly that information to and from a centralized location could be quickly outdated. Conversely, passive control would not permit rapid changes to network loading conditions, nor would it provide for the localized implementation of flow control to avoid network congestion.

Abuse Monitoring. The monitoring of intentional or inadvertent abuses and control of protocol malfunctions can best be implemented with centralized, passive control. The centralized approach can apply an adequate amount of processing power to accumulate needed data on user activity and protocol action. Further, operating passively allows for a degree of anonymity in checking for proper user and network operation. This makes the control system difficult to spoof, since an unauthorized or improper user cannot probe the network looking for acceptable responses from the control elements. In addition, since the abuse and malfunction monitoring is not absolutely essential to the successful delivery of user traffic, a centralized passive control is the most efficient and least costly implementation. The least desirable approach would be an active, decentralized control facility, primarily because of the cost and complexity of collecting all of the data required and processing it.

Cost and Billing. Similar reasoning applies to the collection of cost, usage, and billing information. The most cost-effective and efficient way of doing this is with a passive, centralized facility that collects the information as it is measured and relayed from each of the switches. An active, decentralized facility would be least preferable for the same reasons as those under abuse monitoring.

Engineering Data. The collection of data for planning and engineering has similar needs as abuse monitoring and collection of cost and billing information. However, more active control functions are desirable so that unusual trends or actions can be recognized in real time, and information requests, tracers, or special monitors can be implemented. A centralized, active facility is preferable since use of the engineering data requires considerable manipulation of the information collected.

Cost of Control. The least expensive (and thus most desirable) control structure is a decentralized, passive one. Such an approach would permit the control functions to be incorporated as small, incremental additions to the software (and possibly the hardware) at each network switch. The most expensive approach, in terms of both initial cost and continued operation, is an active, centralized control. It would require sufficient resources to manage network operation in real time, would have to be redundant for reliability, and would use some of the network resources to collect and disseminate the control information.

Adverse Impacts. Similarly, the approach with the fewest adverse impacts is passive, decentralized control. With passive control continued operation of the network is not impaired in real time by a failure of control elements. In addition, decentralized control responsibilities are shared by a large number of facilities and so are not likely to be adversely impacted by the failure of one or a few control elements. The least desirable approach would be a centralized, active approach, which would render the network inoperable, or at least highly degraded, in case of failure.

Choosing an Approach

Now that we have assigned subjective rankings to each of the functions of the packet network control structure, we can look for a measure of preference for one structure compared to the others. We have attempted to do this in Table 9-3 by adding the subjective rankings together. If a particular control structure were best in every category, it would have a summed value of 7, while the least attractive approach would have a value of 28. The total values in Table 9-3 fall in a very narrow range, making it difficult to choose between them. On the other hand, with all approaches virtually equal, we can choose a control structure that is the best for a particular function or that is least expensive. For example, in a military data network, where network integrity might be the most important function of control, we would choose a centralized, active control with many backup facilities that could guarantee integrity if the primary control element failed.

In reality, network designers try to implement a control structure that combines some of the best features of each of the individual structures we have talked about. The most frequently used approach is to combine the features of decentralized, active control with those of centralized, passive control. Active control features are built into each switch, primarily with respect to maintenance of traffic throughput. At the same time, each switch reports measurement data to a centralized, passive network control center for non-real-time processing and corrective actions. This combination is the approach taken in both the ARPANET and the AUTODIN II networks in the United States, and it is also followed in most commercial packet switched networks. In the next section we will take a brief look at the network control center concept as it is used by these two networks.

NETWORK CONTROL CENTERS FOR PACKET NETWORKS

Over the last decade experience has shown that extremely good control and monitoring network performance can be achieved by a combination of a centralized, essentially passive network control center **(NCC)** coupled with a set of active controls applied as part of the protocol and software features of each switch. Putting some of the control functions into the switches themselves achieves a large measure of distributed control, at least in making the basic operation of the network immune to the failure of a single control element. Making the hardware of the NCC identical to that used by the switches themselves adds a further measure of reliability since, by relatively simple reconfiguration of network assets, any one of the switches could assume the functions of the NCC.

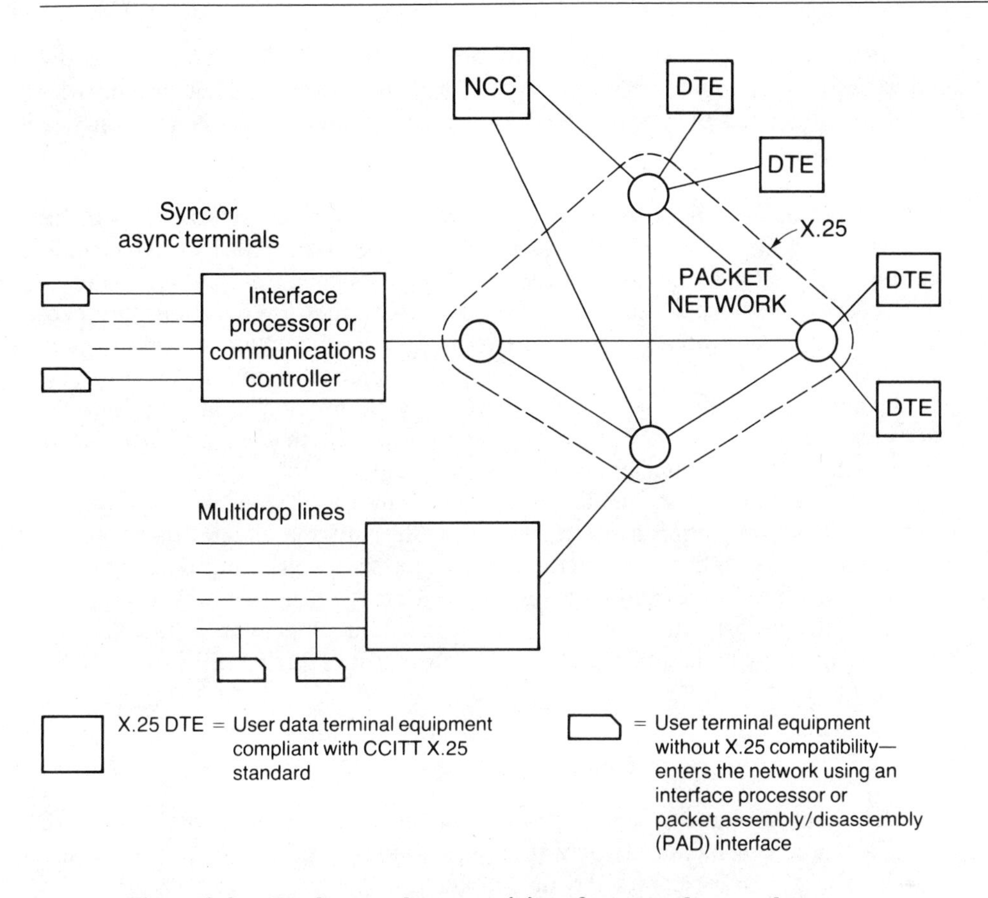

Figure 9-2. Dual network connectivity of a network control center.

A Typical Configuration

Figure 9-2 illustrates a typical configuration of network elements, where the NCC is outside the network boundary. As a passive device with respect to the real-time networkwide flow of information, the NCC is coupled to the network as if it were any other host computer subscriber to the network. For reliability at least two such connections are provided, to two different switches.

Functions. The principal functions carried out by the NCC relate to the information-gathering (abuse monitoring, collection of cost and billing information, collection of engineering data) that is part of all control functions. The NCC provides long-term system control and operational direction to the switches and operators in the network. It provides for long-term controls (such as traffic changes) and correlates and analyzes network disturbance data collected from many network locations to determine the cause of networkwide problems. In addition, because the NCC is a single point with good visibility of the entire network, it can provide current status information to aid staff in dealing with totally unpredicted problems. The NCC and its support staff thus comprise a kind of crisis management capability.

Data Collection. To perform all these functions, the NCC collects disturbance indications, billing and traffic data, and performance data. The disturbance indications include failure of either lines or switches, as detected by other switches, and security/privacy problems, as indicated by excess message rejections, protocol errors, or unusual traffic patterns. The billing and traffic data is useful not only for revenue production but to insure a high degree of network integrity and traffic throughput. Long-term measurements of traffic data indicating that the network is delivering only 10 percent of the traffic for which it was designed would suggest that the distributed control operation was faulty.

The performance data relates to the monitoring of delay, peak load throughput, and call completion rates achieved by the network under varying traffic conditions. This kind of information is part of the engineering data used for sizing network resources and reconfiguring resources when necessary to meet the actual traffic pattern. In addition, since users will judge network service on the basis of the delay their traffic encounters, constant monitoring of this parameter is critical.

Routing and Flow Control

The real-time operations of control, which are distributed to the network switches, are implemented largely through dynamic routing techniques. (These will be discussed more completely in Chapter 12.) The principle employed is that each switch attempts to send its packets over the shortest (in time) path to the destination. The key point is that the shortest path changes as the loading in the network changes. In addition, the shortest path measured in terms of time delay often is not the shortest in terms of distance. The switches estimate the shortest delay path on the basis of information at the switch together with delay information given to each switch by its immediate neighbors. Under most loading conditions this tends to distribute the traffic load and achieves the highest possible network throughput as new traffic is routed around congested areas, over lower delay paths.

The switches also control the traffic originated at each switch by use of the flow control mechanism described in the previous chapter. When more traffic is in the network, response times and acknowledgements become longer. The switches are designed to reduce this loading by reducing the flow control window, thus reducing the rate at which new traffic enters the network. Failures of lines or switches in the network are therefore recognized by the distributed control functions because of the large increase in apparent delay over a path utilizing a failed network element.

Table 9-4. NCC Operator Functions

Set parameters and thresholds
Analyze network problem indications
Initiate performance monitoring
Report production
Respond to user requests
Implement directory changes
Maintain network data base
Modify node level software

Proper operation of the routing and flow control protects the basic integrity of the network. As long as a path exists between two users, it is possible for traffic to move successfully between those users. If individual lines or switches fail, other network integrity controls, such as failure restoration and directory control, are the responsibility of the network control center.

NCC Operator Functions

In most designs the NCC is implemented by hardware that corresponds to the switch hardware. However, it requires specialized control and monitoring software, as well as operators who can deal with unusual conditions. The list of NCC operator functions shown in Table 9-4 summarizes the non-real-time control functions performed by the control structure.

Set Parameters and Thresholds

Various network parameters, such as timeout periods, frequency of routing table changes, or maximum allowable traffic buildup before flow control is applied, have to be changed from time to time, depending on network operating and load conditions. These changes are initiated and executed from the NCC, generally under direction from NCC operations personnel.

Analyze Network Problem Indications

While the NCC software is designed to recognize anomalous conditions as reported by the switches, there is always the possibility of totally unpredictable problems or unusually severe combinations of problems. NCC operators will have to deal with these kinds of problems, taking short-term actions (such as parameter changes) to reduce the impact in real time and performing further analysis to prevent a recurrence.

Initiate Performance Monitoring

From time to time, particularly in response to user complaints, it will be necessary to monitor the performance of some part of the network. The initiation of such monitoring and the collection and analysis of data will normally be done by NCC equipment and personnel.

Report Production

The NCC will produce periodic reports on network performance, loading, and other factors as needed for the administration and engineering of the network.

Respond to User Requests

The NCC will be the focal point for response to user requests, such as a change of service, addition or deletion of capability, or complaints about services rendered.

Implement Directory Changes

The networkwide user directory and the directory of network assets and resources will be maintained by the NCC. Any changes in this directory, particularly in response to user requests, will have to be initiated and directed by the NCC.

Maintain Network Data Base

The master data base, which would include network and user information, as well as information about or copies of the software for each switch, will be maintained by the NCC.

Modify Node (Switch) Software

The NCC can initiate and execute any changes to the software of the switches by performing a downline load via the network. This avoids physically distributing hard-copy versions of the switch software and having personnel available at each switch capable of loading the new software. Thus the packet network is even more flexible since changes can be implemented easily and rapidly across the entire network.

SUMMARY

As we have seen throughout this part of the book, the operation of packet networks is the result of a set of protocols, implemented primarily through software, which permit the orderly flow of information through the network. In this chapter we have outlined the many control functions that are necessary in a packet network, from insuring that the users can communicate, to insuring that information is available for billing users for services they receive from the network.

Control can be implemented through either active or passive elements, and can be either centralized or distributed within the network structure. In present-day packet networks these approaches are commonly combined to utilize the best features of each control technique. As a general rule, it is best to distribute the real-time controls and centralize the more non-real-time controls.

We looked in detail at the network control center (NCC) concept as an example of a centralized, essentially passive approach that is frequently combined with the active controls that are incorporated into the operational protocols of the network. The principal functions of the NCC are gathering information related to abuse monitoring, cost and billing, and engineering requirements. Routing and flow control are implemented by both switch hardware and operators who can deal with unusual conditions. In this way the basic integrity of the network is protected.

SUGGESTED READINGS

COLE, GERALD D. "Performance Measurements on the ARPA Computer Network." Reprinted in *Computer Networks: A Tutorial.* New York: IEEE Press, 1975, pp. 6-12–6-18.

This paper presents detailed information about how the performance of the ARPANET is monitored and measured. Both active and passive measurements are described, including observation of network performance with real traffic and loading of the network with "trace" information. The utility of designing control features as part of the software of each switch is clearly demonstrated.

MATHISON, STUART L. "Commercial, Legal, and International Aspects of Packet Communications." *Proceedings of the IEEE*, vol. 66, no. 11 (November 1978), pp. 1527–1539.

While this paper does not deal directly with the control of packet networks, it provides useful insights into the regulatory environment in which public packet networks will probably continue to operate and the kind of management and information necessary to comply with such regulation. Issues such as degree of standardization and exchange of billing and tariff information are discussed.

MCKENZIE, A. A., COSELL, B. P., MCQUILLAN, J. M., and THORPE, M. J. "The Network Control Center for the ARPA Network." Paper presented at the First International Conference on Computer Communication, Washington, D.C., October 1972. Reprinted in *Computer Networks: A Tutorial.* New York: IEEE Press, 1975, pp. 6-5–6-11.

The authors trace the development of the NCC for the ARPANET from its earliest form to its current implementation as a highly capable, processor control element of the network. The article shows the kinds of information collected and samples of the key network loading and outage information.

PART THREE

Packet Networks in the Real World—Topology, Routing, Robustness, and Some Lucky Guesses

In the first two parts of this book we learned the fundamental principles of operation of packet switched networks. In developing the concepts of the network operation and the description of the network protocols, we frequently used examples of network connectivity and topology where there were a limited number of switching nodes between the source and destination switches. Such topologies are consistent with the objectives of packet switching to insure rapid, low-delay communications among the network subscribers.

In this part of the book we will explore some of the impacts of various topological network designs. Clearly, the objective in the real world is to design a network topology that provides both a high-performance network and economical and efficient use of the network resources. We will look first at the basic principles of topological network design. We will show some tricks and techniques that can be used to make packet switching work better. A case example will illustrate some of the techniques and show that, in the design of moderately large networks, we can safely estimate user requirements. We will also discuss the very important topic of traffic routing in packet networks, where, regardless of the final network topology, clever techniques can insure rapid, efficient delivery of user traffic.

10

Topological Principles—Some Tricks to Make Packet Switching Work Better

THIS CHAPTER:

will look at the topological design of various network structures.

will look at the general formulation and constraints of topological design.

will define the features of network topology that enable packet switching to perform well consistently, and illustrate them through the example of the ARPANET.

When we speak of the **topology** of a network, we are simply referring to the configuration of the network resources. The problem of topological design is by no means unique to packet switching; for many years it has been actively studied in relation to a wide range of transportation and communication problems. Topological design is of equal importance for conventional switching networks for voice communications, for a network of roads connecting a number of communities, or for oil and gas pipelines that carry a fluid product from its source to its destination.

We address topology in this chapter because topology affects a packet network's ability to meet its design objectives consistently. Of particular significance here is the effect that was illustrated by Figure 7-7 (Chapter 7), which showed a rapid drop in throughput as the number of nodes traversed increased. This effect is accompanied by an increase in delay of users' information across the network. As we move through this chapter, we will attempt to define these relationships more fully and will propose an approach that substantially minimizes them.

ALTERNATIVE TOPOLOGICAL STRUCTURES

The topology of a large network is an unusually difficult design problem because of the number of parameters and variables that bear on the solution. The general approach to such problems involves the use of digital computers, which, following a set of procedures or algorithms, can try many possible solutions until the best one is found. Over the last decade a wide variety of design algorithms have evolved.

Approaches to the Problem

For most practical networks the network designer tries to determine the most economical configuration of the network facilities that will meet the users' requirements. This is, in effect, a cost minimization problem. Another approach is to design a network with maximum reliability or survivability under a given attack or failure scenario, subject to a given budget. In this case we are not trying to design the cheapest network but, given a cost limitation, to find the most robust network possible. The latter type of design is encountered in military networks or networks where major failures would have potentially devastating impacts, such as an air-traffic control network or a network of electronic funds transfer terminals.

A Centralized Network. Figure 10-1 contrasts two generic topological structures. **Centralized topology** has a great many applications; it represents the literally thousands of computer installations that support the processes of a large number of user terminals. The central computer, which functions both as the switch and as the computational or data base host, is the principal destination of all communications traffic to and from the individual subscribers. Little, if any, traffic needs to be sent from one user terminal to another. If any such terminal-to-terminal traffic is required, it can usually be passed via the central computer.

The topology problem is thus finding the most economical way to connect the user locations to the central resource, with adequate capacity to handle the users' demands. A direct path from each user to the centralized point, as shown in Figure 10-1, is certainly a possible solution, but it is probably not the most economical one. The solution suggested by Figure 10-2 is likely to be less expensive. Here, users are able to share line capacity, with an arrangement of multiplexors, **concentrators,** and multidrop circuits.

Moving from the basic centralized configuration of Figure 10-1 to the functionally equivalent, but far more economical, structure of Figure 10-2 involves the art, science, and mathematics of topological design. To achieve the best grouping of users, capacity assignment, and type of sharing arrangement for each path, a nearly infinite number of combinations must be searched. The algorithmic and computational procedures to find optimal or nearly optimal arrangements are well documented (see Suggested Readings for this chapter).

Centralized network design can be even more complex if the designer is free to choose the location of the central resource or of some of the terminal devices and communications-sharing devices. Another common variation, known as *paired architecture*, is shown in Figure 10-3. Here, a second computer is tied to the network

Centralized network

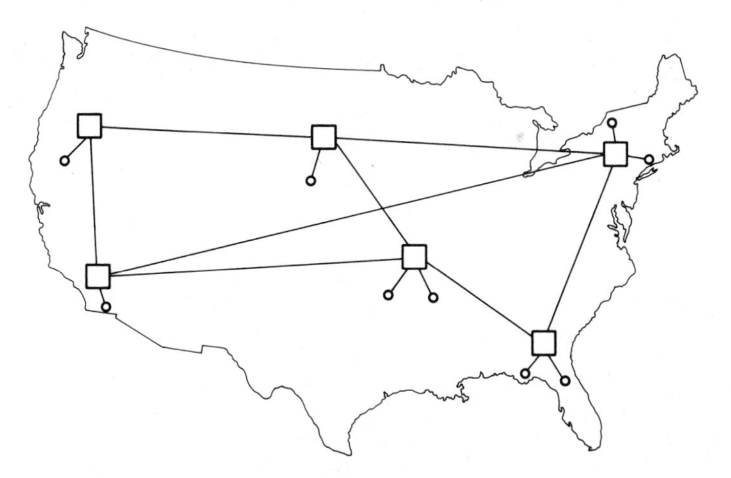

Distributed network

Figure 10-1. Representative topological structures—centralized and distributed.

to provide a level of redundancy and reliability. Additional capacity is added to the network in the form of the connection between the two central sites. The capacity of the remainder of the facilities needs to be shared according to the load on each of the central sites.

A Distributed Network. The distributed network structure illustrated in Figure 10-1 is typified by nationwide common user networks, or large distributed networks serving a substantial number of users. The common telephone network, as well as most of the networks of specialized common carriers, are examples of distributed networks. By distributed network we refer primarily to the fact that the intelligence

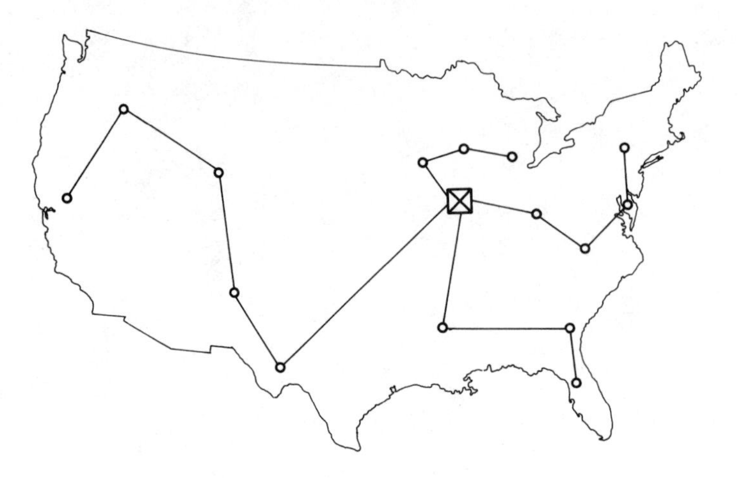

Figure 10-2. Centralized network structure with efficient topological design.

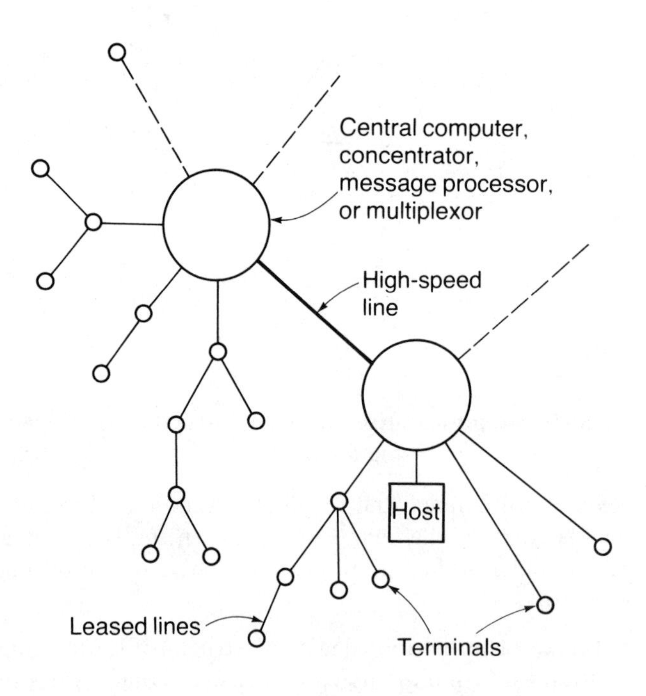

Figure 10-3. A network architecture pairing two large processing centers.

of the network, in the form of switching and processing, is distributed throughout the geographic area being served, thus reducing the apparent traffic load and using the facilities more efficiently.

The apparent traffic load is reduced because traffic destined for a nearby location does not have to flow across the entire network before reaching its destination since the intelligence to route that traffic locally is nearby. Facilities are used more efficiently because techniques of resource sharing (such as packet switching) can be applied at the processing points to achieve high average utilization of the network connections.

An interesting point about distributed networks is that, if we look at one of the switching nodes in isolation, the connection of individual users to that node is a problem of centralized network design. Thus the techniques of centralized network design can be applied as subelements in the design of distributed networks.

Many examples of different topological configurations are found in packet switched network designs around the world. Figure 10-4 illustrates the relatively

Figure 10-4. Network design of the Brazilian REXPAC network. © 1980, ICCC.

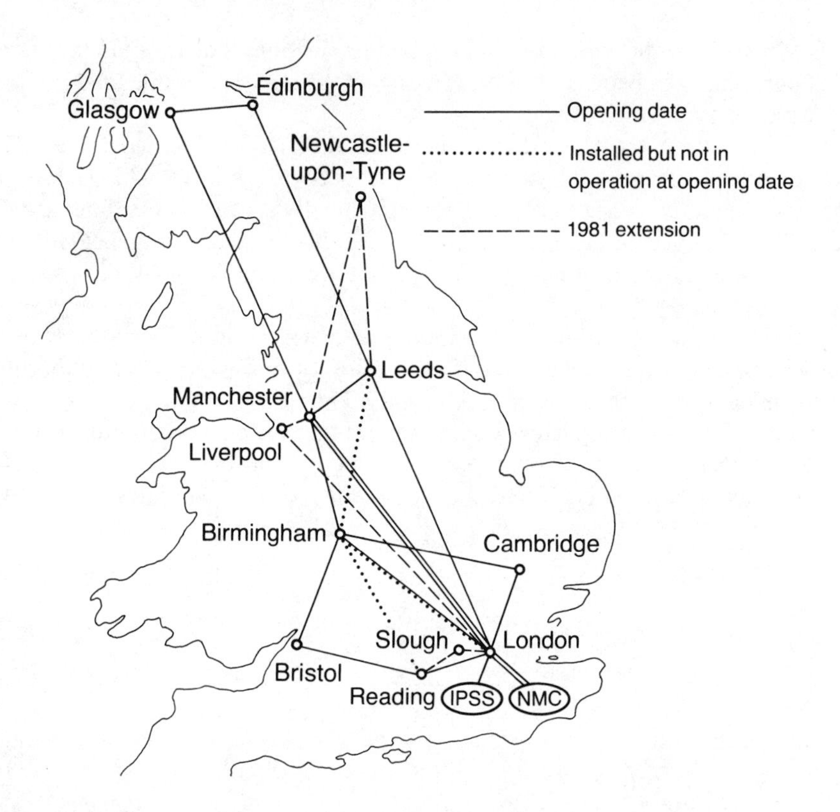

Figure 10-5. Network design of the British national public packet switching network (PSS). © 1980, ICCC.

sparsely connected network designed for Brazil, known as REXPAC. Figure 10-5 illustrates the topology of the British National Packet Switched Service (PSS). Though the latter network covers a rather limited geographical area, it enjoys a richly connected topology.

Hierarchical and Nonhierarchical Networks

Distributed networks can be further divided into hierarchical and nonhierarchical networks (see Figure 10-6). In the basic nonhierarchical distributed structure each switch has a similar function in the network topology and network operation. All of the network switches serve end users, and traffic that originates at one switch may have to pass through several switches en route to its destination.

The **hierarchical** structure employs basically two different types of switches. One type is used to originate and terminate traffic and interface the users to the

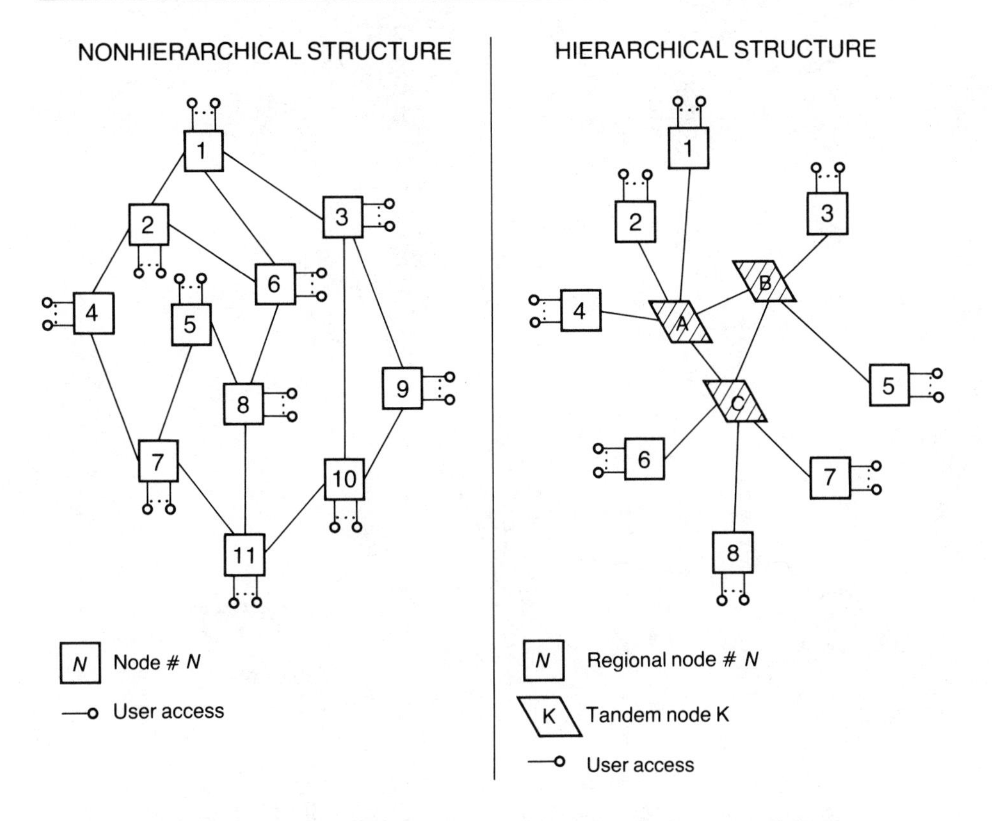

Figure 10-6. Hierarchical and nonhierarchical network structures.

network. The other type passes traffic in a tandem manner between the switches but does not originate, terminate, or directly interface users. The most common form of hierarchical network is the public voice telephone network (illustrated in Figure 4-5, Chapter 4). An excellent example of such a topology applied to a packet switched network is the private network employed by the Nomura Securities Company in Japan (Figure 10-7.).

Because of the separation of functions, hierarchical networks offer advantages in design and operation compared to nonhierarchical designs. Traffic routing is simplified since it is either routed directly or passed up to the next level of the hierarchy. Switches that originate or terminate traffic do not have to get involved with passing and controlling tandem traffic for other switches, and control and network monitoring are simplified. The major disadvantage of hierarchical network structures is the increased vulnerability to failures, since the loss of a single line or switch can isolate some of the users from the network. We will look further at these vulnerabilities later in this chapter.

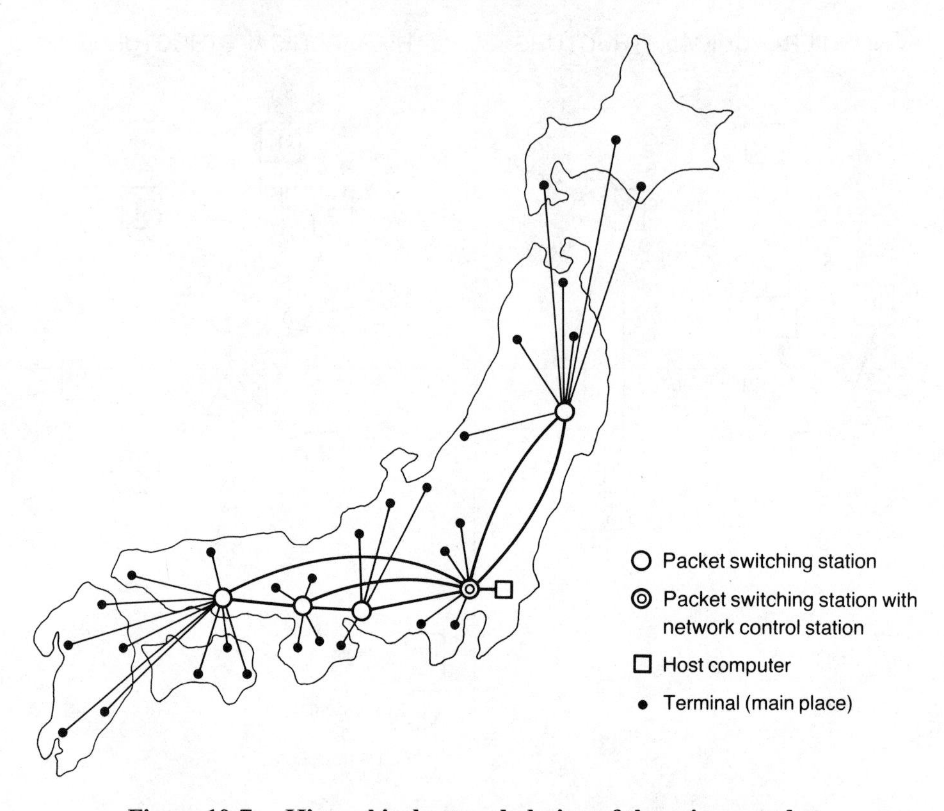

Figure 10-7. Hierarchical network design of the private packet switched network for Nomura Securities Company, Japan. © 1980, ICCC.

THE GENERAL TOPOLOGICAL DESIGN PROBLEM

As we saw earlier, the general network topological design problem is to minimize cost subject to required capacity and performance, or to maximize reliability/survivability subject to a given budget. We will now attempt to define more specifically the nature of the problem and degree of control of the variables, and see how network designs reflect these parameters.

Determining Network Requirements

Establishing the location and traffic demands of the user community is a task that at best is imprecise and at worst involves gross estimates and approximations. In general, we have to characterize user demand in terms of probabilistic traffic demands and user statistics. Next, we have to determine the location, capacity, and connectivity of the network switching nodes, and any other processing power (such as traffic concentrators or multiplexors) that may be used in the network. The capacity and connectivity of the network links (that is, the point-to-point connection between switching nodes) must be determined. Finally, the network performance constraints, such as blocking and delay, have an appreciable impact on the network resources and thus on the cost of the network.

The topological design thus requires:

User locations

User capacity demands and usage statistics

Node (switch) locations

Node (switch) capacities

Node-to-node connectivity

Node-to-node capacity

Network and user response time, throughput, and delay

Network reliability and redundancy

Traffic routing policy and technique

Not all of these quantities are independent variables. For example, the general case does not afford the designer total freedom with respect to node locations, which may be determined by user locations, company locations, access to transmission facilities, availability of real estate, and so on.

A Typical Topology Problem

A typical topology problem may be stated thus:

Given: Average user traffic generation rate, user locations, and allowable node locations.

Minimize: Total network cost by balancing node locations, total line capacity, node connectivity, and routing technique.

Constraints: Achieve a stated average delay for all traffic, and a 90% probability that no traffic will see a delay greater than the stated value. No link will have a capacity greater than that available from the common carriers under a tariffed service.

Even this typical problem, starting with known traffic and locations, is extremely complex. Transmission capacity is a nonlinear function of cost, where, up to a point, additional cost buys increasingly large increments of capacity. If we have N nodes in the network, there are $N!$ different ways to connect the nodes together $[N! = N \times (N - 1) \times (N - 2) \times \cdots \times 2 \times 1]$. To find the best, or "optimal," network design, we would have to try literally thousands of possible network configurations, searching for the best or least expensive solution.

A key aspect of this type of problem is the fact that, even if we are successful in finding an "optimal" solution, the solution does not really exist. That is, a given network design is optimal only for a particular set of user demands. At any given instant of time, the true user demand does not really conform to the estimate used for the design. The design can only be optimal during those rare instants when the actual demand conforms to the estimated demand. Therefore, it is important that the design process insure a degree of flexibility in the configuration and the capacity allocation so that the network can adapt to the true traffic load.

DESIRABLE TOPOLOGICAL FEATURES FOR PACKET SWITCHING

Considering optimal network topologies against estimated and highly variable user demands leads us to define a set of highly desirable design features for the topology of a packet switched network. These features include:

Evolutionary growth

Flexibility in accommodating traffic

Minimized average number of nodes in the principle user-to-user paths

Fail-safe or fail-soft operation

Let us look at these features in somewhat more detail.

Evolutionary Growth

The irony of a networking project is that its very success seals its early doom. The demand for network services tends to rise rapidly as users become dependent on the network to conduct their daily activities. Furthermore, demand growth, especially during the early years of a successful network, tends to be geometric since each new user adds a large number of additional possible traffic flows to the network. Thus networks that are topologically designed against a limited user demand, without full recognition of potential growth, are likely to be stressed to the point of poor operation as the user demands grow. On the other hand, if the flexibility to accommodate growth is designed into the topology, the initial costs are generally higher. In other words, to permit easy growth, the initial network cannot, in general, be optimal in design.

As we saw during our discussions on queueing theory, if the actual loading of the queueing elements of our network—such as the switch processors or transmission channels—begin to approach full (100%) load, the network delays rapidly become excessive. Therefore, spare capacity has to be designed into the nodes and the links connecting the nodes. Node placement has to be a compromise between the length of the distances between switching nodes and the length of the lines accessing the users to the switching nodes. As demand grows, users who are outside the intended service area of the switches become essential users of the network. The remote homing of these users also must be considered in sizing and locating the switching nodes.

The evolutionary growth pattern seen by the ARPANET is illustrated in Figure 10-8. From the first 4 network nodes this network grew, in less than three years, to 30 nodes by August 1972. By the end of 1972 the 38-node configuration shown in Figure 10-9 was in place. Over the next few years growth continued to the point where, in June 1976, the ARPANET consisted of 64 nodes (Figure 10-10).

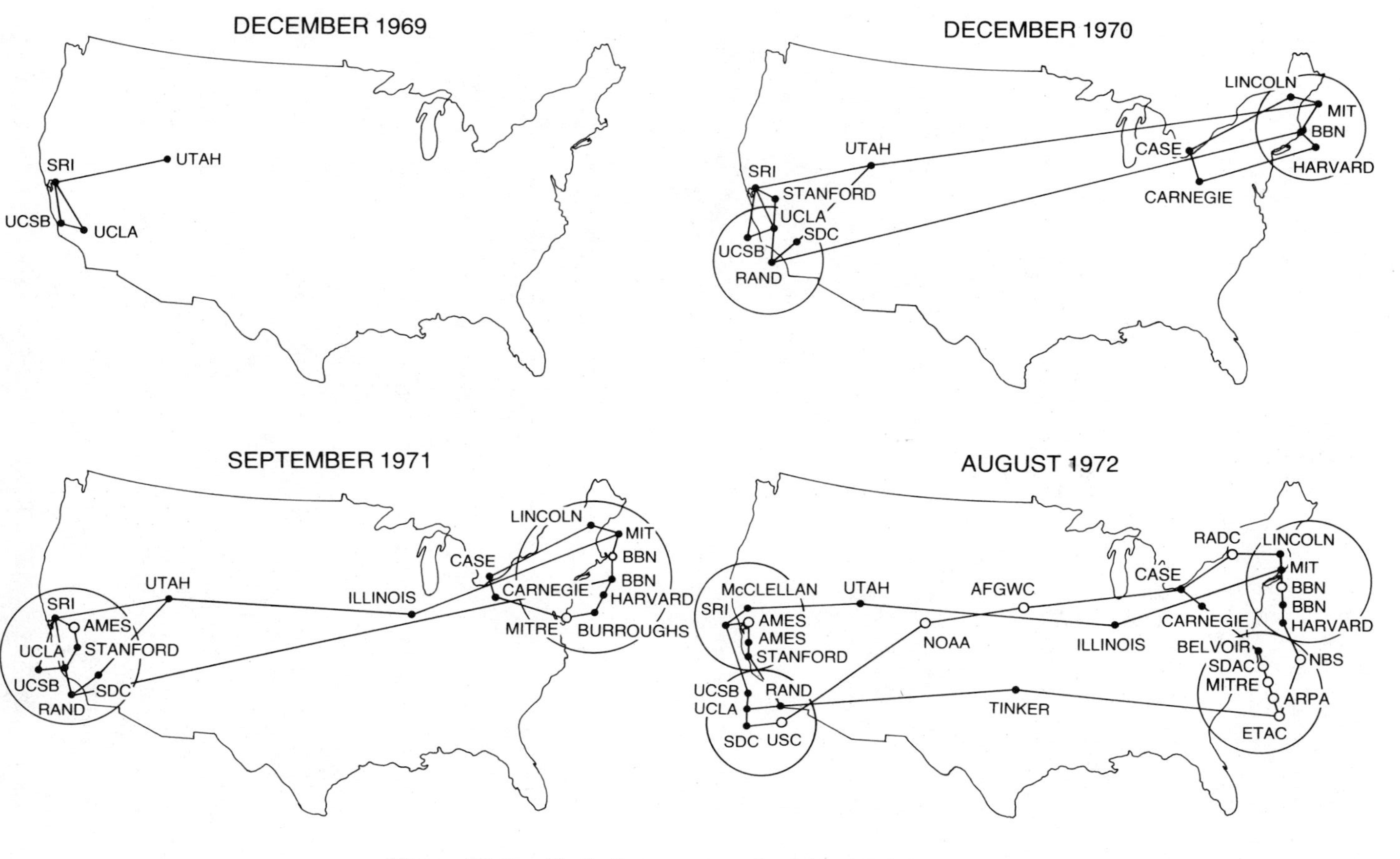

Figure 10-8. Evolutionary growth of the ARPANET.

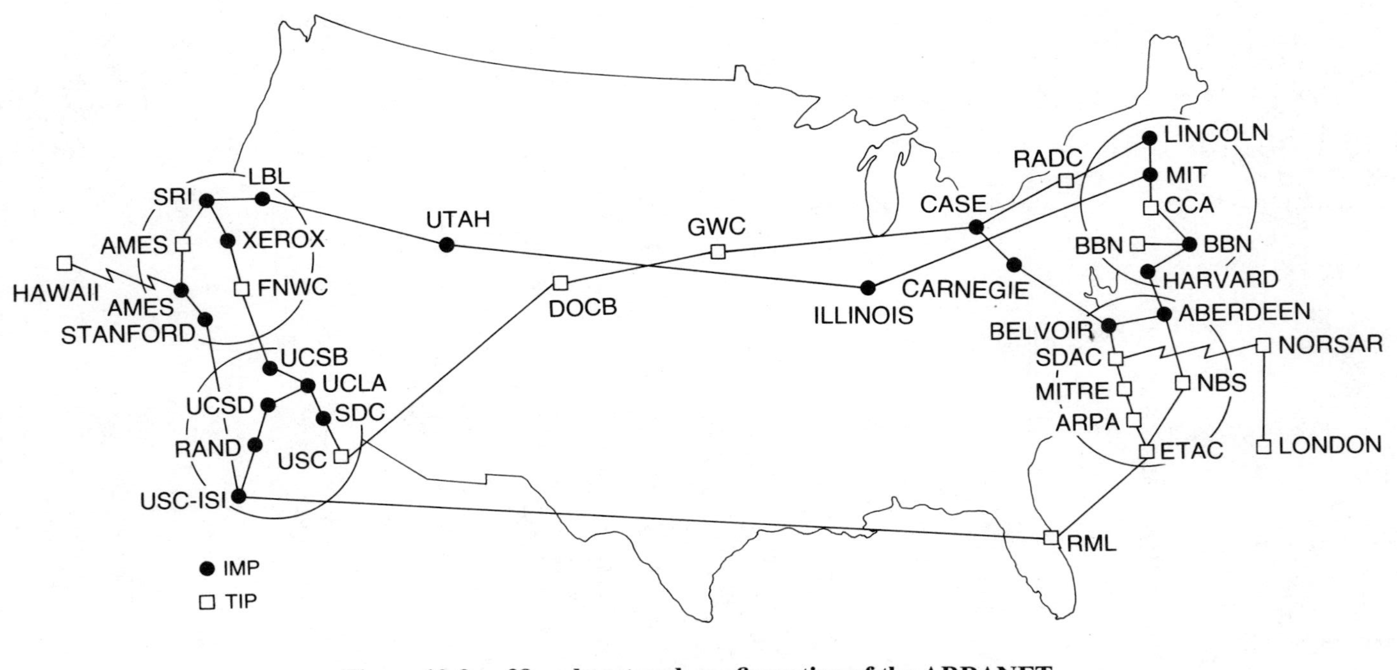

Figure 10-9. 38-node network configuration of the ARPANET.

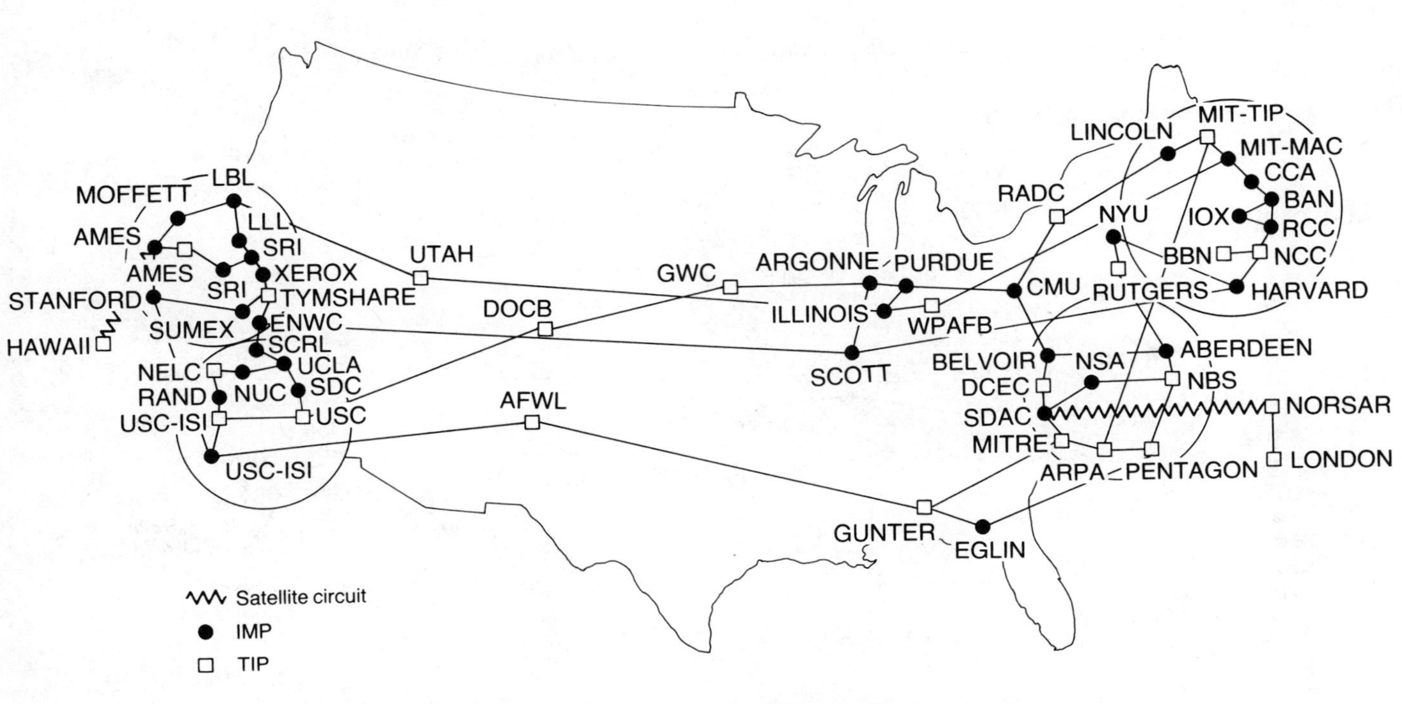

Figure 10-10. Growth limit of the ARPANET using 64 nodes.

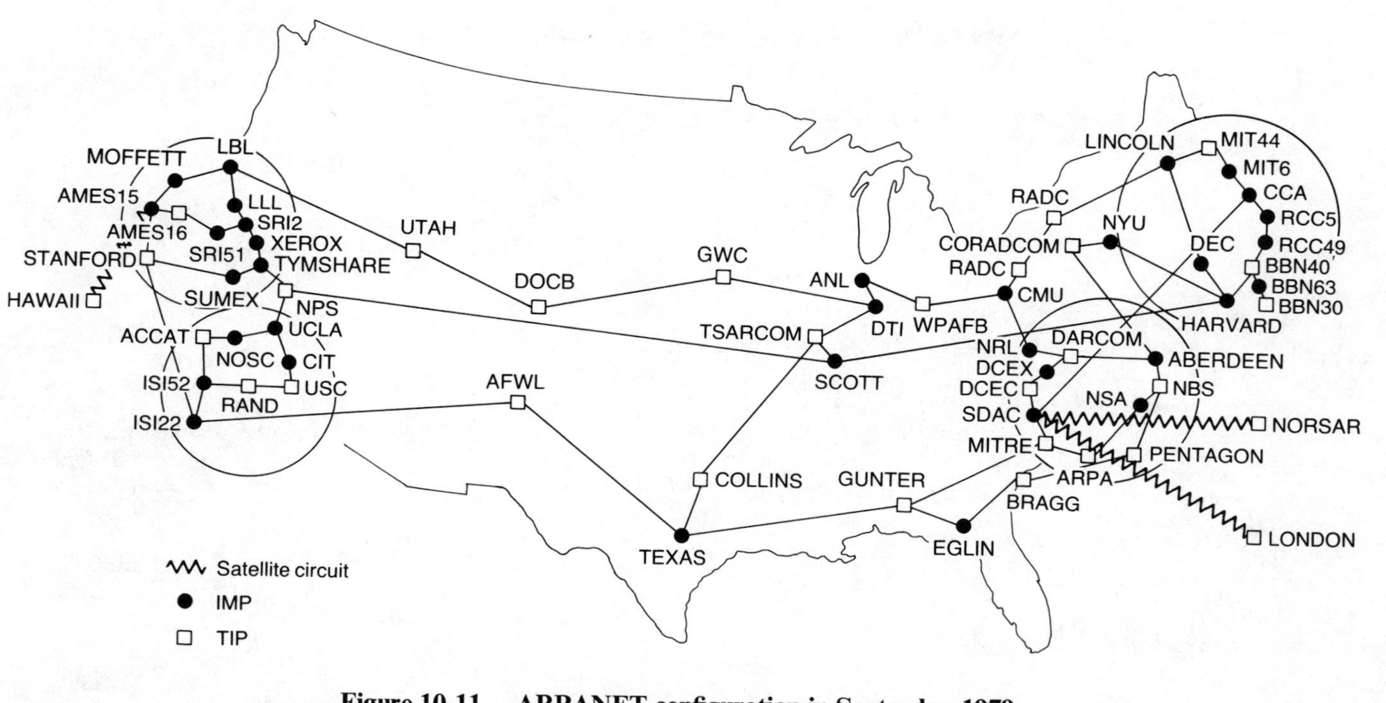

Figure 10-11. ARPANET configuration in September 1979.

At this point a limitation of the initial topological design was apparent. The initial design had set aside 6 bits in the packet header design for the node address. In so doing, it limited the topological design to 2^6 or 64 nodes, unless all of the network software were changed to expand the address capability for the packets. Thus evolutionary growth of the network had proceeded until this address limitation was reached. In a more recent (September 1979) 64-node configuration (Figure 10-11) certain nodes have been relocated to accommodate higher concentrations of users.

Another aspect of design flexibility for evolutionary growth is illustrated in Figure 10-12, where we see a so-called logical map of the ARPANET. This map shows the varied computers that are users and hosts within the ARPANET community.

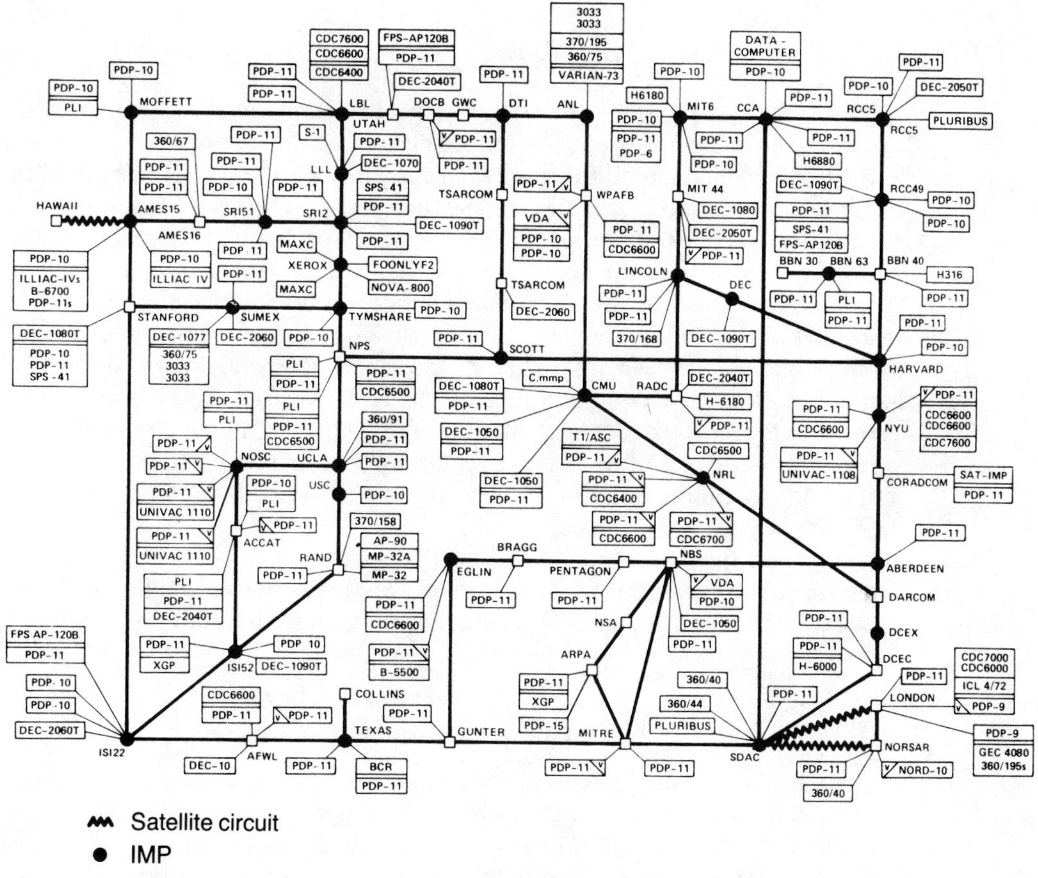

Figure 10-12. Logical map of the ARPANET.

Evolutionary growth in user characteristics, as well as in traffic demand, results both from the topology's ability to grow and from the physical, electrical, and functional protocols' ability to support the many terminal and host capabilities.

Traffic Accommodation

The ability of a network topological design to accommodate varying traffic demands within a fixed structure is established principally by the routing plan and routing technique used within the network. We will discuss routing further in Chapter 12. For the present, we should realize that dynamic traffic routing is essential to the network's ability to handle unusual traffic patterns. The capacity assignment problem and operation of the dynamic routing algorithms to use available capacity effectively are key aspects of the topological design.

User-to-User Path Lengths

As we have seen in our discussions on protocols, packet switching operates best when short, low-delay paths are available between the source switch and the destination switch. As more and more tandem switches are added, the possibilities for the loss, duplication, misordering, and flow control of packets increases.

In the hierarchical structure shown in Figure 10-6 no more than four switches are ever needed in the principal (most direct) path between any pair of users. This is accomplished in the hierarchical design by fully connecting the set of nodes that are designated as part of the upper level of the hierarchy. Fully connecting these nodes means that each of the higher level (tandem) nodes has a direct connection to every other node at this level. Then, when each node that serves users is connected to at least one of the fully connected nodes, the end-to-end path between any pair of users never requires more than four nodes—the originating node, the destination node, and, at most, two nodes within the tandem hierarchy.

However, this configuration tends to be highly vulnerable since loss of a single line in the lower level of the hierarchy would tend to isolate all of the users associated with the failed line. This concern brings us to the question of fail-soft operation.

Fail-Soft Operation

A **fail-soft operation** means that the failure of one or a few network components should not have a catastrophic effect upon network operation. We introduced this problem in Chapter 9, when we discussed making continued network operation possible even if the major control element failed. With fail-soft operation the network will tend to lose capability as components fail, but it should not be inhibited from completing user traffic for all users when single

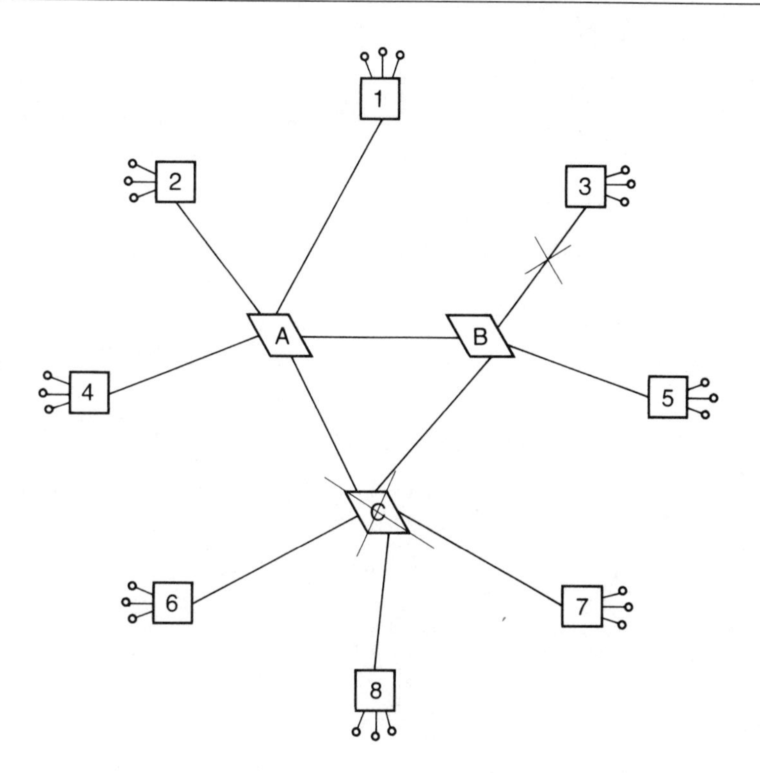

Figure 10-13. Isolation of users resulting from failures in a hierarchical network.

(or a few) components fail. Hierarchical structure does not have this fail-soft property, as we can see in Figure 10-13. For example, the failure of tandem node C caused switches 6, 7, and 8, and all the users connected to these three nodes, to become isolated from the network. If the line from tandem node B fails, node 3 and all of its users become isolated. Network integrity is thus catastrophically affected by the loss of a single node or single line.

The modified hierarchical design shown in Figure 10-14 insures a high degree of invulnerability to single failures in the network topology. Each node that terminates users is connected to one of the higher level nodes, and each of the lower level nodes is connected to at least two others. These additional connections could have a much smaller capacity than the lines from the serving node up to the tandem nodes since they would not be carrying the major traffic load except in case of failure or overload of the tandem nodes.

The topology of this modified hierarchical structure meets the requirements of the design features we have discussed. By contrast, the ARPANET topology in Figure 10-11, as an optimally designed distributed topology, has many very long end-to-end paths between user pairs, often requiring transmission through

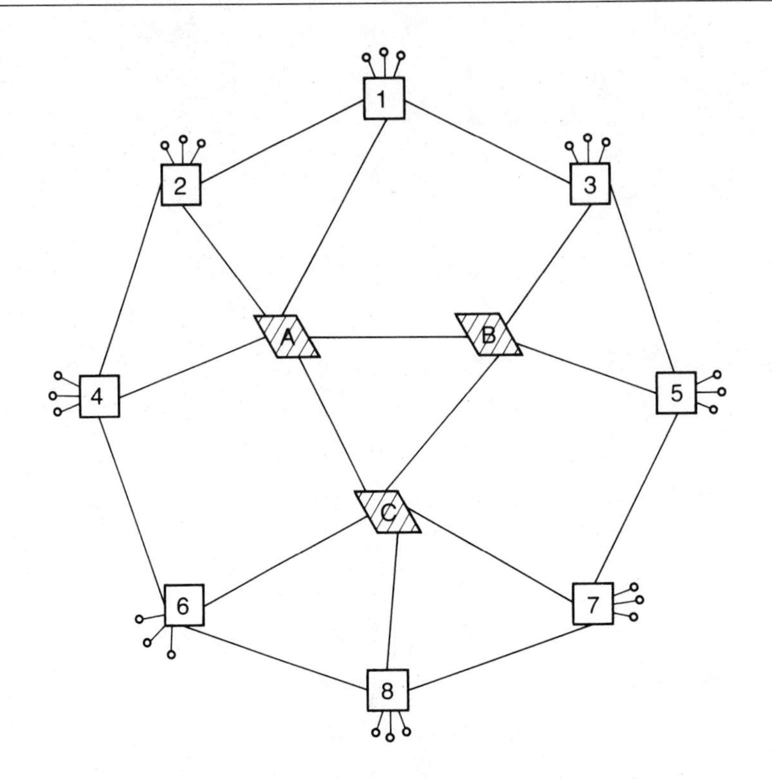

Figure 10-14. A modified hierarchical network structure.

eight or more nodes. Because the modified hierarchical network will require more links and more channel-miles of capacity than an optimally designed distributed, nonhierarchical network, most likely the costs will be higher. However, these costs are really associated with providing the topological features to accommodate growth and performance over a wide range of user loads and user characteristics. We will further examine the implications of these extra resources in Chapter 11, when we look at the results of some specific network designs.

SUMMARY

You should now be familiar with topological design alternatives that are applicable to packet switched networks. In this chapter we contrasted centralized networks with distributed networks and saw that distributed networks could be either hierarchical or nonhierarchical. The general topological design problem is primarily concerned with cost minimization. An "optimal" network design cannot exist in reality since actual traffic patterns will never conform exactly to the patterns estimated for the design.

Four desirable features of the topology of a packet switched network include (1) providing for evolutionary growth, (2) flexible traffic accommodation, (3) short user-to-user paths, and (4) fail-soft operation. The modified hierarchical design, at some additional expense, appears to achieve the desirable attributes of a flexible packet switching topology.

SUGGESTED READINGS

BERGAMO, M. A., and CAMPOS, A. S. "REXPAC—A Brazilian Packet Switching Data Network." *Proceedings of the Fifth International Conference on Computer Communication*, ICCC '80. Atlanta, October 27–30, 1980. Pp. 17–22.

REXPAC, the nationwide experimental packet switching network in Brazil, is described in this paper. The network design uses the CCITT standards family.

BOORSTYN, ROBERT R., and FRANK, HOWARD. "Large Scale Network Topological Optimization." *IEEE Transactions on Communications*, vol. COM-25, no. 1 (January 1977), pp. 29–47.

This paper deals primarily with the design of a large network that is a multilevel hierarchy. The network consists of a distributed backbone network with local access systems that are typically centralized mininetworks. The design can be decomposed and different algorithms applied to each part of it. The result is an overall heuristic algorithm that accomplishes a highly efficient structure.

MEDCRAFT, D. W. F. "Data Network Plans for the UK." *Proceedings of the Fifth International Conference on Computer Communication*, ICCC '80. Atlanta, October 27–30, 1980. Pp. 29–34.

This paper describes the data communications offerings of the British Post Office telecommunications in the United Kingdom and outlines plans for the modernization of those offerings. Details of the National Public Packet Switching Service are presented as a specific example.

SCHWARTZ, MISCHA. *Computer Communication Network Design and Analysis.* Englewood Cliffs, N.J.: Prentice-Hall, 1977, chs. 3, 4, 5, 9, and 10.

These chapters develop the theoretical basis and mathematical formulation of topological design algorithms applicable to both centralized and distributed networks. In some detail they show how to calculate the required capacity along a fixed topology in order to accommodate the required traffic.

SHINOHARA, T., TODA, Y., TAKITA, S. I., and HYODO, T. "Development of a Distributed Computer Network System Nomura-Custom." *Proceedings of the Fifth International Conference on Computer Communication*, ICCC '80. Atlanta, October 27–30, 1980. Pp. 231–236.

A distributed packet switched private network supports five computers and 3400 terminals at 104 separate locations throughout Japan. The network, designed for the Nomura Securities Co., is an excellent example of the application of packet switching to a private network configuration to achieve privacy, security, and economy.

11

Network Design Case Example—Even a Bad Guess Is Better Than None

THIS CHAPTER:

will use an example of the design of a large, nationwide data network to compare hierarchical and nonhierarchical topological structures.

will explain the impact of network data rate and topology on the sizing and complexity of the network.

will show that the lack of detailed user requirements information has only minor impact on initial network design and sizing.

As we saw in Chapter 10, although hierarchical structures offer advantages in terms of network throughput and delay characteristics, they appear to be less reliable and more vulnerable to line or switch failures. A modified hierarchical network, by providing multiple connectivity of the nodes that directly serve the network users, increases network reliability and reduces vulnerability to failure. However, the additional connectivity might increase the resources required enough to make such a network structure impractical. In this chapter we will look at an analytical study that compared the impact of an imposed, modified hierarchical network structure to an optimized distributed network with no imposed connectivity.

THE NETWORK DESIGN REQUIREMENTS

User Characteristics. In this analytical comparison we are dealing with a large, nationwide network of data users who, as a group, generate about 3.3×10^{10} bits of data during the busiest hour of the day, accounting for about 1,600,000 data transactions or messages. Traffic is assumed to emanate from approximately 2400 geographically dispersed locations, with sites varying from single low-speed terminals to those encompassing 10 to 20 computers plus 40 or more high- and low-speed terminals. Seventy potential nodal locations, at or near the concentrations of users, are identified for the network packet switches.

Design Approach. The design approach taken deemphasizes the importance of finding an optimal network design, making that consideration secondary to insuring low user transmission delay, uniformity of performance, and relative insensitivity to changes in user demand. The objective is to design an efficient, survivable, reliable data communications network, starting from what is known to be only a very approximate estimate of the user traffic magnitude and flow distribution.

User Demands. Because only approximate user demands are generally known in the initial design of a network, optimal network designs are fictions of mathematics and cannot occur in practice. The actual network traffic flow, in both quantity and distribution, is a dynamic parameter, which changes from second to second; at most it matches the design optimum for an infinitesimal time. What this analysis will show, however, is that, by imposing a hierarchical structure, we can achieve a network design that is relatively insensitive to the dynamics of a changing user community and user demand pattern, and which is likely to be, at worst, only slightly more expensive than a nonhierarchical network.

The Network Performance Model

In order to quantify the question of network structure, our comparison will be made on the basis of channels, channel-miles, and network delay. The comparison utilizes an analytic store-and-forward network performance model implemented on an IBM 370/155 computer and adapted to packet switching by truncating the maximum message size to the desired packet length (in this comparison, nominally 1000 data bits). The model is based on a specified set of network nodes, at given locations, shown in Figure 11-1. The nodes are connected to each other by a set of links, each of which comprises one or more data channels, with all channels in the network constrained to the same digital data rate. While basic link connectivity is specified as an input to the model, the capacities are sized and optimized to meet the required average delay at the lowest overall resource utilization (channels and channel-miles). All traffic can be assigned to a single user class, or the model can accommodate a two-level priority structure, with the network capacity assigned to meet the constraints of the higher class of users first.

Traffic Flow. The distribution of traffic flow, based on the 70 node locations shown in Figure 11-1, was achieved by prorating the total estimated traffic among the

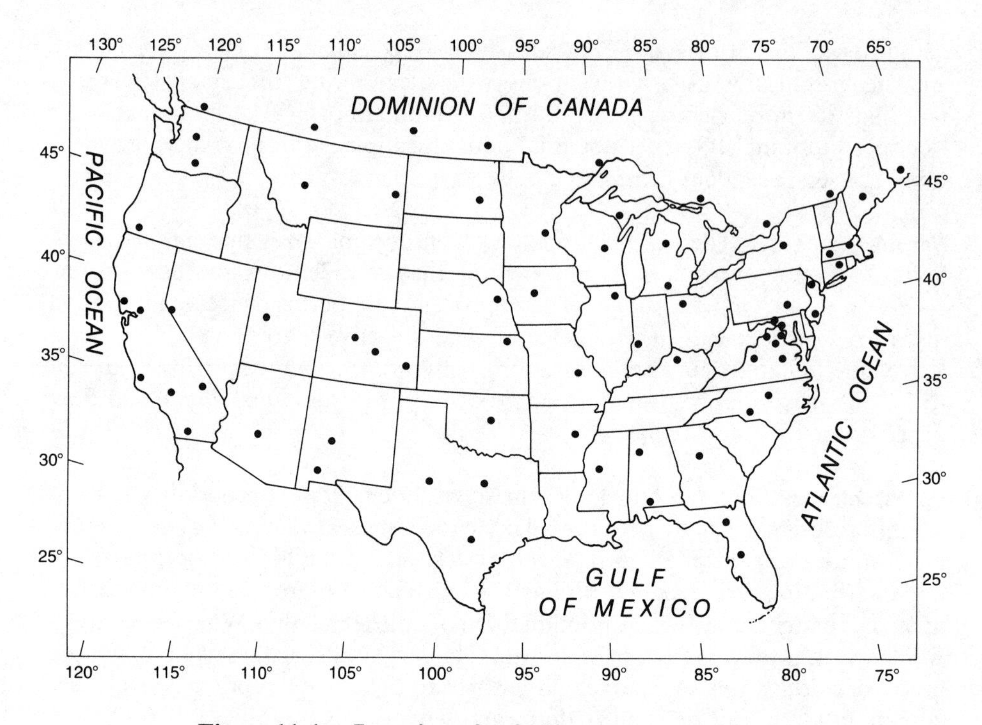

Figure 11-1. Location of switching centers used in the network model.

locations according to the user density in the vicinity of each node. This creates, in effect, a 70 × 70 traffic flow matrix, with 4900 entries, defining the traffic requirements from each node to each of the other 69 nodes. The starting point for the model placed links between all network node pairs with significant traffic between them. The computer model then determined the sizing (capacity) and traffic routing to meet the total traffic demand matrix.

Number of Links. The result was a network configuration employing 362 links to connect the 70 nodal locations to each other. Because of the large number of links relative to the total amount of data traffic, the results of the model run yielded many links with only a few data channels on them. The number of links used was then reduced by eliminating all links with three or fewer channels and rerouting and resizing the network as needed. This produced a more cost-effective network design employing 224 links. Next, all links with five or fewer channels were eliminated, resulting in a 149-link design. Finally, an 84-link connectivity was used, which, as we shall see shortly, represented the minimum connectivity hierarchical network design possible.

Computer-Generated Designs

Computer-generated designs were utilized representing both hierarchical and nonhierarchical configurations, for various bit rates and over the range of topologies. Designs were made using both the projected traffic matrix and a uniform traffic matrix in which the total network traffic was uniformly distributed among all of the network switching nodes. The uniform traffic case, resulting in a traffic matrix where every entry was identical, represented a worst-case situation, where the network designer has no information about the geographic distribution of user traffic.

The basic network connectivity shown in Figure 11-2 illustrates the 362-link, nonhierarchical configuration. The 224-link and 149-link nonhierarchical networks are shown in Figures 11-3 and 11-4, respectively.

The Hierarchical Network. The hierarchical network designs were created by designating the seven busiest nodes in the network as higher level, or tandem, nodes. Following the principles discussed in Chapter 10, each of the remaining 63 nodes was connected directly to one of the tandem nodes, and the tandem nodes were

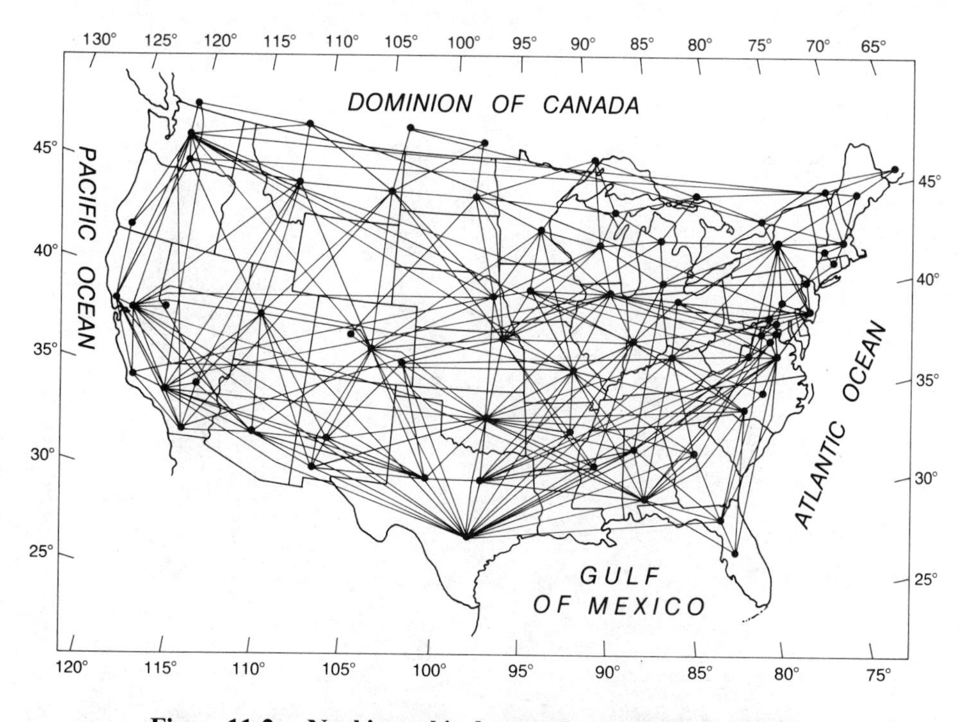

Figure 11-2. Nonhierarchical network employing 362 links. © 1974, ICCC.

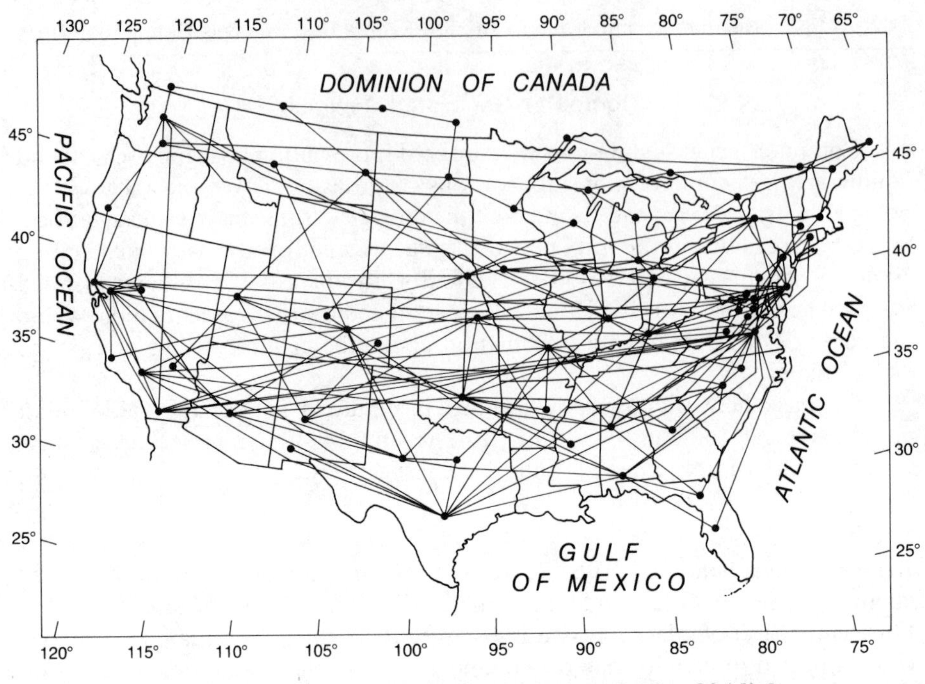

Figure 11-3. Nonhierarchical network employing 224 links.
© 1974, ICCC.

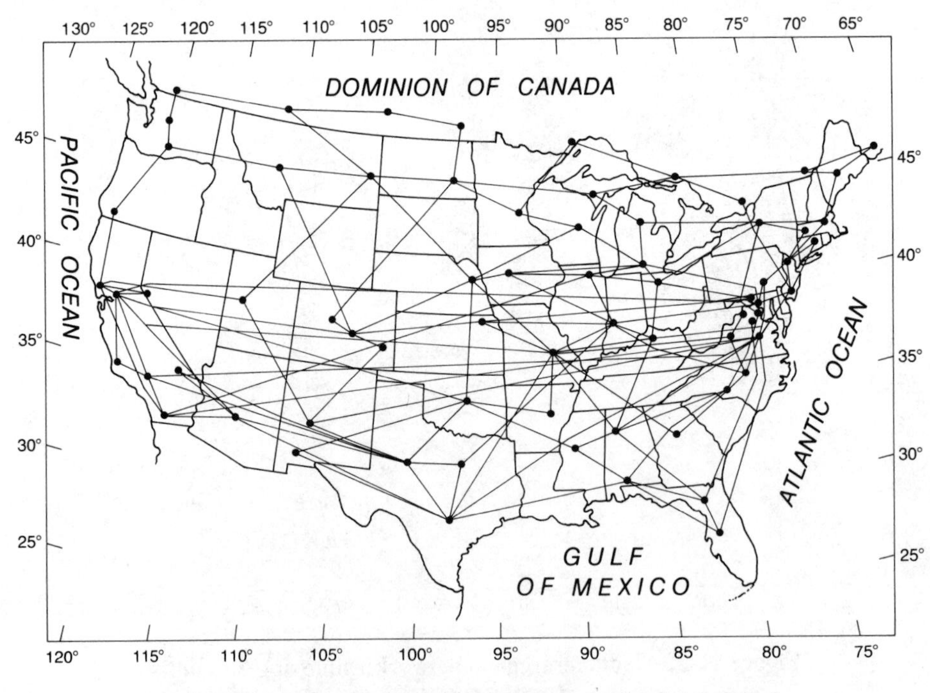

Figure 11-4. Nonhierarchical network employing 149 links.
© 1974, ICCC.

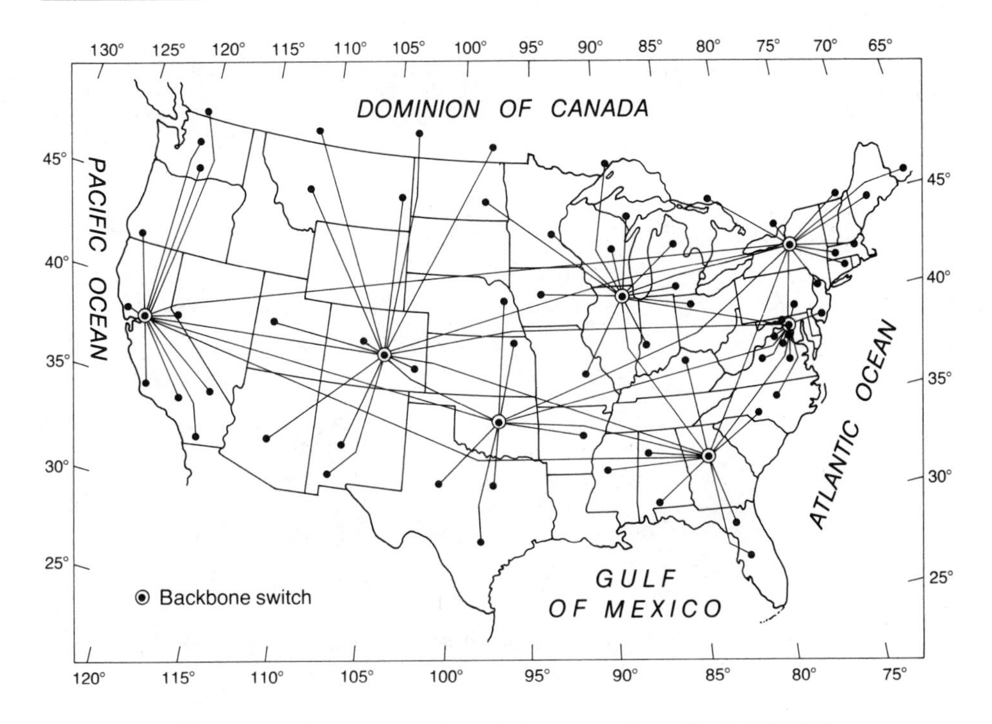

Figure 11-5. Minimum connectivity network employing 84 links.

connected directly to each other. The minimum connectivity network required for the hierarchical case was thus 84 links. This is because it required 21 links to fully connect the seven tandem nodes ($(7 \times 6)/2 = 21$), plus 63 links to connect each of the remaining nodes to the nearest tandem node. The seven tandem nodes were nominally in the vicinity of New York, Washington, Atlanta, Chicago, Oklahoma City, Denver, and San Francisco.

The 84-link hierarchical structure, shown in Figure 11-5, illustrates the principles discussed in Chapter 10 quite graphically. Here we can see that any pair of users, wherever they are located in the network, can communicate with each other with, at most, four nodes in the end-to-end path. However, the vulnerability of that connectivity to single-node or single-link failures is also readily apparent.

The Modified Hierarchical Network. By keeping the hierarchical structure but adding links, we can reduce the vulnerabilities, as shown in the 149- and 224-link designs in Figures 11-6 and 11-7, respectively. In these modified hierarchical network designs each lower level node is connected directly to one of the tandem (higher level) nodes and at least two other lower level nodes. This results in much richer connectivity, many more alternate paths through the network, and thus

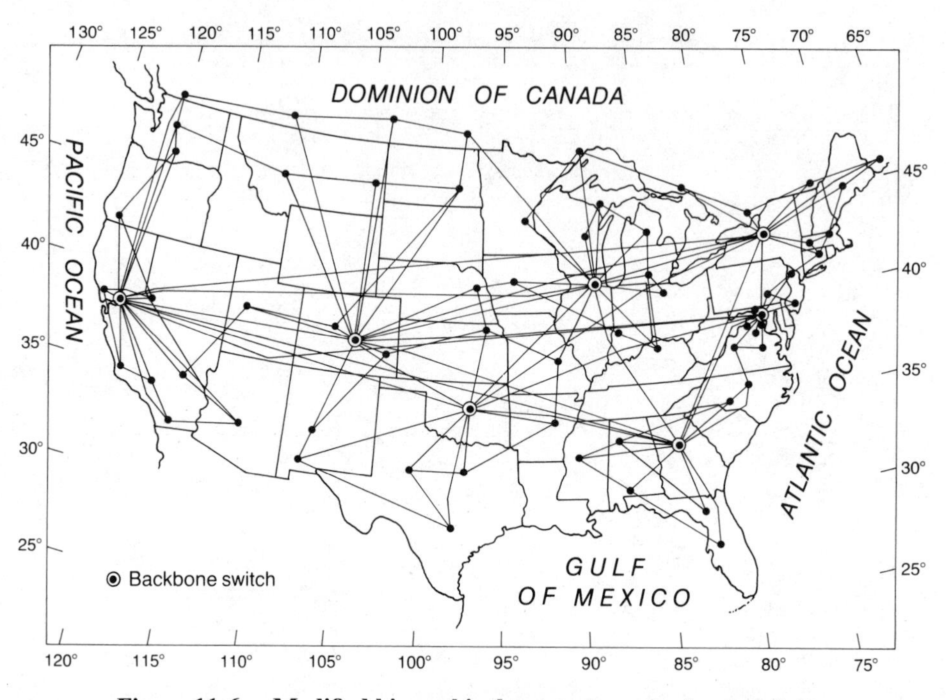

Figure 11-6. Modified hierarchical network employing 149 links. © 1974, ICCC.

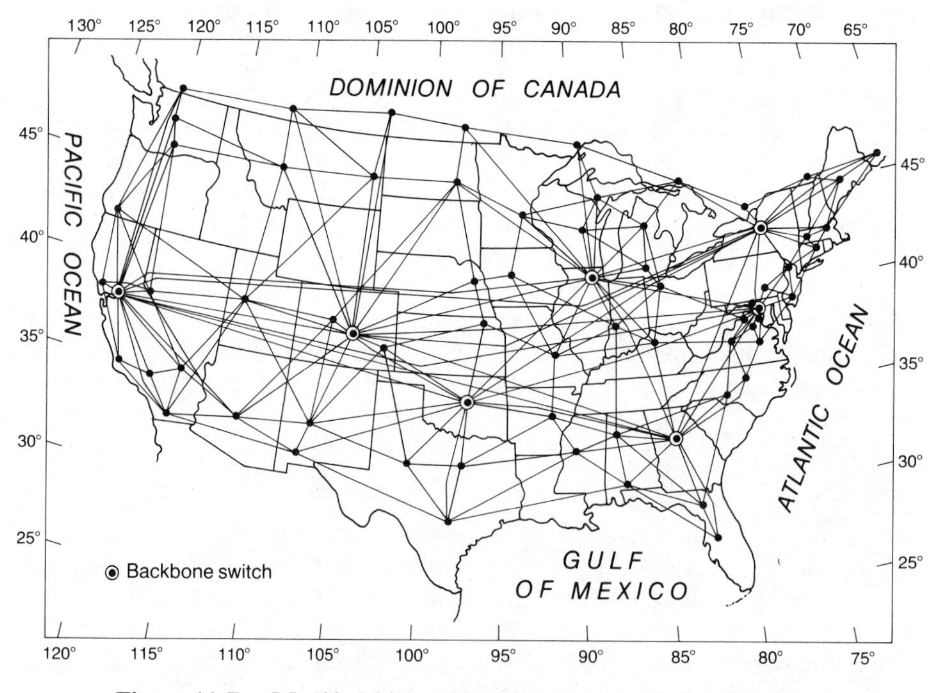

Figure 11-7. Modified hierarchical network employing 224 links. © 1974, ICCC.

higher network reliability. However, under normal operation the network still provides a primary path between all users employing four switches or less.

THE IMPACT OF TOPOLOGY AND DATA RATE ON NETWORK RESOURCES

When this range of network structures, topology, and connectivity were defined, the results of the optimizing computer runs were summarized graphically to see the impact of the variations.

Number of Channels and Channel-Miles

The cost of the network design, in terms of the resources used, is highly correlated to the number of channels and channel-miles of transmission facilities. Figures 11-8 and 11-9 show the number of channels and channel-miles plotted

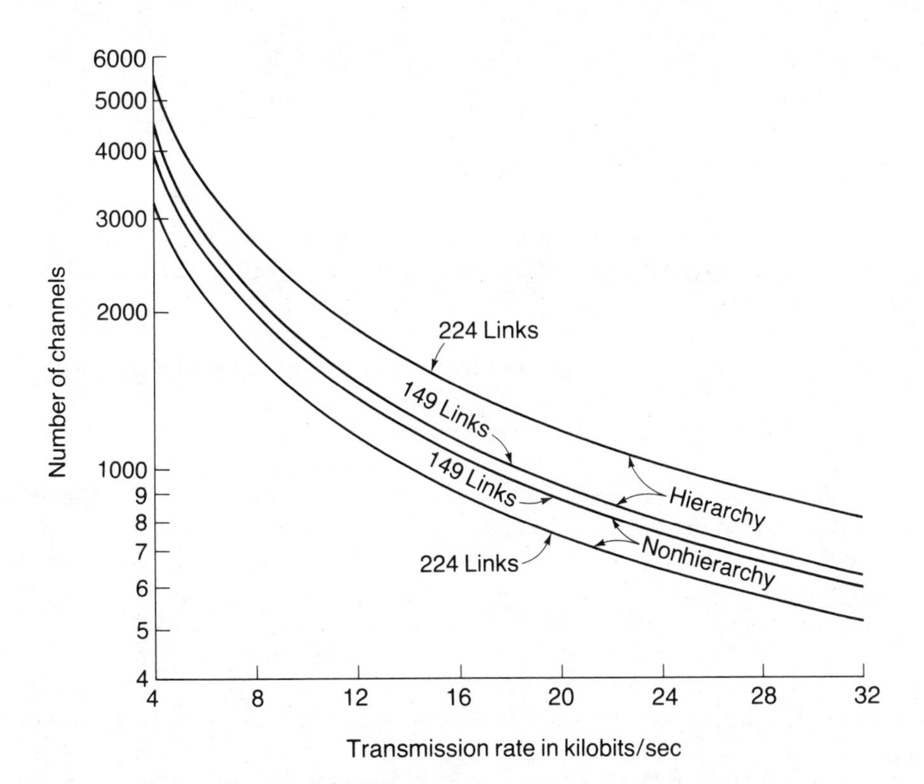

Figure 11-8. Plot of the number of channels versus line bit rate for nonhierarchical and hierarchical networks. © 1974, ICCC.

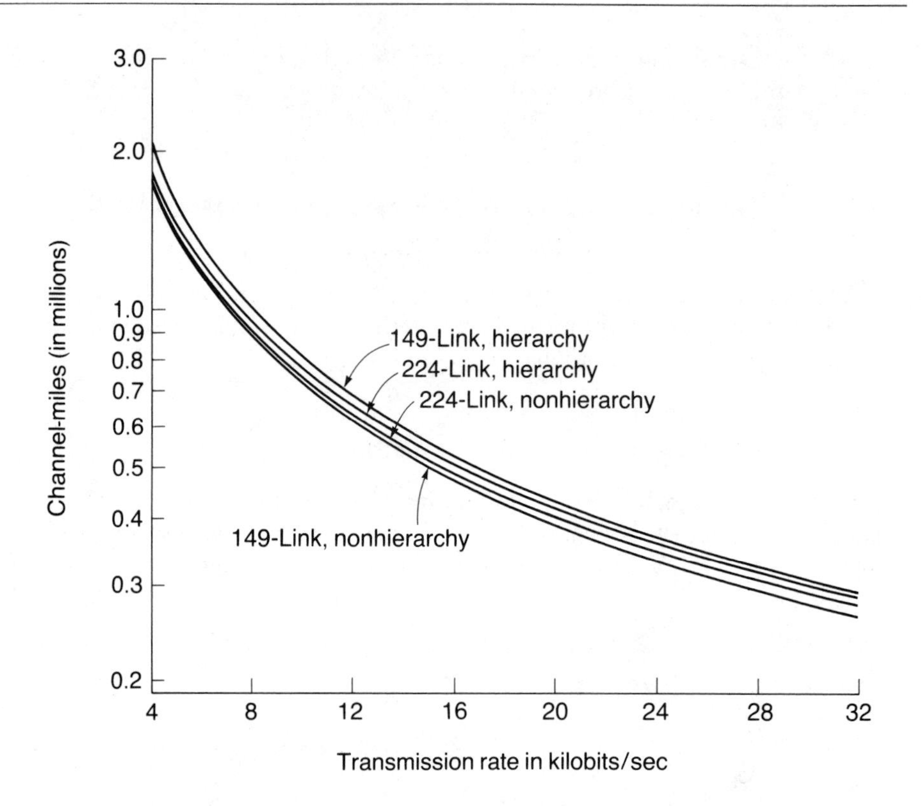

Figure 11-9. Plot of the number of channel-miles versus line bit rate for nonhierarchical and hierarchical networks. © 1974, ICCC.

against channel bit rate for both the hierarchical and the nonhierarchical network designs.

It can be seen that, at any bit rate, there is a fairly large spread in the number of channels between the alternative designs. On the other hand, the number of channel-miles for the 149- and 224-link cases are tightly grouped. As the connectivity gets thinner, the number of channels for the nonhierarchical and hierarchical networks tend to converge. This is further illustrated in Figure 11-10 and Figure 11-11, in which the number of channels and number of channel-miles are plotted against the number of links used in the network. These plots happen to be at a channel bit rate of 16 Kb/sec, but similar results are obtainable at other bit rates.

In Figure 11-10 the number of channels rapidly diverge as the number of links increase from the 84-link case, with the nonhierarchical network showing a maximum number of channels in the area of 150 links. The number of channels required by the hierarchical network increase rather significantly as the number of links increase, and there appears to be a maximum value outside of the range of topologies examined.

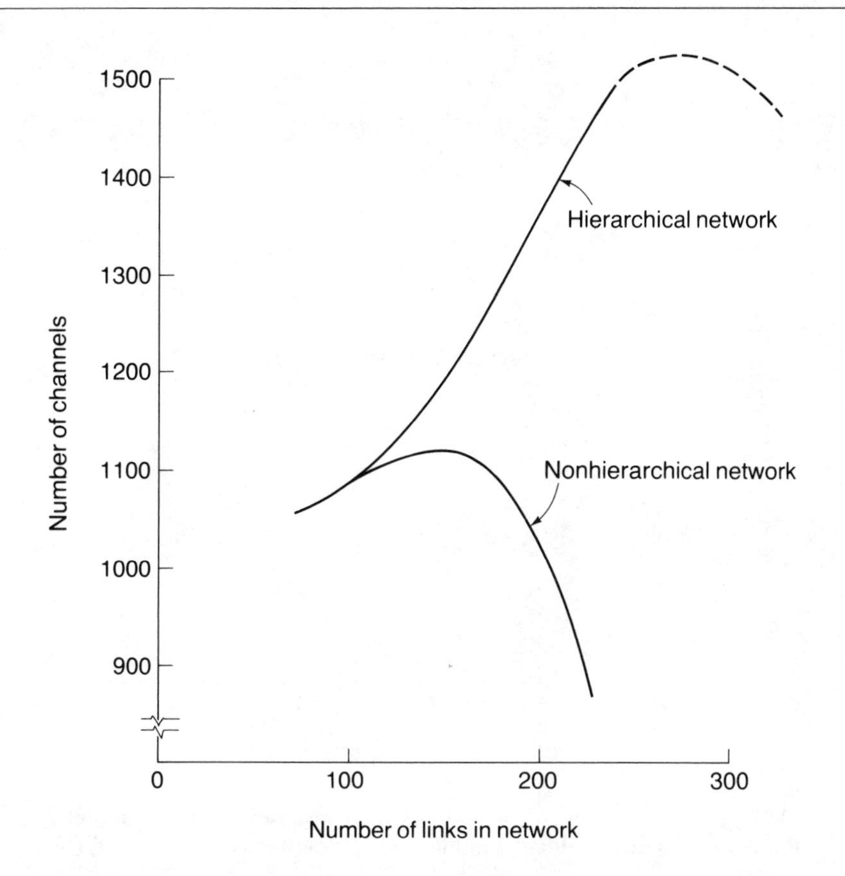

Figure 11-10. Plot of the number of channels versus the number of links for nonhierarchical and hierarchical networks. © 1974, ICCC.

Remember that the presence of additional links indicates a richer network connectivity, more alternate routes between user pairs, and less likelihood of user isolation due to single network line or switch failures. However, as links are added to the network, the number of channels tend to increase since more of the traffic is routed over circuitous routes. Finally, enough links are present so that there are many direct paths for most of the traffic through the network. Traffic is now routed more efficiently, ultimately resulting in more efficient network flows and a reduction in the number of channels required. This result is shown by the nonhierarchical network reaching a maximum value and then seeing a reduction in the number of channels required (Figure 11-10). A similar effect is expected for the hierarchical case, although the maximum occurs at a higher value of both links and channels because the imposed hierarchical structure does not allow the additional links to be added in the optimum locations.

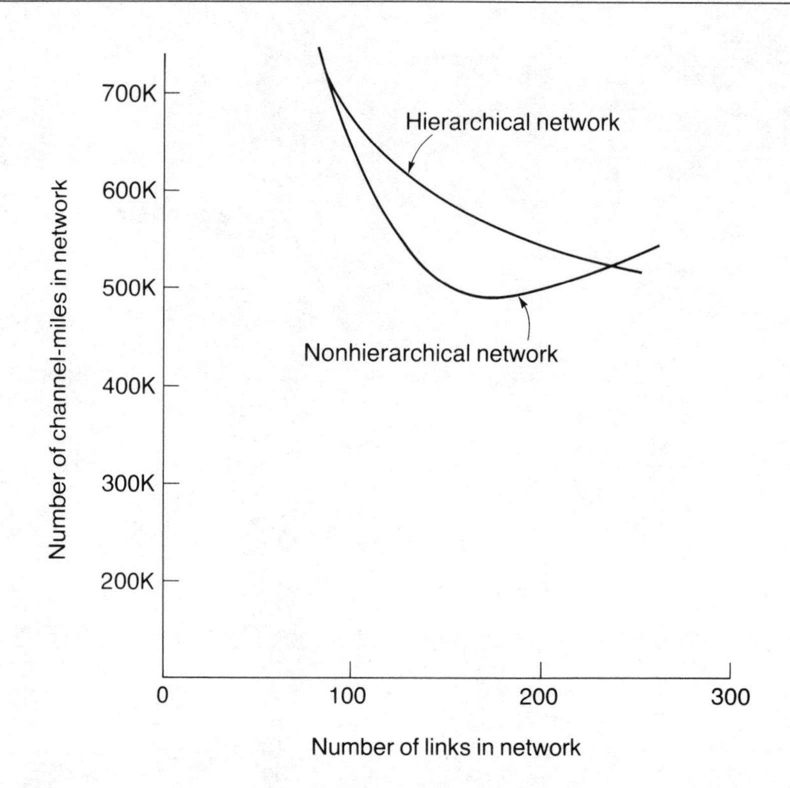

Figure 11-11. Plot of the number of channel-miles versus the number of links for nonhierarchical and hierarchical networks. © 1974, ICCC.

Figure 11-11 shows a much closer and regular comparison of channel-miles for the hierarchical and nonhierarchical network configurations: The relative difference is generally within 10%. Here again, the nonhierarchical network reaches a minimum value of channel-miles at a lower number of links than the hierarchical network because the imposed structure of the hierarchy tends to make less efficient use of each added increment of network capacity.

The significant point of this comparison is that the degree of inefficiency is relatively minor, in that the hierarchical design requires only slightly more resources than the "resource-efficient" nonhierarchical network. However, the advantages of the modified hierarchical network structure in network routing, delay, delay variance, system management, control, and implementation phasing probably balance the nominal increase in the number of channels and channel-miles used in the network. In the design range that provides the best balance between network robustness and network efficiency (that is, between 100 and 200 links) the hierarchical network appears to require about 15% more channels and less than 10% more channel-miles than the nonhierarchical network.

Actual Performance

Now that we have defined the modified hierarchical structure as desirable, it is of interest to look at the actual performance achieved with such a network structure. A key measure of performance is the end-to-end delay, averaged over all of the user pairs in the network. Results of such an investigation are shown in Figure 11-12 for a 149-link hierarchical network, where the achieved average network delay is plotted versus the channel bit rate.

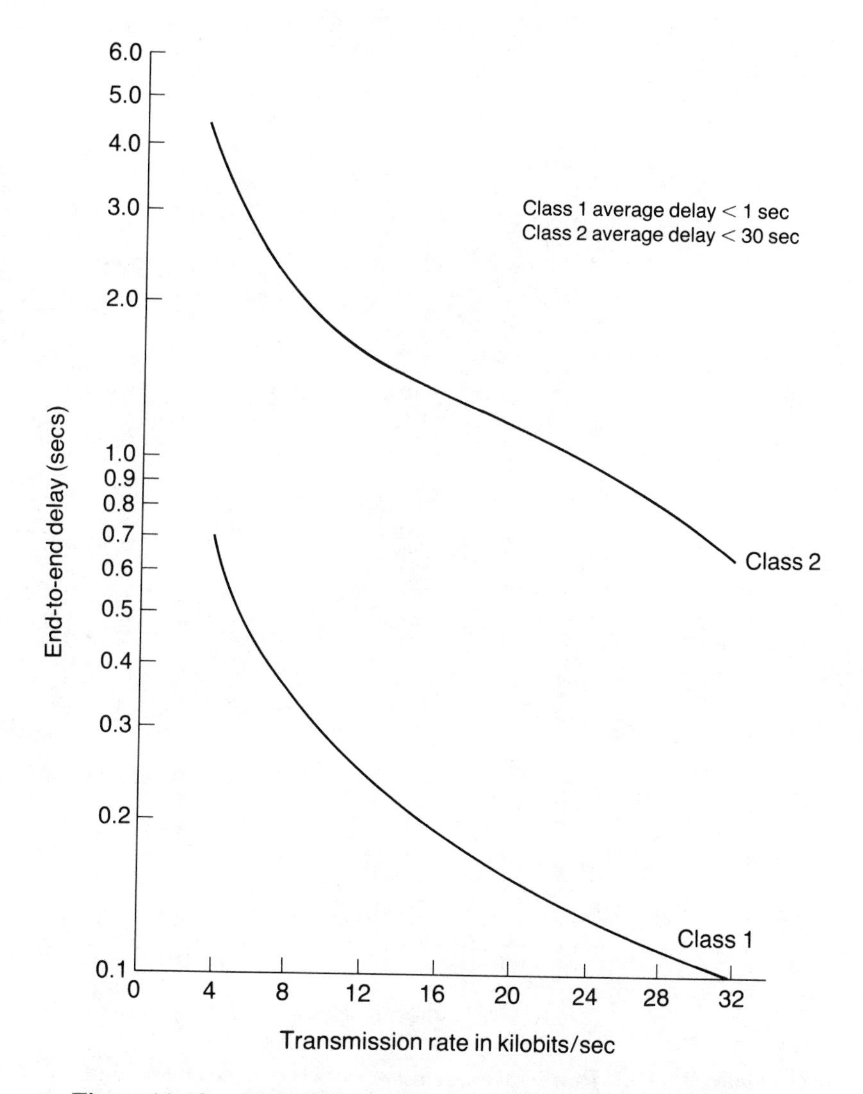

Figure 11-12. Plot of the end-to-end packet delay versus line bit rate for a seven-switch modified hierarchical network. © 1974, ICCC.

One further capability was added—the ability to group the traffic into two classes, with different priorities. Class 1, comprising one-third of the total traffic, is handled ahead of class 2. The network design constraint required that all class 1 traffic be delivered with an average delay of less than 1 second, while the average

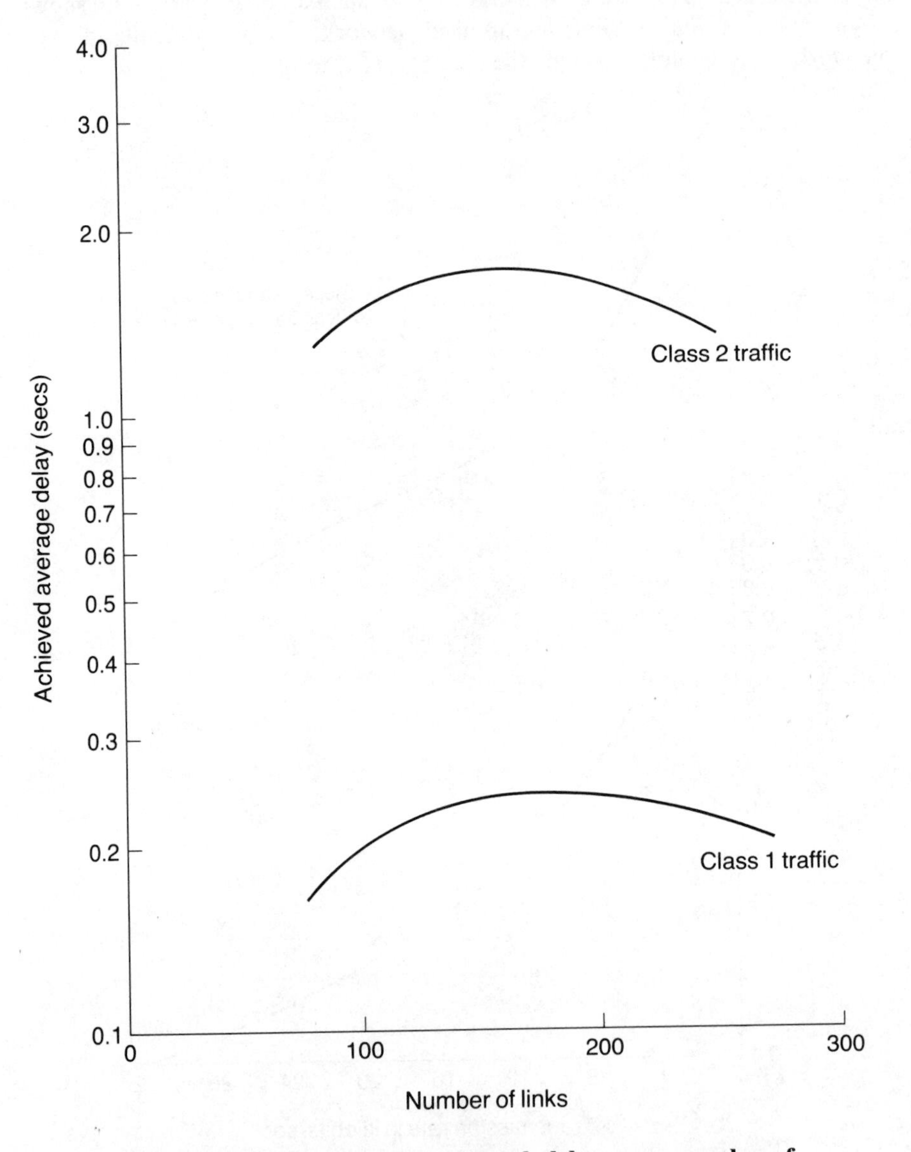

Figure 11-13. Plot of average network delay versus number of network links in a modified hierarchical network. © 1974, ICCC.

delay for class 2 traffic was 30 seconds. From Figure 11-12 we see that these constraints were significantly exceeded, and with channel bit rates above about 20 Kb/sec, average delays of less than 1 second were achieved for all traffic.

The excellent delay characteristics are attributable in part to the operation of the queueing type of network, which requires sufficient capacity to meet the basic demand and shows a rapid improvement in performance as extra capacity is added. However, the delay performance is also attributable to the fact that the hierarchical network design insures direct paths between all user pairs employing four or fewer nodes, thus reducing the likelihood of excessively large delays.

Figure 11-13 illustrates the network average delay in a somewhat different way. The achieved average delay for class 1 and class 2 traffic is plotted as a function of the number of links in the network. This particular plot is for the case of individual channel rates of 16 Kb/sec. We can see that the delay performance is rather constant, with only minor variation over the range of topological designs. The main reason for this constancy of performance is the fact that the hierarchical network provides direct, low-delay paths through the network for all user traffic. The additional links added to the network increase the richness of connectivity to protect against user isolation, but the principal route through the network still remains the shortest direct path through the higher level, tandem switches.

DETAILED REQUIREMENTS AND BAD GUESSES

We have now seen the impact of various topological designs and network structures both on the amount of network resources required to meet a fixed total user demand and on network performance. We should also remember from queueing theory that, once we have sized the network capacity to meet the estimated demand, the network performance can change very rapidly if the traffic carried on a facility begins to approach 100% of rated capacity.

One other interesting facet of demand estimation and network design is illustrated in Figures 11-14 and 11-15. Here, some of the previous plots of channels and channel-miles are repeated, but curves are shown for both uniform and nonuniform (actual) traffic flow. The uniform traffic case handles the same total volume of traffic as the nonuniform case, but assumes no knowledge of the distribution of the traffic among the various nodes. In most cases the network resources required to meet the total demand are remarkably close in the uniform and the actual traffic patterns. Agreement between two random estimated nonuniform traffic cases would likely be even closer.

It is valuable to establish a nominal distribution of traffic in order to study varying traffic flows and the impact on network design. However, the fact that changing user demand and network loading factors are likely to alter the node-to-node traffic patterns is a secondary consideration in sizing and tradeoff studies

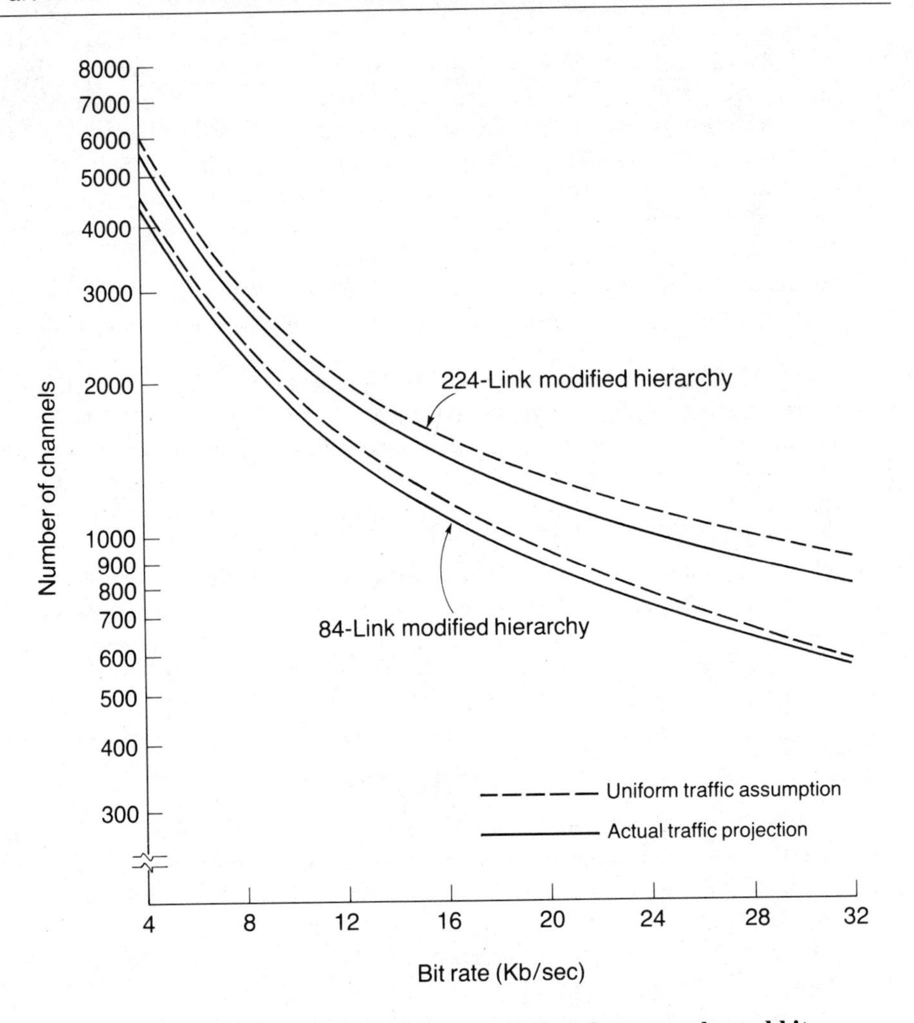

Figure 11-14. Plot of the number of channels versus channel bit rate for different network structures under different assumed distributions of user traffic. © 1974, ICCC.

for a large data network. What is really essential, as we will see further in the next chapter, is the ability of the network to adapt to changing user patterns, primarily through dynamic routing algorithms. Nevertheless, it is reassuring to know that a rough estimate of the initial network loading will give a usable network design that will not differ substantially in resource utilization and overall sizing from a more precise design done with full knowledge of the actual traffic demand. So, as we said in the title of this chapter, when it comes to the topological design of the network, even a bad guess is better than no guess at all.

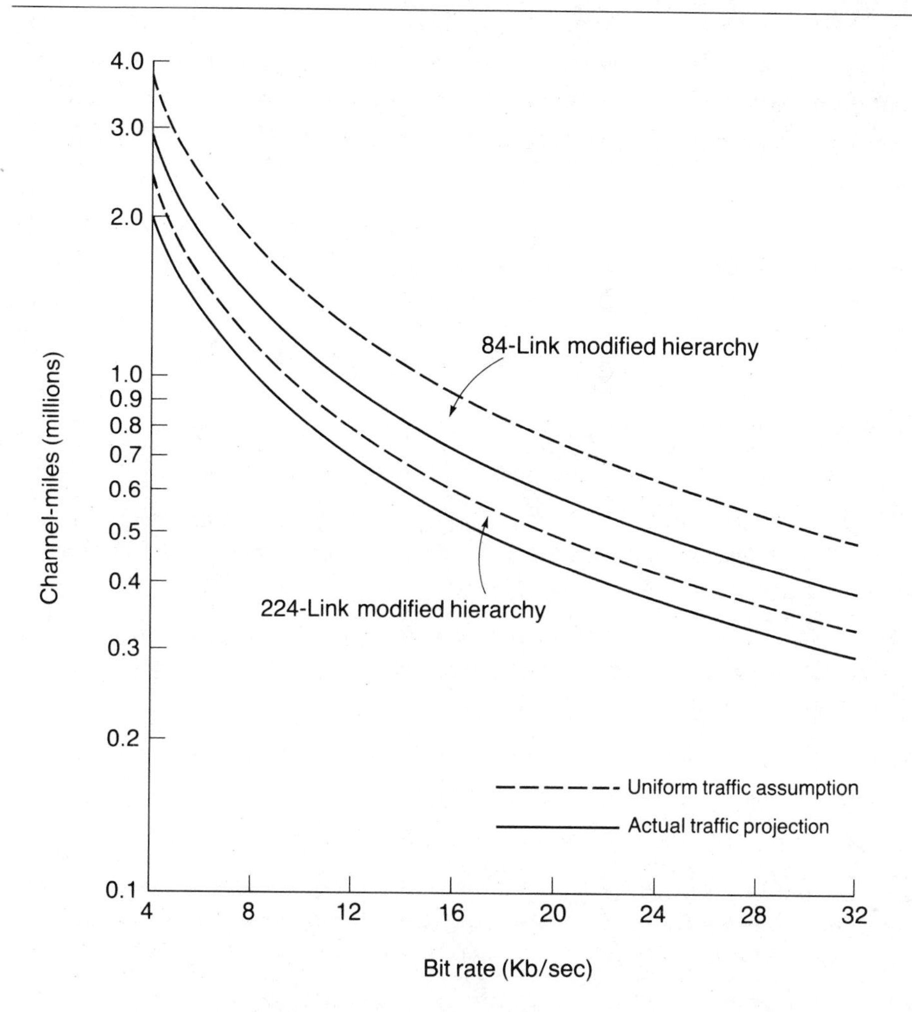

Figure 11-15. Plot of the number of channel-miles versus channel bit rate for different network structures under different assumed distributions of user traffic. © 1974, ICCC.

SUMMARY

In this chapter we used the example of a nationwide network to obtain a quantitative comparison between hierarchical and nonhierarchical network topologies.

For a packet-switched network a hierarchical network topology offered advantages because we could, under normal operating conditions, limit the total number of nodes in the end-to-end path between two users. Consequently, we could insure the delay performance of the packet network with relatively simple network protocols.

Our analysis of many different packet switched topological designs revealed that the network resources used to meet performance objectives were only slightly greater with a modified hierarchical network than with the lowest cost non-hierarchical network designs. In addition, the resources required for a network designed against a very gross estimate of user requirements were only slightly greater than for a network designed against a precise estimate. These results enable us to have a measure of confidence when embarking on the design of a new network with only very imprecise estimates of the potential demand.

SUGGESTED READINGS

GERLA, MARIO, and KLEINROCK, LEONARD. "On the Topological Design of Distributed Computer Networks." *IEEE Transactions on Communications*, vol. Com-25, no. 1 (January 1977), pp. 48–60.

This article presents an excellent survey of formal algorithms and computer-based design techniques useful in modeling, analysis, and design of network topologies. The general design problem is stated and decomposed into simpler subproblems for which exact solutions can be found. A heuristic topological design procedure attempts to solve the more complex original problem. The paper attempts to illustrate the relationship between performance measures, network cost, and effort in the design process.

ROSNER, R. D. "Large Scale Network Design Considerations." *Proceedings of the International Conference on Computer Communications*, ICCC '74. Stockholm, August 1974. Pp. 189–197.

This paper provides more details on the network design model used to obtain the results described in this chapter. The resources needed to meet network requirements of different topological structures are analyzed in some detail. References to the technical literature describe various approaches to the topological design of networks.

12

Routing in Packet Networks

THIS CHAPTER:

will introduce several ways to route packets through a network from source to destination.

will discuss the advantages and disadvantages of each routing method.

will describe the technique of adaptive directory routing and various ways it has been implemented.

In this part of the book we have found that one of the key topological attributes of practical and reliable packet networks is the presence of multiple paths or routes between the source and destination of the packets. Consequently, the operation of the network must include a method by which the packets learn the proper path.

Suppose we have to select the best route to drive between two cities. We can look at a map, select the most direct route (probably based on the shortest distance), and follow that path. Alternatively, after looking at the map we may choose a route that follows the interstate limited-access highways. Though possibly longer in distance, this path will take less time because we usually encounter less delay along such a route. Thus we may choose a route based on shortest distance or one based on lowest expected delay.

There are other possible ways to approach this problem. We can totally disregard any map and simply set out along the road, following the countryside, watching the scenery, and randomly wandering along until we end up at our destination. We can set out on the shortest path but listen to traffic reports on the radio and make path changes as new information is available about traffic delays. Instead of using a map, we can ask local residents along the way for their suggestions about the best way to reach the next milestone on the route to the ultimate destination.

Finding the best route through a packet network is quite analogous to finding our way between cities. Information has to be disseminated about the availability of different paths. Selection criteria have to be defined. And, finally, the route over which to transmit the packet must be decided on.

ALTERNATIVE ROUTING METHODS

There are many alternative routing methods, starting with the basic vehicular traffic analogy, and various implementations within each method. The four basic methods we shall investigate are packet flooding, random routing, directory routing, and adaptive directory routing. Each method is capable of achieving low **delay** through the network, which is the essential feature that packet switching tries to insure to the user community. However, the ability to achieve low delay is very dependent on the complexities of each technique, as well as on the relative amount of traffic in the network at any given time. Let's start by defining the basic operating techniques of each routing method.

Packet Flooding

Packet **flooding,** in effect, tries every possible path between the source and destination. Consequently, it can insure that traffic will eventually get through the network, even if the network is severely damaged or disrupted, as long as a physical path exists.

With the flooding method the packet is sent by the originating node to each of its neighbors. Each node that receives a packet checks to see whether it has ever received this particular packet. If it has, it simply discards the packet. If it hasn't, this node then sends the packet to all of its neighbors except for the one the packet came from. The process is repeated at each succeeding node until the destination is finally reached.

Since, in principle, every possible path through the network is tried, the first copy of the packet to reach the destination would have had to arrive over the path with the lowest total delay. Any copies that eventually arrive after the first one have to be discarded by the destination node.

Random Routing

Random routing has a basic concept similar to flooding, except that the packet is not sent out to every neighbor connected to each node. Instead, one of the routes from a node is randomly selected, and the packet is sent over that path only. The random selection of possible paths has to include the path over which the packet was received.

In effect, the packet "wanders" around through the network, eventually reaching the destination. The probability of selecting a particular line from a node can be biased—based on traffic loading, line capacity, or other network conditions—but in general the path between source and destination is unpredictable.

Like flooding, random routing has the advantage that, as long as a path exists through the network, the traffic will eventually arrive at the destination.

Directory Routing

Directory routing employs a **routing table** (directory) at each node that indicates the path to choose for any destination in the network from the present node. The directory is fixed in the switch memory and is developed "off-line" according to any of a number of possible criteria, such as shortest path, lowest expected delay path, highest capacity path, etc.

To operate reliably, directory routing has to include not only the primary choice path but also secondary choices to use under specified conditions, such as line failures, known switch failures, or severe traffic overload.

Adaptive Directory Routing

Adaptive directory routing is quite similar to directory routing in that each switch has a routing table in memory that indicates the best route for any destination in the network. However, the entries in the routing table can be changed in real time, "on-line," depending on changes in network operating conditions, such as traffic congestion or line or switch failures. Therefore, secondary routing choices are not needed. Current entries in the routing table, in principle, always represent the best possible current route to the destination according to the network selection criteria.

Each of these techniques has certain qualities that assure the reliable and accurate delivery of packets across the network. At the same time, each has inherent problems and complexities that limit its utility in practical networks. As we will see, the relative advantages and disadvantages have led most existing and planned packet networks to use some form of adaptive directory routing.

ADVANTAGES AND DISADVANTAGES OF ROUTING METHODS

Flooding

To understand the operation of the flooding technique, let us look at the simple network illustrated in Figure 12-1. Notice that each of the six nodes in this network is connected to three other nodes. (This is said to be a three-connected network.)

A packet originating at node 1 is destined for node 6. A copy of this packet departs from node 1 to each of its directly connected nodes—node 2, node 4, and

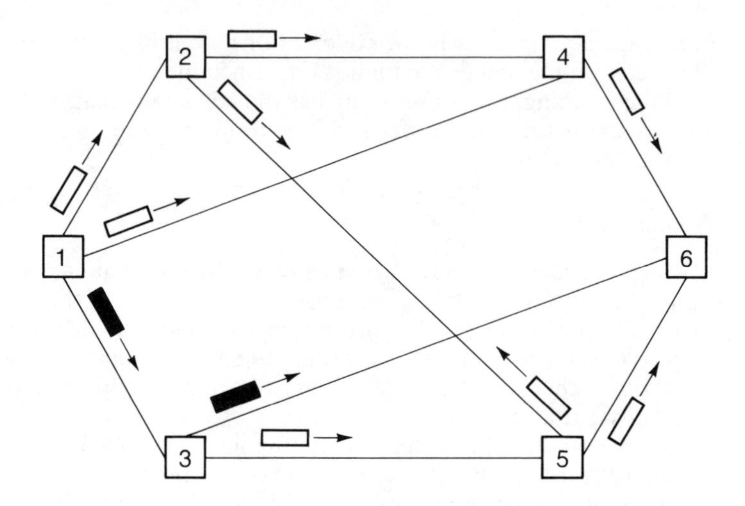

Figure 12-1. Packet flooding routing technique.

node 3. After the packet is successfully received at node 2, copies are sent to nodes 4 and 5. At node 4, however, the copy of the packet received from node 2 duplicates the copy received directly from node 1; therefore, it will be disregarded. At node 3 copies of the packet received from node 1 are sent to node 5 and node 6. Of course, the packet that arrives at node 6 completes the transaction, since this is the defined destination of the originating packet. In normal circumstances the fastest path from node 1 to node 6 is either the route node 1–node 4–node 6 or the route node 1–node 3–node 6. In this case we assume that the first copy of the packet to arrive at the destination—node 6—arrives from node 3.

Notice that, after the packet has arrived at its destination, there are still many copies of this packet circulating throughout the network. For example, the copy of the packet that arrives at node 4 from node 2 presumably duplicates the copy of the packet that arrived at node 4 from node 1 directly. Since node 4 recognizes that it has received this packet before, it does not send out any additional copies of the packet. Additional copies of previously delivered packets will slowly disappear from the network as they arrive at nodes that have previously received and relayed them. Furthermore, several additional copies of the packet are likely to arrive at the destination. Duplicate packets must be recognized by the destination switch and discarded. In our example copies of the packet are received from node 4 and node 5, in addition to the first one that arrived from node 3.

We can see that the disadvantages of the flooding technique are based on the multiplication of the traffic load, which is directly proportional to the average connectivity of the network. The increase in traffic intensity in the network increases the queueing delay, which generally leads to an increase in end-to-end delay, even though each packet is delivered via the currently fastest route. Secondly, each packet must contain complete address and identification information.

Each switch has to keep a complete record of all the packets it has seen over a sufficiently long time period to insure that duplicate copies of packets received and relayed earlier will be properly recognized and discarded. Similarly, all switches have to be constantly alert to the possibility that copies of packets will arrive much later, after the first copy has been successfully delivered to the destination subscriber.

Random Routing

Random routing reduces the problem of traffic multiplication by sending the packets over only one route emanating from each node. In Figure 12-2 we see that each route leaving each node is marked with the probability of its selection in the random routing process.

If all links and nodes were identical, we might assign the same probability to each route. However, differences in link capacities or possibly historical measurements that indicate the most probable destination of traffic emanating from each node lead to nonuniform probabilities. In this case one route at each node will be selected with 50% probability and the other two routes with 25% probability each. Figure 12-2 shows the packet, headed from node 1 to node 6 following the most probable route, always selecting the 50% probability path from each node. The path thus goes from node 1 to node 4, to node 2, to node 5, to node 3, to node 6—a rather roundabout route indeed.

This is admittedly an extreme case, but it illustrates the primary disadvantage of random routing. That is, while traffic, on the average, is distributed around the

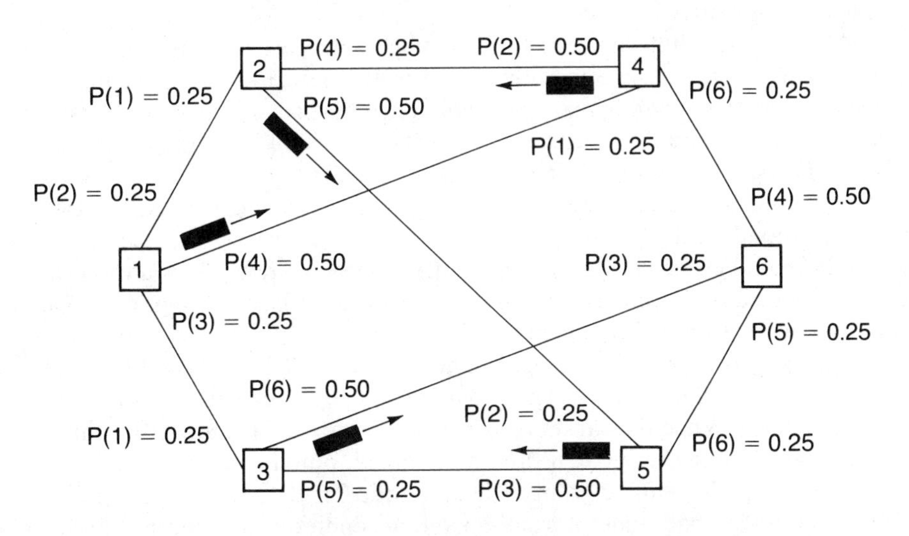

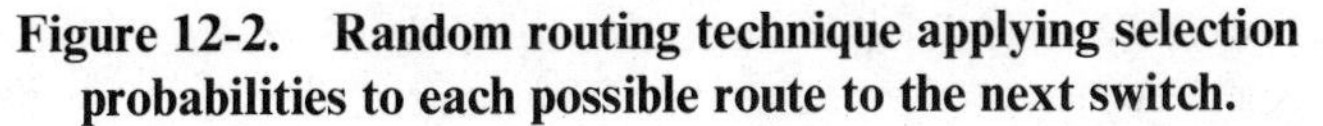

Figure 12-2. Random routing technique applying selection probabilities to each possible route to the next switch.

network according to the route selection probabilities, the average path length from source to destination will tend to be considerably longer than the most direct possible path. Consequently, the packets will incur end-to-end delay that is considerably longer than the lowest delay path. In addition, since the same packet flows through more than the minimum number of nodes, there is a traffic multiplication effect, though it is nowhere near as bad as the traffic multiplication effect of the flooding method.

Directory Routing

Directory routing has the advantage that its operation is completely deterministic; that is, every packet flowing between a particular source/destination pair will follow the same route according to the routing table or directory. Once the routes are set up according to some criterion, such as shortest path or lowest expected delay, they are faithfully followed unless the routing directory is changed by the operators or the network control center.

The major disadvantages of directory routing are (1) that it is so highly structured, and (2) that it is unable to adapt to changes in network configuration or operation resulting from line failures, node failures, or severe traffic overloads. In Figure 12-3 we again see our sample network, this time under the control of directory routing tables. A master—the shortest path routing matrix—contains 36 entries for the six-node network. Each entry in the matrix shows the next node along the shortest path from any node to any other node in the network. For example, packets going to node 2 that are presently at node 6 should be routed to node 4 next. (This is shown by the shaded entry in the sixth row, second column of the routing matrix.)

Although the full shortest path routing matrix exists conceptually, and probably would be constructed and stored in the network control center, only the single row of the routing matrix related to that particular switch need be stored at each switch. Figure 12-3 illustrates the local routing table for each switch, which simply shows, for each possible destination node, the proper node to which a packet should be routed. The local routing table is identical to the horizontal row corresponding to that node in the overall routing matrix. Note that in this example several entries in the routing matrix show two possible routes between certain node pairs. In practice, to insure truly deterministic routing, only one of the two choices would actually be put in the matrix.

For a packet flowing from node 1 to node 6, the node 1 routing table indicates that we should send the packet to node 3 or node 4. In this case we show the packet going to node 3. When the packet reaches node 3, the local routing table indicates that packets destined for node 6 should be routed to node 6 directly.

Thus the path followed from node 1 to node 6 is the shortest possible path. We would expect that, under normal operating conditions, the shortest path will result in the lowest average delay from source to destination. In addition, with this routing no excess copies of the packet are generated.

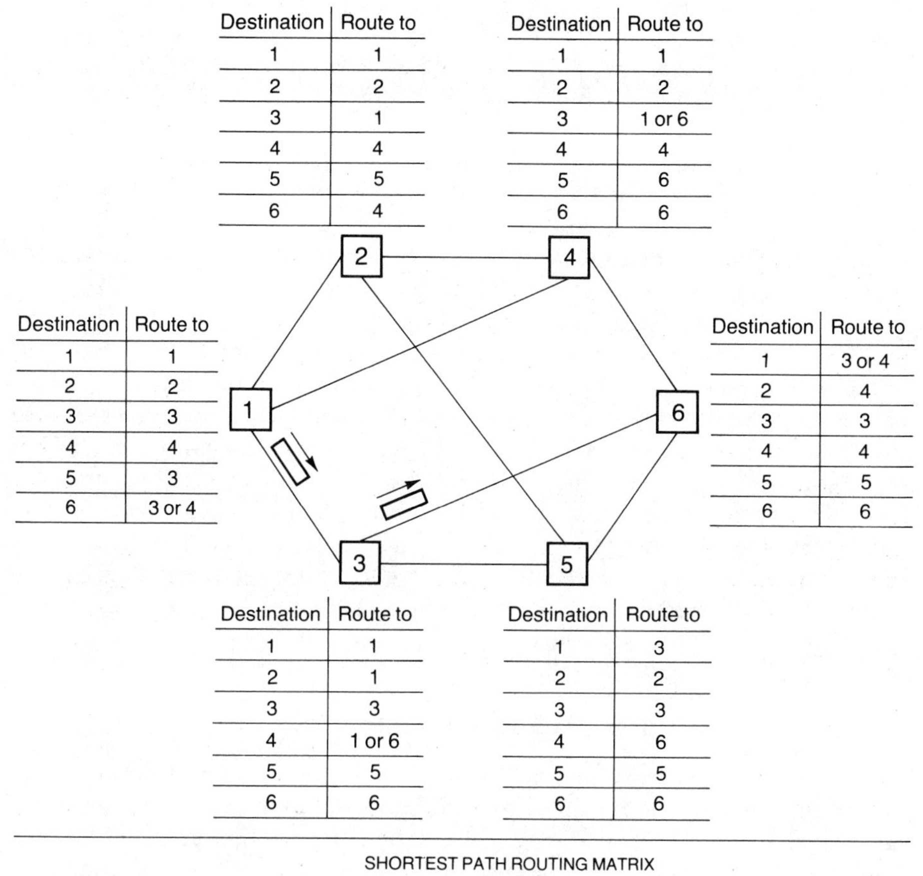

Destination	Route to
1	1
2	2
3	1
4	4
5	5
6	4

Destination	Route to
1	1
2	2
3	1 or 6
4	4
5	6
6	6

Destination	Route to
1	1
2	2
3	3
4	4
5	3
6	3 or 4

Destination	Route to
1	3 or 4
2	4
3	3
4	4
5	5
6	6

Destination	Route to
1	1
2	1
3	3
4	1 or 6
5	5
6	6

Destination	Route to
1	3
2	2
3	3
4	6
5	5
6	6

SHORTEST PATH ROUTING MATRIX

To node

From node	1	2	3	4	5	6
1	—	2	3	4	3	3 or 4
2	1	—	1	4	5	4
3	1	1	—	1 or 6	5	6
4	1	2	1 or 6	—	6	6
5	3	2	3	6	—	6
6	3 or 4	4	3	4	5	—

Figure 12-3. Directory routing technique using a routing table at each switch.

Adaptive Directory Routing

The adaptive directory routing method attempts to capture the advantages of the directory routing method while eliminating its disadvantages. The major difference is that, in the adaptive method, the routing matrix can be changed rapidly and frequently as conditions in the network change.

The major disadvantage of this routing technique is the complexity of performing the routing matrix formulation in real time. In computing the routing tables, the switch processors must be able to detect network conditions, provide the processing power to perform the routing calculations, and permit the routing tables to be dynamically updated. In addition, since the routing is neither deterministic (like fixed directory routing) nor highly robust (like the flooding method), problems in the routing could cause failures in the network's ability to deliver user traffic properly. However, complexity is part of the tradeoff of computer processing power in return for higher transmission efficiency that was discussed in Chapter 1.

Because of this transmission efficiency most actual packet networks use some form of adaptive directory routing. The method has had numerous different implementations in order to protect against anomalies that could interfere with timely traffic delivery. However, they all share the basic characteristic that each switch uses routing tables that can be changed readily in response to changes in the network operating conditions.

APPROACHES TO ADAPTIVE DIRECTORY ROUTING

As adaptive directory routing must be able to assess current network conditions in order to route traffic most efficiently, all approaches to this routing method entail some combination of information, criteria, and decision making.

First of all, information is needed on the status of the various network facilities—particularly lines and nodes—together with traffic loading and delay information along the various paths and routes through the network. Secondly, there must be criteria on which to base routing decisions. In most cases the shortest possible delay is the principal routing criterion. However, under various conditions, such as very heavy local loading, even distribution of traffic may be an alternative routing criterion. Finally, there must be a process to combine the information with the criteria and make a final decision for entry into the routing tables. In addition, a protocol must allow for quick and efficient distribution of routing and status information to all switches.

Delay as a Criterion

Since a primary objective of packet switched networks is the rapid delivery of computer-based information, the most widely used routing criterion is generally the lowest current end-to-end delay. Under light loading conditions this will often mean choosing the route with the shortest total length or the fewest number of total switches in the end-to-end path. However, as network loading increases so that there are appreciable queueing delays at intervening switches, the lowest delay path may turn out to be long and circuitous, attempting to bypass points of congestion and high queueing delay.

A routing plan based on minimizing delay can be readily implemented because each packet switch in the network is sufficiently intelligent to estimate the length of time it will take a new packet to reach each of its nearest neighbors. (A **nearest neighbor** is a switch directly connected to the present switch.) Each switch can make an accurate estimate if it knows, for each line connected to it, the bit rate, the error rate, and the number of bits of data that are queued up waiting for transmission over that line. The delay over that line is simply the line rate divided by the number of bits awaiting transmission, times a factor that represents the probability that a certain fraction of the data will have to be transmitted more than once due to line errors.

Similarly, each node can make a good estimate of the delay to its nearest neighbors by the same process. Any node can thus estimate the delay to a node two hops away by combining its estimate of delay with its neighbors' estimates. Therefore, if all nodes in the network exchange delay information with their nearest neighbors, each node can make an overall estimate of delay to all possible network destinations.

Delay Tables. To illustrate the principle of delay tables, let us refer again to the example network (Figure 12-4). In addition to the network connectivity, the bit rates are indicated for each of the network lines. For purposes of illustration we have assumed that each packet has 1000 bits of length. Figure 12-4 also shows the number of packets queued on each line leaving each switch. At each node we see a delay table, which is calculated by each node on the basis of the line rate and the amount of traffic queued for each line. At this point in the development each delay table contains entries for only those destinations to which each switch has a direct line, so the estimates can be considered quite accurate.

Developing a Routing Table

To further understand the way the routing algorithm operates, let's focus on the development of the routing table for node 5 (Figure 12-5). The starting point is the delay table that node 5 constructs for itself from its knowledge of its own line rates and queued traffic. The delay table (Figure 12-5a) which is the same as the one in Figure 12-4, has delay estimates entered only for those nodes to which node 5 has a direct line: From node 5 the estimated delay is 71 ms to node 2, 104 ms to node 3, and 312 ms to node 6.

Having made its own estimate of delay to its nearest neighbors, each node then exchanges its delay table for a copy of the delay table calculated by each of its

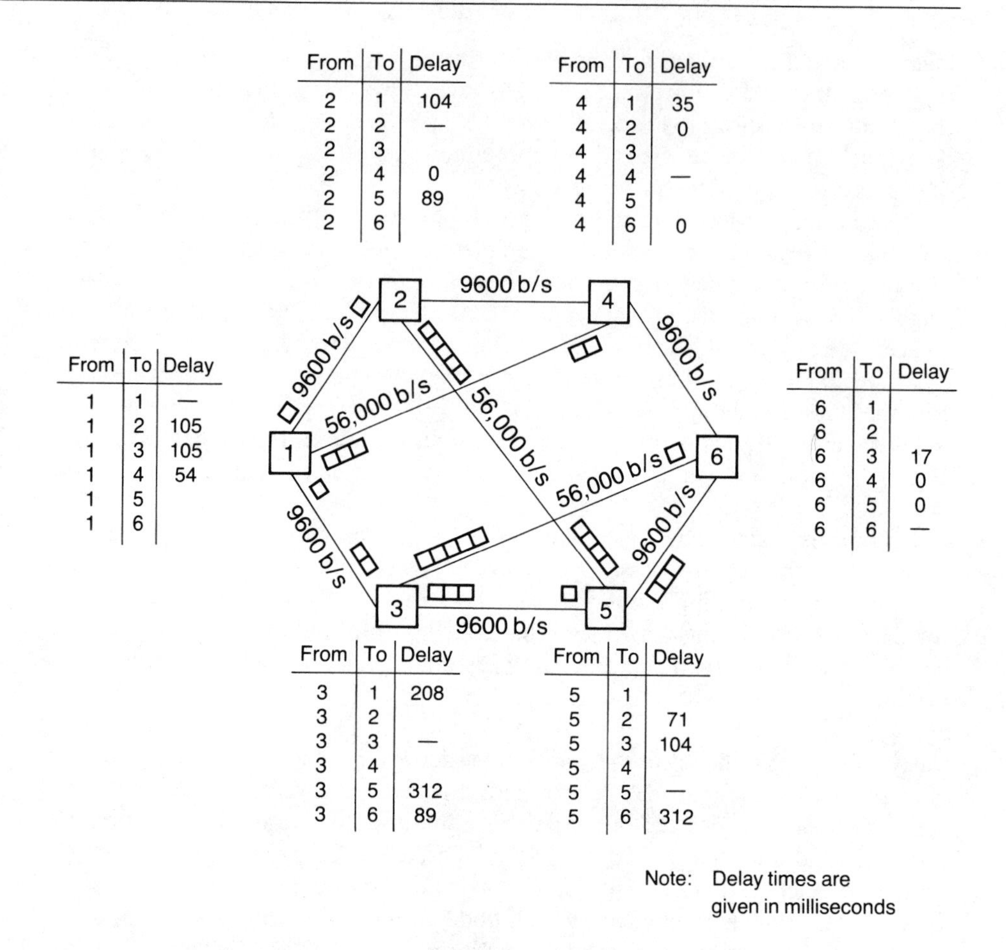

From	To	Delay
2	1	104
2	2	—
2	3	
2	4	0
2	5	89
2	6	

From	To	Delay
4	1	35
4	2	0
4	3	
4	4	—
4	5	
4	6	0

From	To	Delay
1	1	—
1	2	105
1	3	105
1	4	54
1	5	
1	6	

From	To	Delay
6	1	
6	2	
6	3	17
6	4	0
6	5	0
6	6	—

From	To	Delay
3	1	208
3	2	
3	3	—
3	4	
3	5	312
3	6	89

From	To	Delay
5	1	
5	2	71
5	3	104
5	4	
5	5	—
5	6	312

Figure 12-4. Adaptive directory routing using a delay table at each switch.

nearest neighbors. In our example node 5 exchanges delay tables with nodes 2, 3, and 6. As a result of this information exchange node 5 has its own delay table plus the three copies of its neighbors' delay tables (Figure 12-5b). The processor of node 5 now combines the information from these delay tables and calculates a routing table to use until updated information becomes available. If we assume that the processing time for a packet arriving at a node is 25 ms, in addition to the line delay entered into each node's own delay table there would be an additional delay of 25 ms to transmit data via an intermediate node.

Let us consider how information is routed from node 5 to node 1. Node 5's own delay table has no entry for delay to node 1 since node 5 does not have a direct line to node 1. However, two of the delay tables received from its neighbors have entries for transmission to node 1. Node 5 selects the lower of these estimated

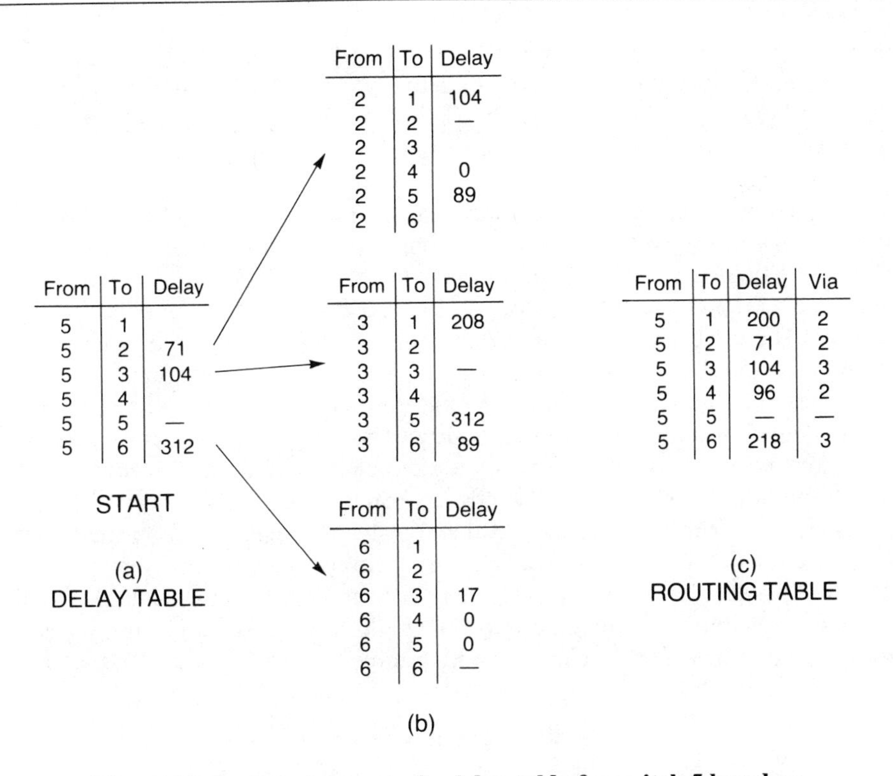

From	To	Delay
5	1	
5	2	71
5	3	104
5	4	
5	5	—
5	6	312

START

(a)
DELAY TABLE

From	To	Delay
2	1	104
2	2	—
2	3	
2	4	0
2	5	89
2	6	

From	To	Delay
3	1	208
3	2	
3	3	—
3	4	
3	5	312
3	6	89

From	To	Delay
6	1	
6	2	
6	3	17
6	4	0
6	5	0
6	6	—

(b)

From	To	Delay	Via
5	1	200	2
5	2	71	2
5	3	104	3
5	4	96	2
5	5	—	—
5	6	218	3

(c)
ROUTING TABLE

Figure 12-5. Development of a delay table for switch 5 based on local information and information obtained from switch 5's nearest neighbors.

delays—in this case 104 ms achieved by routing via node 2—and adds 25 ms processing time and its own estimated delay to node 2 (currently 71 ms). This information is then entered into its own routing table (Figure 12-5c): To reach node 1, node 5 will route via node 2, with an estimated delay of 200 ms (104 + 25 + 71 = 200 ms).

As node 5 is directly connected to node 2, the second entry in the routing table shows to route from node 5 to node 2 directly, with an estimated delay of 71 ms. Similarly, the third entry in the routing table shows that the route to node 3 is direct, with an estimated delay of 104 ms.

The next entry is interesting. As node 5 has no direct route to node 4, there is no entry in its own delay table. However, both node 2 and node 6 have entries in their delay tables, in both cases showing zero delay. In order to choose the better of these, node 5 has to combine its own delay estimates to reach the intermediate nodes. Reaching node 4 via node 2 will take 71 ms to reach node 2, plus 25 ms processing at node 2, plus zero delay from node 2 to node 4—a total of 96 ms. Reaching node 4 via node 6, however, will take 312 ms to reach node 6, plus 25 ms processing at node 6, plus zero delay from node 6 to node 4—a total of 337 ms. The next entry in the routing table, therefore, is to route to node 4 via node 2, with an estimated delay of 96 ms.

The last entry in the routing table illustrates the adaptivity of the routing process. Node 5 has a direct line to node 6, as shown by the 312 ms delay estimate entered in node 5's delay table. However, node 3 also has a direct line to node 6, with a delay of only 89 ms. The delay from node 5 to node 3 would be 104 ms, processing time at node 3 is 25 ms, and the delay from node 3 to node 6 is 89 ms. This totals 218 ms delay, compared to 312 ms delay to go from node 5 to node 6 directly. Therefore, to reach node 6 from node 5, it would be advantageous to route via node 3, and the last entry in the routing table reflects this routing.

The final routing table for node 5 indicates that, at the present time, all traffic will be routed over the lines to either node 2 or node 3. It is also clear that within a very short period of time—200 ms or less—the information in this routing table will be outdated. For instance, all traffic has been routed via nodes 2 and 3 because the delay to node 6 was much greater than the delay to the other connected nodes. But since no additional traffic is being sent via node 6, the queue over this line to node 6 is being reduced. As a result, after the routing table is used for 200 ms, the delay over the line from node 5 to node 6 would be reduced to 112 ms, which by itself would induce a change in the routing table.

Conclusions about Routing Tables

In order for an adaptive routing technique of this type to operate well, either the routing updates have to take place quite frequently, or else the marginal difference between two routes has to be quite large before a circuitous route would be chosen above the most direct route.

Furthermore, we built the routing table up sequentially, beginning with the delay table of each node, for the purpose of illustration. In reality, however, the overall routing table is retained and periodically exchanged between each switch and its adjoining neighbors. The node then combines its locally measured delay information with the routing tables received from its neighbors to calculate its own current best estimate of network delay and routing. This then becomes the routing table to be transmitted to its neighbors in the next time interval.

The algorithm we have described here is the basis for many variations applied to a wide variety of message and packet switched networks. Early in the ARPANET development, for instance, the algorithm was implemented with synchronous exchange of routing tables between adjoining switches between three and five times per second.

Experience revealed that synchronous exchanges can lead to instabilities, where two nodes simultaneously tell each other that the best route to a point in the network is via the other node. If the routing tables are exchanged asynchronously, each node has an opportunity to utilize the latest information available in its own routing table before passing the aggregated information onto its adjoining nodes.

It was also found that very frequent exchange of routing tables could cause network instabilities, creation of packet loops, and shuttling packets back and forth between a pair of nodes. Routing table updates on the order of every second or longer were found to be sufficient for effective utilization of the adaptive process. Latest implementations of the ARPANET routing algorithms update the routing tables about every ten seconds.

It can be seen from the description of the routing algorithm and routing table update process that information about line and node failures will gradually spread through the network. A line failure would be indicated by a very large (infinite) delay entry for the nodes connected to the failed line. If a failed node's nearest neighbors didn't receive the updated routing table as expected, they would realize that the node had failed. Routes including the failed node could be avoided by the inclusion of large (infinite) delay estimates associated with routes passing through the failed node. When the failed node became operational, it could be returned to the network simply by beginning to transmit routing tables to its immediate neighbors.

Variations of Adaptive Directory Routing

As we have described the algorithm so far, the only true connectivity information exists between each node and its nearest neighbors. The connectivity beyond a particular node's nearest neighbor can be deduced only from the information obtained from its neighbors. There are many variations and methods for implementing adaptive directory routing methods.

A more accurate routing table process would trade memory and storage for increased accuracy. A complete connectivity matrix of the network, together with the delay status on all lines, would be stored at every node. Current delay information measured directly at each node would be sent periodically to every node in the network (not just the nearest neighbors). Each node would then calculate its own minimum delay routing table based on all the information received from the other nodes of the network. Experience has demonstrated that, although this algorithm takes about three times as much memory at each switch as the local update–only algorithm, it has improved performance and reliability. If the table update rate is reduced to about once every 10 seconds, the network overhead is no greater than that for local exchanges. The global algorithm responds quickly to line and node failures and to congestion due to traffic buildup. Because of the extended measurement period the process is quite stable, and packet loops or packet shuttling are extremely unlikely.

Other approaches to adaptive directory routing employ centralized information processing, where the routing plan is computed "off-line" by the network control center on the basis of traffic information received from each node. Periodic changes to the routing plan are sent by the NCC to each node in the network. This approach minimizes the processing and storage burden on each switch to individually process routing table updates. However, it tends to make the routing updates somewhat less current than those obtained from the distributed process. An even simpler approach to the table update process would be to restrict updates to only major network changes, such as line or switch failures.

SUMMARY

An essential element in achieving consistently low delays for the subscribers of a packet network is the routing algorithm, which is used to find the shortest or lowest delay path through the network. A good routing technique must assure the reliability of the network so as to prevent the isolation of users due to the failure of lines or nodes in the network. Both flooding and random routing will insure the reliability of network connectivity, but the overall delay will suffer because of both traffic multiplication and the choice of nonoptimum paths. Shortest path routing using predetermined routing directories can assure good delay performance under normal design operating conditions, but it cannot adapt to changes in traffic patterns or problems with network facilities.

Adaptive directory routing, which can be implemented with a wide range of detailed algorithms, provides a sufficient degree of flexibility both to achieve good delay performance and to adapt to changes in network operating conditions. Of course, this means that network resources in the form of switch memory, switch processing capacity, and transmission line capacity must be dedicated to the routing process. However, experience with existing networks has shown that both processing and line overhead associated with the routing function are very small and well worth the effort to insure low network delay and high network reliability.

SUGGESTED READINGS

McQuillan, John M., Falk, Gilbert, and Richer, Ira. "A Review of the Development and Performance of the ARPANET Routing Algorithm." *IEEE Transactions on Communications*, vol. COM-26, no. 12 (December 1978), pp. 1802–1811.

In this article the specific ARPANET routing algorithm used as the basis for the discussion in this chapter is reviewed. The ARPANET algorithm is traced from its original implementation through several major changes that resulted from practical experience.

McQuillan, John M., Richer, Ira, and Rosen, Eric C. "The New Routing Algorithm for the ARPANET." *IEEE Transactions on Communications*, vol. COM-28, no. 5 (May 1980), pp. 711–719.

This paper describes the change of the ARPANET routing algorithm from the nearest neighbor driven technique to a networkwide information driven algorithm. Because each node computes its routing table from information gathered from every other node in the network, routing performance tends to be more accurate and more stable. The actual performance of the new algorithm is compared to the earlier implementations.

SCHWARTZ, MISCHA, and STERN, THOMAS E. "Routing Techniques Used in Computer Communication Networks." *IEEE Transactions on Communications*, vol. COM-28, no. 4 (April 1980), pp. 539–552.

This is an overview article of a number of different adaptive routing algorithms used in several operating networks, including TYMNET, ARPANET, TRANSPAC, SNA, and DECNET. After a general introduction to shortest path and minimum delay routing algorithms, detailed implementations, together with their advantages and pitfalls, are described.

PART FOUR

Packet Switching without Packet Switches

In the first three parts of this book we dealt with packet switched networks that were geographically well defined; had switches and transmission lines that were discrete elements of the network; and operated within this structure with a set of carefully defined protocols to insure rapid, efficient interchange of information among the subscribers to the network.

In Part Four we are going to deal with several different approaches to packet communications, using packet-based demand assignment techniques and the notion of packet broadcasting. Synchronous communications satellites are often a central element of these approaches. Because of their location more than 22,000 miles above the equator, satellites can be seen by essentially all the network subscribers at the same time.

Although the unique characteristics of satellite communications will provide the starting point for our understanding of packet broadcasting, we will find that demand assignment and broadcasting techniques can be applied to purely terrestrial systems as well. We will also be able to estimate quite accurately the best structure and configuration of distributed packet networks that utilize satellite communications in combination with terrestrial resources.

As we proceed through this part of the book, notice the interesting effect that packet broadcasting has on our network approaches. We will see that packet broadcasting is the ultimate in distributed packet switched networks: It achieves packet switching without packet switches!

13

Resource Sharing and Multiple Access Techniques

THIS CHAPTER:

will review the general question of resource sharing and the typical application of shared communications among a large community of low duty-cycle users.

will describe the basic operation of the ALOHA technique.

will see how packet switching functions are achieved without packet switches over broadcast satellite channels, and estimate the capacity, throughput, and delay over such channels.

The words *LIVE VIA SATELLITE* flashed across millions of television screens nearly every day are probably the most vivid reminder of how satellite-based communications has added multiple dimensions to worldwide media coverage. Not quite as visible is the huge growth in high-quality international communications service. It is particularly significant that this service is available to all lesser developed countries as well as to highly developed, highly industrialized nations.

Despite its various applications satellite-based communications has functionally been viewed as a giant "cable in the sky." By and large, satellite-based communications has been carried out on a point-to-point basis, even though many pairs of points can be served simultaneously by a single satellite. Only in the very late 1970s was it recognized that satellites provide capabilities that go far beyond the cable in the sky. Furthermore, as these capabilities have been expanded, their costs have been reduced impressively.

Before we get into the application of specific satellite technology to packet communications, let us briefly revisit the general problems of sharing communications.

THE GENERAL PROBLEM OF COMMUNICATIONS RESOURCE SHARING

Figure 13-1 shows the generalized switched network that we introduced in Chapter 1, where a single switch permits an arbitrarily large community of N users to interchange information. Although the center of this network has been conceptualized as a switch, it could very well be a central shared computer, particularly in computer-based information systems. Figure 13-2 makes the same situation slightly more general by indicating a centralized resource, M, and a number of individual terminals, each designated by T. However, contrary to the case in Figure 13-1—where each user was connected to the central point by his own physical, point-to-point line—the connection between M and each T is presumed to be a shared medium. The medium could be a single multidrop circuit, a shared terrestrial radio channel, or a common user broadcast satellite channel.

Regardless of what the shared medium is, a problem arises from the fact that the communication between the users and the central point is not symmetrical. Since, in general, M is an intelligent, processor-based device, communication of messages or packets from M to the various T's over the shared medium is not difficult. By use of temporary packet buffering, or storage at M, the various packets going from M to the group of T's can simply be transmitted sequentially over the shared medium. The problem is the transmission of information from the various independent T's to the central point, M. At first, it appears that we have to provide a formalized structure or transmission protocol so that the resources of

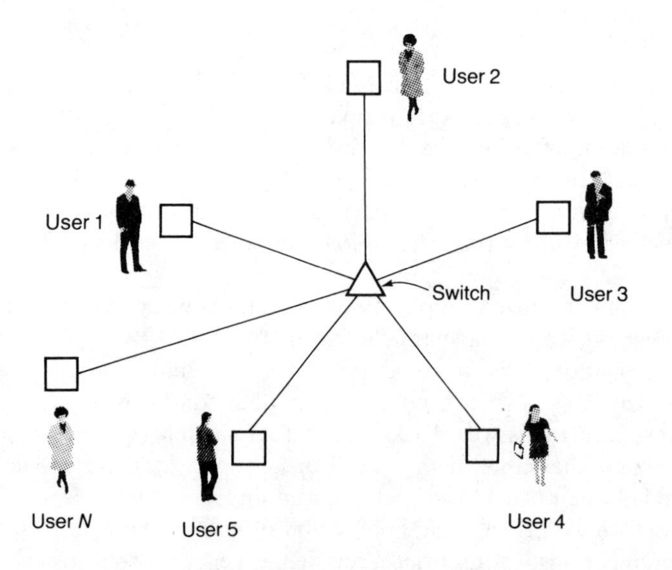

Figure 13-1. Generalized switched network of N users.

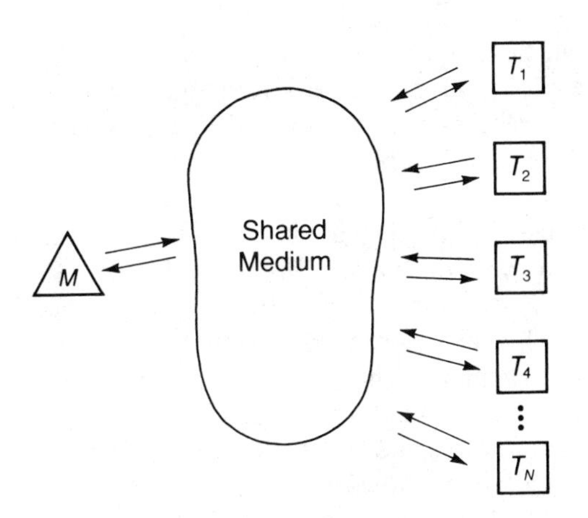

Figure 13-2. Model of independent user terminals accessing a central location by using a shared medium.

the shared medium can be fairly allocated among the large group of potentially competing users.

The sharing can be achieved using highly structured capabilities, such as frequency-division and time-division multiplexing, or less structured techniques, such as polling or multidrop circuits. Finally, essentially unstructured random-access demand-assignment techniques, such as the ALOHA technique that we will study in detail, can provide surprisingly good results for communications medium sharing. In any case, the utility of sharing a resource is most evident with **low duty-cycle** communications users. By low duty-cycle, we mean that the user sends occasional bursts of information, and the time between transmissions is long compared to the length of each individual transmission.

Table 13-1 summarizes the peak and average communications usage of several classes of low duty-cycle, or bursty, communications users. In each case, because of processing time and other protocol and network functions, as well as

Table 13-1. Low Duty-Cycle (Bursty) Communications Users (in bits per second)

	Average Rate	*Peak Rate*
Terminal-to-computer	1 B/S	100 B/S
Computer-to-terminal	10 B/S	10,000 B/S
Remote job-entry	100 B/S	10,000 B/S
Computer-to-computer	10,000 B/S	1,000,000 B/S

the time for human thought process between network actions, there is a 100 to 1 or 1000 to 1 difference between the peak rate and the average rate of usage. For example, an airline reservationist may enter 20 characters via a terminal device to query a data base about flight and seat availability. A few seconds later, the computer responds with a screen full of data, possibly 1000 characters. After a minute or two discussion with the customer, several more characters are entered, to get additional information, or possibly to confirm a reservation. Thus over a period of several minutes only a few characters are entered by the reservations clerk, resulting in a low average utilization of the line.

In any case, particularly where people are operating a terminal tied to a computer, inputs typically come as a few hundred bits every few minutes (averaging out to between 1 and 5 bits per second). Responses from the computer typically come at the same interval. However, because they have greater information content, they have higher average data rates. Consequently, resource sharing approaches that use fixed allocation of capacity tend to be wasteful and inefficient.

Let us now turn to some specific applications of satellite-based techniques to shared communications, beginning with the ALOHA technique.

THE SATELLITE-BASED ALOHA TECHNIQUE

What Is ALOHA?

First of all, **ALOHA** is not an acronym. Used in the sense of its Hawaiian meaning as a greeting both of arrival and of departure, it is the name of a communications protocol and technique that was developed during the late 1960s at the University of Hawaii. The word is also used in the operation of the protocol: Whenever a user has something to transmit, he simply says "Aloha" into the channel and begins transmitting the message.

Imagine the situation illustrated in Figure 13-3. Here we see a group of users, each communicating with a central processor over a commonly available channel. The communication takes places via a synchronous satellite 22,300 miles above the equator. All users can receive the communications of all other users, and can also hear their own transmissions. The satellite is more than just a mirror in the sky. It receives signals transmitted on the **uplink** frequency, amplifies them, and retransmits them on a different **downlink** frequency. Because of the long distance the transmissions must travel, it takes about one-fourth of a second for a user's signal to reach the satellite and be returned to earth.

Typical Operation of the Basic ALOHA Channel

Figure 13-4 shows the typical occurrences in the channel. Although the ALOHA technique is not confined to any particular data rate, it is explained most easily in terms of a specific data rate. We will assume that the channel is operating

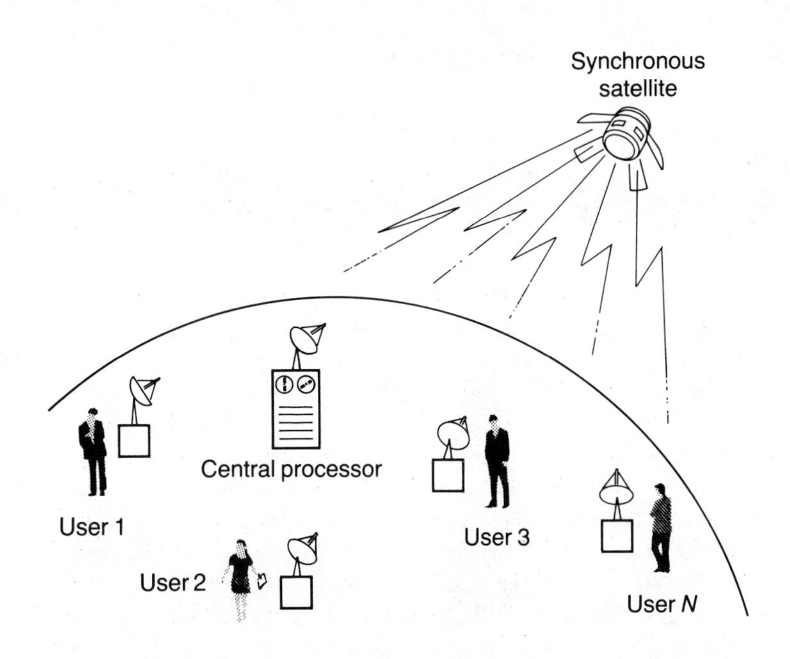

Figure 13-3. Distributed network using a satellite channel for common access.

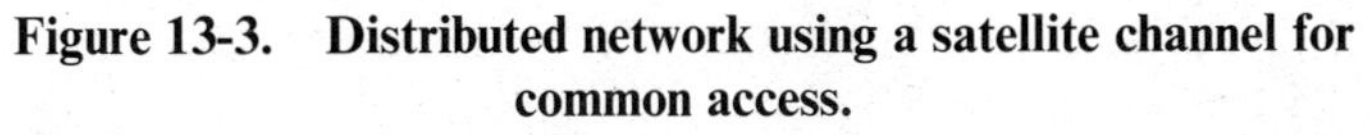

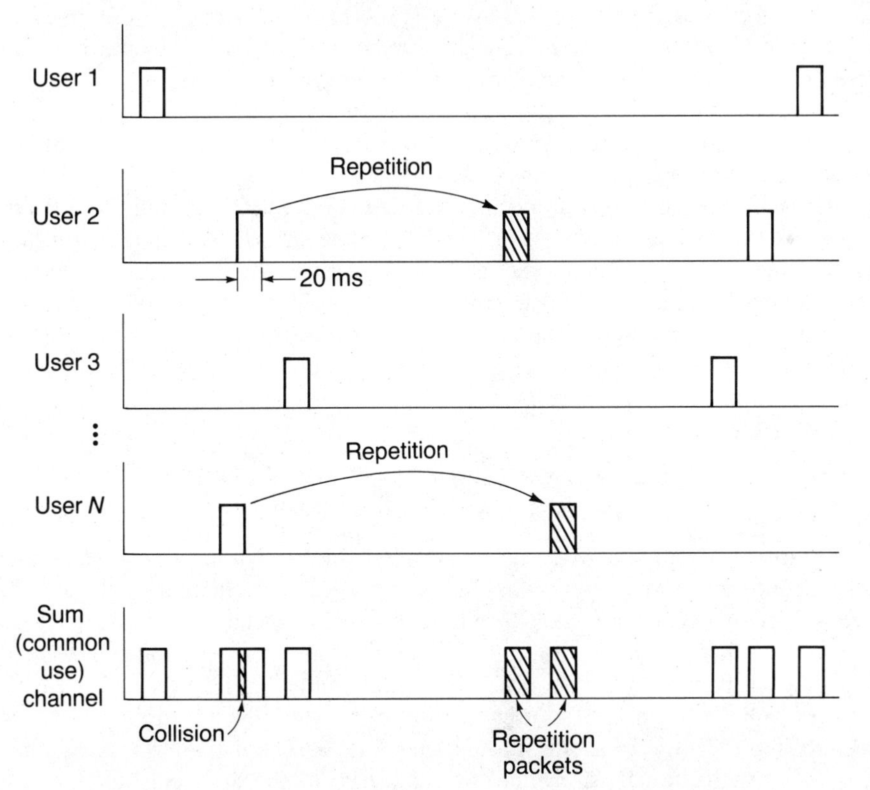

Figure 13-4. Summary operation of the basic ALOHA channel.

at a rate of 50,000 bits per second, and that each user sends a data packet consisting of 1000 bits or less. A full packet will thus have a duration on the channel of one-fiftieth of a second (20 ms), which is relatively short compared to the quarter-second (250 ms) that the packet takes to travel up to the satellite and back.

Figure 13-4 shows each individual user's transmissions separately and the sum of all the user transmissions. User 1 transmits a packet for 20 ms first. Shortly thereafter user N transmits a packet, but before this packet is completed, user 2 begins to transmit a packet. As we see in the sum channel, this results in a **collision,** or overlap, of these two packets, making both of them unintelligible at the destination.

One-quarter second after the initial transmission, user 2 and user N, by listening to the satellite downlink, hear that their packets have been involved in a collision. They each transmit a repetition packet to replace the packet that was damaged in the collision. (We assume that, if any part of a packet is damaged, the entire packet has to be repeated.) If both users act immediately, however, they are likely to collide again. In order to avoid this, both users wait for a randomly selected delay time to elapse before attempting the retransmission. As we see in Figure 13-4, the procedure is successful: The repetitions of the collided packets are retransmitted without interference.

In order to portray the packet overlap process, we have had to use a time scale for Figure 13-4 that is quite inaccurate. The entire width of the figure represents less than 1 second of real time. In reality it is very unlikely—for most users quite impossible—that any user will generate more than one packet during a 1-second interval. (To do so from a keyboard-type device would require a typing speed in excess of 1500 words per minute.)

In any case, collisions do occur, causing delays and retransmissions. As the number of active users increases, or the frequency with which each user transmits packets increases, the likelihood of collisions increases. As the collisions increase, the channel becomes even busier because each collision generates at least two attempted retransmissions. Therefore, it is important to see if we can predict the behavior of the ALOHA channel and determine how much traffic the channel is actually capable of delivering.

THE NETWORK ASPECT OF THE PACKET BROADCAST CHANNEL

Before analyzing the performance of the ALOHA channel we should emphasize some of the unique operational features of the technique that established the network environment of packet switching without the need for switches.

In this discussion we will consider the user and the terminal device to be the same entity. Also note that, when we described the ALOHA technique earlier, we spoke of the need for the user to listen for a possible collision one-quarter second after the packet was transmitted. In reality, the channel interface equipment associated with the user's terminal would carry out the protocol functions and perform any necessary retransmissions. The person at the terminal would not be required to take any actions at all different from the processes and procedures he might follow even if he had a dedicated line into the network.

Network Functions

The network functions generally associated with a packet switch are automatically absorbed into the basic operation of the ALOHA channel. In a network employing terrestrial switches, the switch acts as the user connection point, performing the packet routing function and the capacity allocation of the lines to which it is connected. It allocates capacity by buffering (storing) packets until each has its turn for transmission.

Figure 13-5 depicts a general application, where users may want to send packets to other users directly, as well as to any one of many computers in the network. Three users—user 3A, user 3B, and user 3C—are in close proximity to each other and can share a common ground station equipment set.

Network access is achieved by any user who has connection to an authorized satellite earth terminal. He may be directly connected to such an earth station locally, or he may use a short piece of dedicated terrestrial transmission to achieve

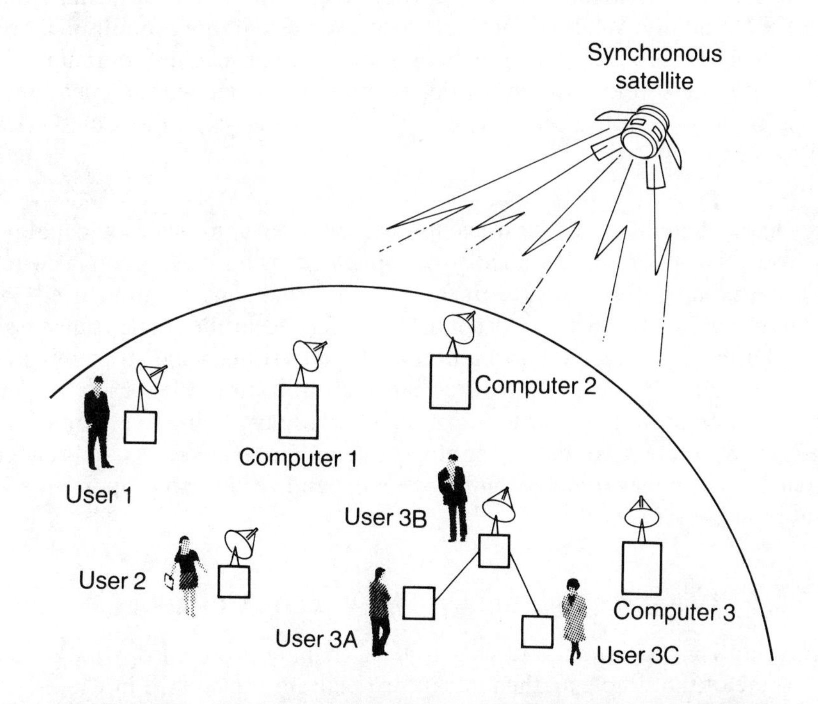

Figure 13-5. Serving a distributed network of users over a shared satellite channel using dedicated and shared earth stations.

connectivity to an earth station. Routing of traffic is not needed since all users transmit directly to each possible destination via the universal broadcast coverage of the satellite. All required capacity, instead of being distributed into a large number of point-to-point links, is aggregated into the common capacity available through the satellite. If capacity is available when a user transmits, he gets through successfully on the first try. If the channel is busy, and capacity is not available, collisions and retransmissions occur until traffic delivery is successful.

A destination and source address, as well as a repetition counter to prevent confusion between original packets and their repeats, are included in each packet transmitted. Each packet then contains all the needed information to insure its arrival at its ultimate destination. Consequently, all the networking functions are achieved using a satellite broadcast technique—without the need for network switches.

Disadvantages. There are, of course, some disadvantages to broadcast network techniques. Since a single satellite can cover only about one-third of the earth's surface, user communities that span very large distances may not be reachable via a single satellite broadcast network. Another problem is privacy. Not only is every user in the network able to hear the traffic of any other user, but anyone, anywhere, within visible range of the satellite can monitor all transmissions of the user community. While the problem seems somewhat more serious in a broadcast network, all forms of common user or common carrier communications services are vulnerable to interception. Users of electronic communications are moving rapidly to utilize security and privacy devices, regardless of the transmission medium.

Advantages. Satellite broadcast techniques provide a number of capabilities that would be much more difficult to implement using conventional switched network techniques. Packets destined for more than one location need to be transmitted only once, with all destinations indicated in the single transmission. Since all users can hear the information at the same time, single transmission of multiple destination messages reduces capacity utilization. New users can enter the network or existing users can move around without having to rewire facilities or inform any address, routing, or control tables of the changes. Thus a broadcast-based network can be rapidly reconfigured or expanded, and can support mobile or movable users very easily.

CAPACITY AND DELAY OF THE ALOHA CHANNEL

Having learned how the basic ALOHA channel protocol operates, we can apply some of the simple mathematical formulations introduced in Chapter 5 to estimate the capacity and user delay that will be achieved by a broadcast channel used in this way. The notation is summarized in Box 13-1.

$$\lambda = \text{messages (packets) to be delivered (in packets per second)}$$
$$\lambda' = \lambda + \text{repetition packets}$$

therefore:

$$\lambda' > \lambda$$

$$K = \text{length of packet (in bits)}$$
$$R = \text{channel rate (in bits per second)}$$
$$\tau = \text{packet length (in seconds)} = \frac{K}{R}$$

then

$$\lambda \times K = \text{traffic intensity in the channel (in bits per second)}$$

for convenience, define:

$$S = \lambda K = \text{channel throughput}$$
$$G = \lambda' K = \text{channel traffic}$$

or:

$$s = \frac{\lambda K}{R} = \text{normalized channel throughput}$$

$$g = \frac{\lambda' K}{R} = \text{normalized channel traffic}$$

or

$$\boxed{s = \lambda\left(\frac{K}{R}\right) = \lambda\tau}$$

$$\boxed{g = \lambda'\left(\frac{K}{R}\right) = \lambda'\tau}$$

Box 13-1

The Notation

We assume that the aggregate community of users has a total of λ messages or packets per second to be delivered at the present time. This is the *current packet demand* on the channel, measured in packets per second. However, because of the way the ALOHA channel operates, some of the delivered packets will have been involved in collisions and will be repetition packets. The total traffic actually flowing in the channel is thus defined as λ' (which is the sum of the packets delivered on the first attempt, the packets delivered as a result of one or more repetitions, and the packets that were damaged by collision). In other words,

$$\lambda' = \lambda + \text{repetition packets}$$

Therefore, although we don't know the actual value yet, λ' must be larger than λ.

We will denote the length of each packet, in bits, as K, and the rate of the channel as R bits per second. The amount of traffic that the users require to be delivered is thus $(\lambda \times K)$ bits per second.

For convenience, we will define:

$$S = \lambda K = \text{channel } \textit{throughput}$$

$$G = \lambda' K = \text{channel } \textit{traffic}$$

and since the channel bit rate is R bits per second,

$$s = \frac{\lambda K}{R} = \text{normalized channel } \textit{throughput}$$

$$g = \frac{\lambda' K}{R} = \text{normalized channel } \textit{traffic}$$

By *normalized* throughput and traffic we mean simply that both of these measures are expressed as a fraction of the total capacity of the channel. The values of s and g are thus fractions that can range from 0.0 to 1.0. Furthermore, by expressing the results in a normalized measure, the analysis is completely independent of the actual channel rate we are considering. However, for purposes of the example we will continue to use a 50,000-bits per second channel and 1000-bit packets.

Finally, if the packet length is K bits per packet, and the channel rate is R bits per second, then the duration of each packet in time on the channel is:

$$\tau = \frac{K}{R} \text{ seconds per packet}$$

By substituting this relationship into the expressions above, we find:

$$s = \lambda \tau$$

and

$$g = \lambda' \tau$$

It is interesting to note that these normalized measures of throughput and traffic on the channel are in the same form as the definition of Erlangs (traffic intensity) we worked with in Chapter 5.

An Example Transmission

Now that we have all the basic notation and parameters we need to understand the operation of the ALOHA channel, let us look at a specific case, that of user V.

User V's packets are transmitted at random times, each packet lasting τ seconds. However, for a packet to be successful, that is, to avoid collisions, no other user can start to transmit a packet beginning τ seconds ahead of user V, or any time during the interval up to the end of the transmission of user V's packet (Figure 13-6). In other words, user V's packet will be successful only if no other user transmits during a time interval (the vulnerable period) that is twice as long as user V's packet.

In Chapter 5 we learned about the Poisson arrival process for statistically unrelated users. For such users the probability of having exactly K new packets arrive over a time interval of τ seconds was given as:

$$P(K) = \frac{(\lambda\tau)^K e^{-\lambda\tau}}{K!}$$

where λ is the average packet arrival rate.

We can use this expression because the probability of user V's packet being successful is the same as the probability that exactly zero other packets are transmitted during the time interval of length 2τ seconds. If we substitute these values into the Poisson expression, the probability of zero arrivals ($K = 0$)—which is also the probability that user V's packet is successful—is then:

$$P_s = P(K = 0) = \frac{(\lambda' 2\tau)^0 e^{-2\tau\lambda'}}{0!} = e^{-2\lambda'\tau}$$

We must use the value of λ' in this formula since the traffic seen by the channel includes all the likely repetition packets. The expression $P_s = e^{-2\lambda'\tau}$ is helpful, but it does not fully describe the utilization of the ALOHA channel.

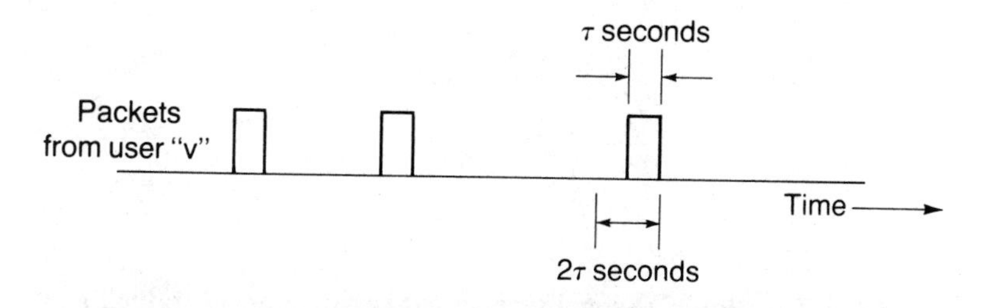

Figure 13-6. Vulnerable period for packet interference in an ALOHA channel.

What this expression shows is the probability that a particular user's packet will be successful when it is transmitted. In the case of our example the average successful traffic is what we defined as the channel throughput, and the total number of trials is what was defined as the channel traffic. By the basic definition of probability, we thus have:

$$P_s = \frac{\text{channel throughput}}{\text{channel traffic}} = \frac{\lambda}{\lambda'}$$

Since, on the average, both measures of probability of success for the same channel must be equal, we can equate these two expressions, yielding:

$$\frac{\lambda}{\lambda'} = e^{-2\lambda'\tau}$$

or

$$\lambda = \lambda' e^{-2\lambda'\tau}$$

By multiplying both sides of the equation by τ, we change nothing but the notation, yielding

$$\lambda\tau = \lambda'\tau e^{-2\lambda'\tau}$$

or, since $\lambda\tau$ and $\lambda'\tau$ define s and g, respectively,

$$\boxed{s = ge^{-2g}}$$

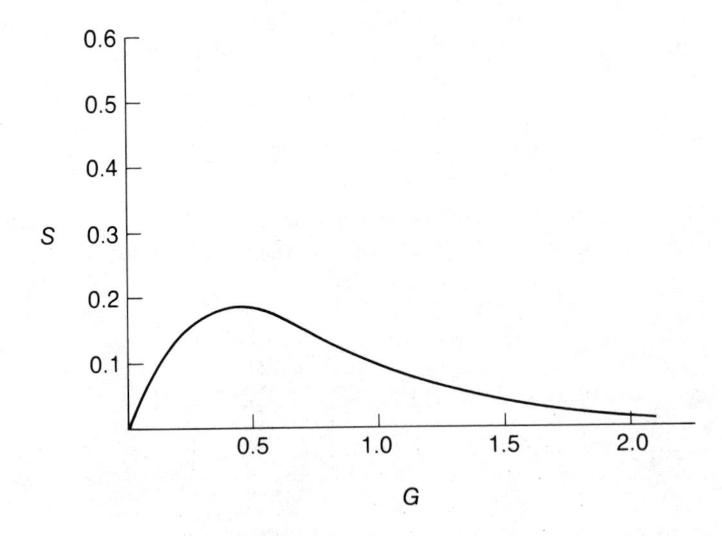

Figure 13-7. Plot of channel throughput versus channel traffic in an ALOHA channel.

This expression relates the useful, delivered throughput of the ALOHA channel, s, to the total traffic flowing on the channel. An approximate plot of this relationship is shown in Figure 13-7.

As the channel traffic begins to increase, the useful throughput begins increasing relatively quickly. However, the probability of collisions also increases, resulting in a lower probability of successful transmission. A point is finally reached—at a value of g equal to one-half—where any further increase in traffic creates collisions with such a high probability that the useful throughput is actually reduced. The point of maximum useful throughput, known as the ALOHA *channel capacity*, occurs at a value of channel traffic of $g = 0.5$. The useful channel throughput is $s = \frac{1}{2e} = 0.184$. Remember that these were relative or normalized measures. Therefore, the maximum useful throughput of the ALOHA channel is only 18.4% of the original channel basic bit rate and occurs when the channel is filled to 50% of its bit rate with transmitted traffic.

Implications of the Technique

Although the useful throughput is "only" 18.4% of the basic channel bit rate, that capacity is usable in a very flexible way. Since users have no multiplex equipment and no allocated share of the capacity, the ALOHA channel throughput is composed only of truly useful delivered information. In the case of the earlier example our 50,000-bits per second channel would have a useful throughput of about 9200 bits per second. However, if the true, *average* demand of the users was only a few bits per second, the channel could support possibly as many as 5000 users.

The capability is achieved because none of the channel capacity is used unless the user is currently active and ready to transmit. There is no control in the channel other than the repeat protocol associated with the collided packets. Users require no multiplex or timing equipment. The only basic network overhead is the destination address of each packet since no other network overhead, routing, or sequencing information is needed. The channel capacity is thus used very efficiently, as long as operation is maintained below the channel traffic level of $g = 0.5$.

However, if demand exceeds capacity such that g begins to exceed a value of 0.5, the additional demand creates more collisions, which in turn create more demand in the form of additional retransmissions, which creates more collisions, driving the useful throughput to zero (the value of s gets smaller as g increases beyond 0.5). As a result, control has to be imposed on the users' ability to transmit to insure that the operation of the channel always remains in the region below (or immediately around) a value of $g = 0.5$.

The easiest way to achieve this is for a control facility to monitor the channel performance and, as the load becomes heavy, command the user terminals to increase the time delay before retransmitting a packet that has been involved in a collision. This will reduce the apparent demand and keep the operation in the region of high channel throughput. Many other approaches are possible. The ultimate doomsday approach, for example, assumes that, as the channel performance deteriorates, many users will defer their use to a later time, thus reducing the load to a more acceptable region of operation.

ALOHA CHANNEL DELAY CHARACTERISTICS

The (human) user of the ALOHA channel sees the collision and retransmission process in terms of delay between the transmission of the packet and confirmation that it has been delivered to its destination. By using the satellite channel, the user has a minimum delay of 0.25 second to transmit the packet from origin to destination. If an immediate reply or confirmation is expected from the destination, the "round-trip" delay will be a minimum of 0.5 second, plus the processing time associated with the distant end. If both the user and the destination are using the same ALOHA channel, then the collision/retransmission process will affect both directions of communication.

The Delay Process

Figure 13-8 depicts the elements associated with the delay of a single packet message through the ALOHA channel. To make the discussion general, all times will be measured in terms of packet lengths, where τ denotes the length of the packet in seconds. The propagation time up to the satellite and back is given as $N\tau$. For example, since the satellite delay is about 250 ms for packet lengths of 20 ms, N would have a value of 12.5.

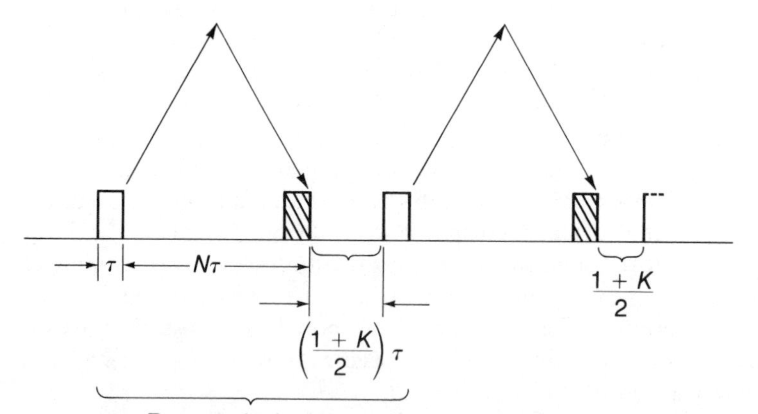

Figure 13-8. Factors contributing to packet delay in an ALOHA channel.

The process illustrated by Figure 13-8 is as follows. The packet is initially transmitted, which takes τ seconds. After $N\tau$ seconds of transmission delay the packet is received, possibly with interference. If the packet was received without interference the first time, the added delay of the channel is simply $N\tau$ seconds. If, however, a collision occurred, the protocol requires that the sender wait for a random period of time—between zero and K packet lengths (the length of time to transmit one full packet)—before retransmitting, in order to minimize the probability of a second collision. On the average, then, users will have to wait $\frac{(1 + K)\tau}{2}$ seconds before attempting retransmission. The retransmitted packet will then proceed over the channel, experiencing a delay of $N\tau$ seconds, and possibly repeating the retransmission process again.

The number of times this process has to repeat can be determined from the basic definitions of channel traffic (g) and channel throughput (s). The value of s is the amount of traffic that is ultimately delivered successfully, regardless of how many times it has to be repeated. The value of g is the total traffic in the channel—the successful throughput plus all the previously unsuccessful collisions. The ratio of g/s, then, represents the average number of times each packet has to be retransmitted before being successfully delivered.

For example, at maximum ALOHA channel capacity the channel of 50,000 bits per second of gross data rate would have a value of $g = 25{,}000$ bits per second and $s = 9200$ bits per second. The value of g/s would thus be $25{,}000/9200 = 2.72$. This means that each bit (and consequently each packet) has to be transmitted an average of 2.72 times. Clearly, some packets will make it on the first try, some on the second, some on the third, and so forth. But, on the average, it takes g/s transmissions for success. In Figure 13-7 the brackets indicate that the actions are repeated an average of g/s times. We can now write the following expression for the total average delay through the ALOHA channel, including the time to transmit the packet initially:

$$\text{total average delay} = \tau + \left(N\tau + \frac{(1 + K)\tau}{2}\right)\frac{g}{s} - \frac{(1 + K)\tau}{2}$$

The last term in this expression is needed to subtract the random waiting time following the successful retransmission which is embedded in the second term of this equation. Box 13-2 portrays the mathematical steps needed to put the expression into a somewhat more convenient form:

$$\text{total average delay} = \frac{\tau}{2}[1 + e^{2g}(1 + 2N) + K(e^{2g} - 1)]$$

where the first parenthetical term relates to the satellite delay, and the second parenthetical term relates to the retransmission protocol waiting delay.

These expressions perhaps will be easier to visualize by graphical presentation, as shown in Figure 13-9, and by some examples.

$$\text{TAD} = \text{total average delay} = \tau + \left\{N\tau + \frac{(1+K)\tau}{2}\right\}\frac{g}{s} - \left(\frac{1+K}{2}\right)\tau$$

from the ALOHA capacity equation:

$$s = ge^{-2g}$$

therefore

$$\frac{g}{s} = \frac{1}{e^{-2g}} = e^{2g}$$

$$\text{TAD} = \tau + \left\{N\tau + \frac{(1+K)\tau}{2}\right\}e^{2g} - \left(\frac{1+K}{2}\right)\tau$$

$$= \tau + N\tau e^{2g} + \frac{K\tau}{2}e^{2g} + \frac{\tau}{2}e^{2g} - \frac{K\tau}{2} - \frac{\tau}{2}$$

$$= \tau\left[\frac{1}{2} + \frac{e^{2g}}{2} + \frac{K}{2}(e^{2g} - 1) + Ne^{2g}\right]$$

$$\boxed{\text{total average delay} = \frac{\tau}{2}[1 + e^{2g}(1 + 2N) + K(e^{2g} - 1)]}$$

where

τ = packet length (in seconds)
g = channel traffic factor
N = satellite propagation delay (in packet lengths)
K = retransmission protocol delay (in packet lengths)
e = base of natural logarithms = 2.718

Box 13-2

Some Examples of Delay

The plot in Figure 13-9 is based on the parameters we have been using, that is, a basic channel rate of 50,000 bits per second and a packet length of 1000 bits. The value of τ is thus 20 ms, and the satellite round-trip delay is thus $N = 12.5$. The figure plots the user-observed average delay (in packet lengths) as a function of the channel throughput. The curves are plotted only up to the value of $s = 0.184$, which corresponds to the value of maximum channel throughput.

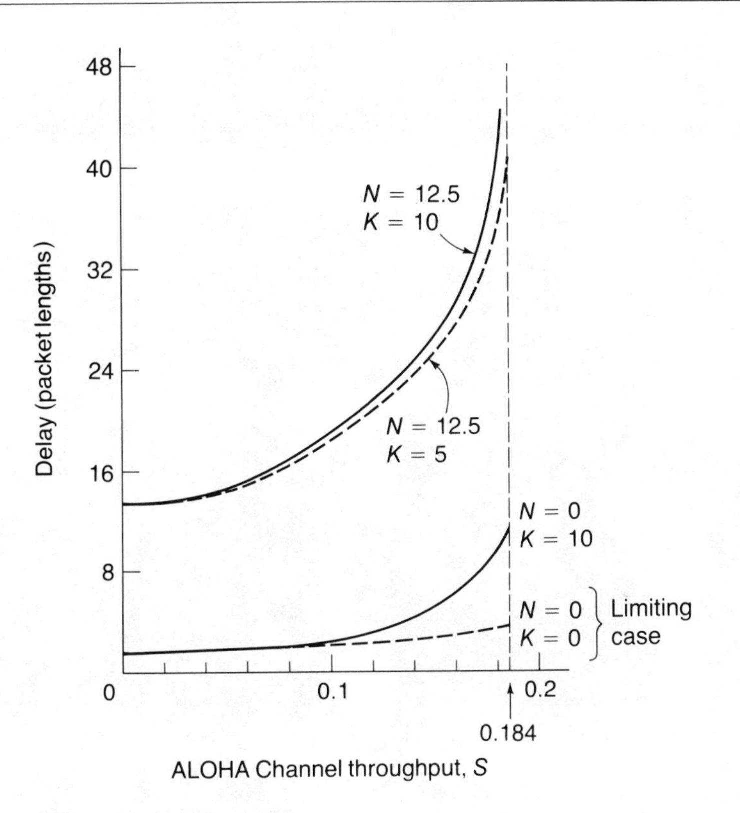

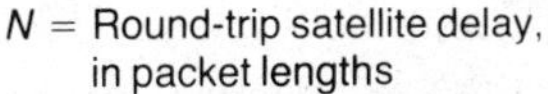

N = Round-trip satellite delay, in packet lengths

K = Protocol retransmission delay (maximum waiting time before retransmitting a collided packet), in packet lengths

Figure 13-9. Plot of delay versus channel throughput for the basic ALOHA channel.

When the throughput is low, the traffic is light in the channel and there are very few collisions, resulting in low values of delay. As traffic throughput increases, the channel is more heavily loaded, resulting in more collisions, more retransmissions, and more average delay. The upper curve is based on values of 12.5 packet lengths satellite propagation delay and 10 packet lengths maximum protocol retransmission waiting. The middle curve shortens the retransmission waiting time to a maximum of 5 packet lengths. The lower curve treats the theoretical case of $N = 0$, or zero propagation delay, which would be the case of a terrestrial repeater system (discussed in more detail in Chapter 15).

$$\text{TAD} = \text{total average delay} = \frac{\tau}{2}[1 + e^{2g}(1 + 2N) + K(e^{2g} - 1)]$$

where:

$$\tau = 20 \text{ ms} = \frac{1000 \text{ bits}}{50{,}000 \text{ b/s}} = \text{packet length}$$

$$g = 0.5 \text{ at maximum capacity}$$
$$N = 12.5 \text{ packet lengths}$$
$$K = 10 \text{ packet lengths}$$

$$\text{TAD} = \frac{20}{2}[1 + e^{2(0.5)}(1 + 2(12.5)) + 10(e^{2(0.5)} - 1)]$$
$$= 10[1 + 2.718(1 + 25) + 10(2.718 - 1)]$$
$$= 10[1 + 70.67 + 17.18] = 10[88.85]$$
$$= 888 \text{ ms} \approx 0.9 \text{ second}$$

Box 13-3

Box 13-3 shows the specific calculation for the average delay of a 50,000-bits per second ALOHA channel, operating at capacity ($s = 0.184$; $g = 0.5$), with 1000-bit packets ($\tau = 20$ ms). We can see that the delay of the ALOHA channel operating at maximum capacity is less than 0.9 second for a 1000-bit packet in the 50,000-bits per second channel. We can also recall that, at this level of performance, the actual throughput is about 9200 bits per second. If, for example, the average user demand, measured over time, is a full packet every 2 minutes, the channel could support more than 1100 users, with each user seeing an average delay of about 0.9 second for every transmission. If the demand, or average usage, were smaller, then the delay would decrease, and the number of users who could be accommodated would increase. At an overall average demand of, say, 2 bits per second per user, the channel could accommodate nearly 5000 users.

SUMMARY

This chapter has introduced the operation of a random-access, demand-assigned channel, based on the notion that the users are permitted to transmit whenever they care to. Conflicts of demand for the common capacity are resolved by each user repeating his transmission until it is successful. Because the process uses a satellite channel in broadcast operation, each user can hear the transmissions of every other user and can thus determine if there is any interference with his own transmission. However, since there is a quarter of a second delay inherent in the satellite channel, the fact that the channel is quiet now does not necessarily mean that it is really free of other traffic.

The nature of the satellite broadcast channel, with universal access from all user locations, allows most of the features and services of a packet switched network to be attained without packet switches. Destination and source address information incorporated in each packet permit user data to be delivered to the destination and an acknowledgement returned to the source.

The overall capacity of the ALOHA channel was found to be only about 18% of the basic channel rate. However, because of the simple allocation protocol and the fact that only *active* users consumed any of the channel capacity, the channel was able to support a much larger number of users than a more structured allocation of the same basic channel rate. In addition, the overall average delay through the channel was quite acceptable (less than 1 second for typical cases) as long as channel operation was maintained below full capacity. As we shall see in the next chapter, the channel capacity can be substantially increased at the expense of some additional complexity in user equipment and operational protocol.

SUGGESTED READINGS

ABRAMSON, NORMAN. "The Throughput of Packet Broadcasting Channels." *IEEE Transactions on Communications*, vol. COM-25, no. 1 (January 1977), pp. 117–128.

This paper brings together the results of a number of researchers by presenting a unified development of the various aspects of packet broadcasting. The unified presentation treats not only the basic ALOHA channel but several variations in the ALOHA protocol and the application of packet broadcasting to terrestrial networks as well. The paper concludes with a summary of a number of practical applications of packet broadcasting.

ABRAMSON, NORMAN, and CACCIAMANI, EUGENE R., JR. "Satellites: Not Just a Big Cable in the Sky." *IEEE Spectrum*, vol. 12, no. 9 (September 1975), pp. 36–40.

This paper traces the evolution of satellite capabilities, but emphasizes services and techniques that are more comprehensive than the traditional use of satellites as a high-capacity point-to-point medium. Applications based on broadcast characteristics are described from a functional (application) viewpoint rather than a technical viewpoint. Future applications, many of which have actually occurred since the article was published, are introduced and described.

BARGELLINI, PIER L. "Commercial U.S. Satellites." *IEEE Spectrum*, vol. 16, no. 10 (October 1979), pp. 30–37.

This article provides a survey of communications satellites, their effect on domestic networks, and a look into the near and long-term future of nationwide satellite communications. Bargellini emphasizes the availability of relatively inexpensive, user-premises earth stations, as well as the ability of satellites to serve mobile users.

KLEINROCK, LEONARD. *Queueing Systems Volume II: Computer Applications.* New York: John Wiley, 1976, pp. 360–407.

This part of Kleinrock's book presents a different approach to the analysis of packet broadcasting channels, placing particular emphasis on some of the techniques that can increase the apparent channel capacity. The discussion of ground radio packet switching techniques provides an excellent overview of the Carrier Sense Multiple Access techniques, which achieve high channel capacity and utilization.

14

Improvements on the Basic ALOHA Channel—Slots and Reservations

THIS CHAPTER:

will investigate some methods for increasing the basic capacity of the ALOHA channel by applying more discipline to the users' operation.

will look at a technique known as *slotted ALOHA*, which allows users to transmit only at discrete time intervals.

will find that channel capacity improves if one of two users who collide on the channel actually "captures" the channel and gets through successfully.

will discuss a number of possible reservation techniques that can achieve very high channel capacities and throughputs.

The operation of a satellite packet broadcast channel using the basic ALOHA protocol permits users to transmit packets whenever they care to, in a purely random fashion. As a result, there is a relatively high probability that at least some portion of any given packet will be interfered with by some other user's packet. In analyzing the channel operation, we found that the frequency of packet collisions results in an overall channel capacity of about 18% of the basic channel rate. If users attempt to use the channel at a rate above this capacity, the probability of collision becomes so great that the throughput on the channel quickly tends toward zero.

INCREASING THE CAPACITY OF THE PACKET BROADCAST CHANNEL

The apparent lack of efficiency of the basic ALOHA channel has stimulated a great deal of research into methods for achieving channel operation with greater capacity while retaining much of the inherent simplicity of the random broadcast protocol. Many possible techniques have been suggested, but each involves increased complexity of both the access protocol and the user terminal device. Nevertheless, the very low incremental cost of adding additional intelligence to the user devices by application of small processors within the terminals makes many of these approaches very attractive.

The first technique we will discuss is known as the slotted ALOHA channel. This technique decreases the probability of interference between packets by requiring that users transmit only at the beginning of discrete time intervals. A user's packet is "guaranteed" no interference since no other user is permitted to transmit until the beginning of the next packet interval. As we shall see in the next section, this technique can effectively double the channel capacity at only a small increase in average delay.

Another apparent improvement in channel capacity occurs if each user transmits at a slightly different power level. If two packets collide at the satellite, but have a different signal level, there is a good chance that the stronger of the two signals will "capture" the receiver and be transmitted by the satellite without error. Thus, when a collision occurs, only one of the two colliding packets will need to be retransmitted. Such a protocol could be implemented randomly, with power levels being adjusted upward or downward on a random basis. Or the power levels could be determined on a priority basis, with more important or higher priority users being given higher powered terminals, thus establishing a multiclass priority packet broadcast system. In any case, this technique will result in a channel capacity that is more than three times as large as the basic ALOHA technique.

Finally, much of the randomness of the packet broadcast channel can be eliminated by requiring users to make advance reservations for the capacity they intend to use. So-called reservation techniques can greatly increase the capacity of the packet broadcast channel when some users consistently have more information to transmit than will fit into a single packet. Reservation techniques are thus very useful for situations where many low-capacity or low-utilization terminals are trying to transmit to a central point, while the central processor or computer is sending a relatively large amount of data back to the individual users. Use of slot reservations in the broadcast channel permit these two kinds of traffic to be mixed together, without mutual interference, in such a way as to achieve very high channel capacity. A number of different reservation techniques will be discussed later in this chapter.

THE SLOTTED ALOHA CHANNEL

The basic, or unslotted, ALOHA channel is a marvel of simplicity (neglecting, of course, the technological achievement of placing the satellite in orbit to begin with). Users transmit whenever they desire, without regard for the state of the channel. Although the maximum channel throughput can reach only about 18% of the basic channel rate, that capacity is able to serve a large number of users since none of it is assigned to inactive users. However, the basic ALOHA tech-

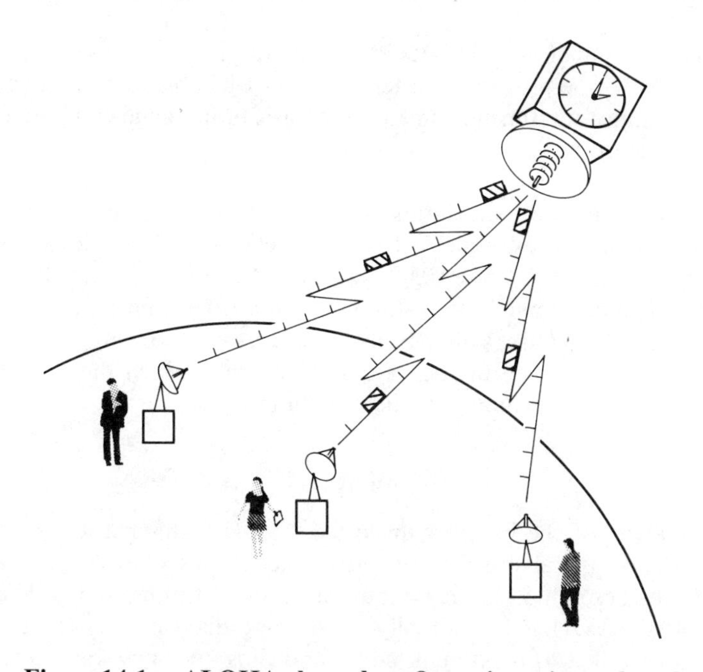

Figure 14-1. ALOHA channel configuration using a slotted (clocked) channel.

nique entails a period of time over which each packet is vulnerable to interference. By establishing a **slotted channel**—that is, a channel with discrete time slots in which users may transmit their packets—we can considerably reduce this vulnerable time.

Figure 14-1 is a pictorial representation of the slotted ALOHA system. Imagine that a clock on the satellite ticks at intervals exactly corresponding to the length of a packet. Each user terminal can hear the clock ticking by listening to the channel and synchronizes his transmitter to the clock. The transmission protocol requires that the user must wait until the beginning of a time interval—that is, until the next clock tick—before he can transmit a packet. By following this discipline, packets overlap completely or not at all. Packets are not destroyed or damaged if just a few bits at the beginning or end overlap another user's earlier or later packet, as can happen in the unslotted ALOHA case. In Figure 14-1 we see the two packets closest to the satellite about to collide, but the remaining packets in the system are safe from collision.

Disadvantages

The slotted ALOHA channel protocol has two disadvantages. One is the potential complexity of the terminals, not only in order to synchronize the time reference for all users, but to allow for the slight variation in actual distance between each terminal and the satellite. Since the time reference (exact time of the "ticks") has to be precise, each terminal must be able to determine range or distance and adjust its transmit time on the basis of its range compared to a networkwide standard.

Another disadvantage is that the packet length, and the resultant time between clock ticks, represents the maximum amount of data a single user can transmit at any given time. For example, a 1000-bit packet, which would correspond to 20-millisecond time slots on a 50,000-bit/second channel, represents about 125 characters (or approximately two standard typewritten lines). However, typical user data terminal operations often transmit only a small percentage of this amount of data at any given time. In such a case the time between the end of a user's transmission and the beginning of the next time slot is wasted.

Analysis

Our analysis of the basic or unslotted ALOHA channel was a worst case analysis, assuming that every transmitted packet was of maximum allowable length and thus presented the maximum likelihood of interference. The analysis for the slotted ALOHA case is really a best case analysis. We will assume that every packet is filled to maximum length and thus does not waste any capacity between the end of one packet and the beginning of the next allowable time slot.

Box 14-1 shows the analytical situation for the slotted ALOHA channel. We begin with the definitions:

s_i = the probability that user i successfully transmits a packet

g_i = the probability that user i transmits any packet (that is, a successful or unsuccessful packet)

For user i to transmit a successful packet, since this is a slotted case, he must transmit his packet while no other user transmits one. Mathematically, this can be stated:

$$s_i = g_i(1 - g_1)(1 - g_2)(1 - g_3) \cdots (1 - g_n)/(1 - g_i)$$

where all terms are included for all n possible users of the channel. The expression is divided by $(1 - g_i)$ because we want to exclude the term involving user i, since it is this user's probability of success that we are attempting to compute.

Using the symbol $\prod$ to indicate the product of terms, this expression can be written:

$$s_i = g_i \prod_{\substack{j=1 \\ j \neq i}}^{j=n} (1 - g_j)$$

It is difficult to proceed beyond this point without making an assumption that was inherent in the analysis of the unslotted ALOHA channel. If we assume that all of the users of the channel are statistically equal—that is, they share the

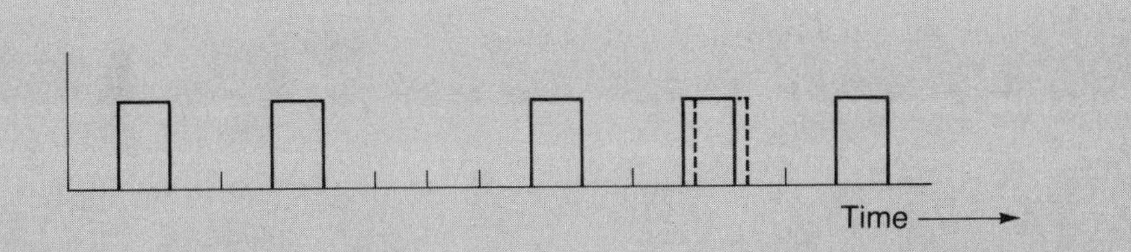

n = total number of users in the system

s_i = probability of a successful packet by the ith user

g_i = probability that the ith user transmits a packet

thus

$(1 - g_j)$ = probability that the jth user does not transmit a packet

$$s_i = g_i \prod_{\substack{j=1 \\ j \neq i}}^{j=n} (1 - g_j)$$

for equal users

$$s_i = \frac{S}{n} \quad \text{and} \quad g_i = \frac{G}{n}$$

$$\frac{S}{n} = \frac{G}{n} \prod_{\substack{j=1 \\ j \neq i}}^{j=n} \left(1 - \frac{G}{n}\right) = \frac{G}{n}\left(1 - \frac{G}{n}\right)^{n-1}$$

$$S = G\left(1 - \frac{G}{n}\right)^{n-1}$$

if n is large,

$$\left(1 - \frac{G}{n}\right)^{n-1} \approx e^{-G}$$

thus

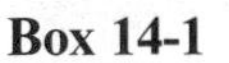

$$S = Ge^{-G}$$

Box 14-1

capacity equally and have an equally likely probability of transmitting at any given instant of time—we can write:

$$s_i = S/n$$

and

$$g_i = G/n$$

where S and G are the channel throughput and channel traffic, respectively, with the same meaning they had in the unslotted analysis.

These expressions say simply that a given user's probability of a successful packet, s_i, is one-n^{th} of the total normalized throughput of the channel. Similarly, his probability of making a transmission is just one-n^{th} of total traffic transmitted on the channel.

By substituting these expressions above, we find:

$$S/n = G/n \prod_{\substack{j=1 \\ j \neq i}}^{j=n} (1 - G/n) = G/n(1 - G/n)^{n-1}$$

or

$$S = G(1 - G/n)^{n-1}$$

This last expression results because we have the product of $n - 1$ identical terms, each of the form $(1 - G/n)$.

If the number of users is large, we can employ a limit property of the expression $(1 - x/n)^{n-1}$, which is approximately equal to e^{-x} as n tends toward infinity. As a result, we then have:

$$S = Ge^{-G}$$

Recall that the result for the unslotted case is:

$$S = Ge^{-2G}$$

Capacity and Delay Results

The results of both of these equations is shown in Figure 14-2, where the relationship of channel throughput to channel traffic is plotted for both the slotted and the unslotted ALOHA case. Notice that, for any given channel traffic, the slotted case is better than the unslotted case by a small amount, up to the point where the unslotted channel reaches maximum capacity ($G = \frac{1}{2}$). At this point the unslotted throughput begins to decline, while the throughput of the slotted channel continues to increase. The curve for the slotted channel ultimately reaches a value of $1/e$, or about 0.37, representing a normalized throughput of about 37% of the basic channel bit rate. This throughput is twice the achievable throughput of the unslotted ALOHA channel.

We must recall, at this point, that the impression that the channel is achieving twice the throughput is based on each packet's being full. If, on the average, the

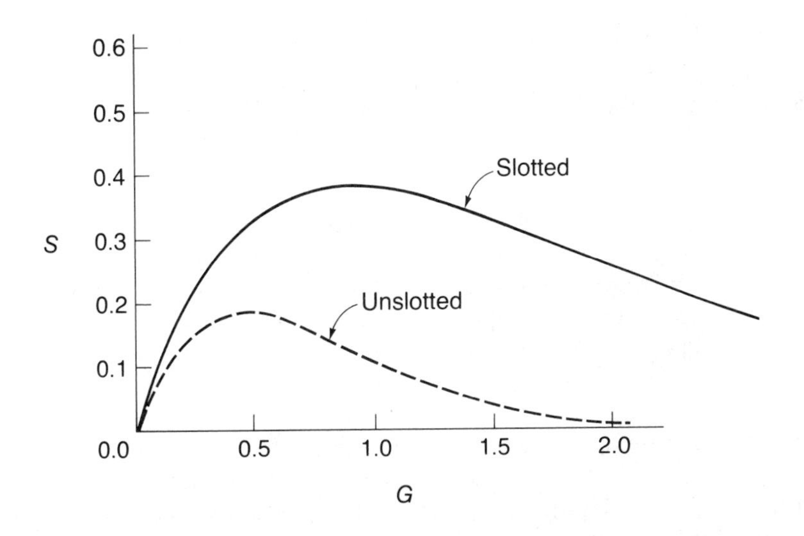

Figure 14-2. Plot of channel throughput versus channel traffic for slotted and unslotted ALOHA channels.

packets are only half full, the actual throughput is back to the same value as that found on the unslotted channel since no other user can transmit during the empty interval.

The delay associated with the slotted ALOHA channel can be derived in a fashion analogous to the unslotted delay. The only difference is that, on the average, each time a user is ready to make a transmission he has to wait one-half of a packet time until the beginning of the next slot interval before he can actually transmit. At peak capacity $G = 1$ and $S = 0.37$, which means that each successful packet is transmitted an average of 2.7 times, and each retransmission results in an average of one-half packet time delay. Thus the maximum excess delay for the slotted channel is about 1.4 packet times compared to the unslotted channel, operating at capacity. For the 20-millisecond packet time this excess delay—less than 30 milliseconds—is negligible compared to the approximately 900-millisecond delay encountered by the average packet when the channel is operating at maximum capacity.

In summary, the slotted ALOHA packet broadcasting channel can achieve a twofold increase in channel throughput capacity at a slight increase in average delay, assuming that all packets are filled to their maximum allowable length. The slotted technique requires some additional complexity in user equipment, but, with some additional processing within the user terminals, the incremental cost should be low. The slotted ALOHA approach would be ideally suited for a highly homogeneous user community, so that all transmissions of all users were identical in length even though they occurred at random times. The packet length could then be set to the exact length of each user's transmission, and no capacity would be wasted with partially filled packets. For example, a network of credit validation terminals or electronic funds transfer terminals, where every transmission represents the same amount of data, would meet this description.

SLOTTED ALOHA CHANNEL WITH CAPTURE

The analysis of the ALOHA packet broadcast channel made the assumption that, when any part of two or more packets overlap, all packets involved in the collision must be retransmitted. In reality, it is possible that one of the packets will be sufficiently strong to **capture** the receiver and be received accurately. If this were the case, not every packet involved in a collision would have to be retransmitted. As a result the apparent interference would be reduced and the channel throughput at any level of traffic increased. It would be possible to create a priority based system, where users with higher need to communicate (that is, higher priority in the system) could be alloted greater transmitter power, giving them a substantially higher probability of being received correctly even in the presence of interfering packets.

We can analyze the situation initially by simply assuming sufficient random fluctuations in received signal levels that, for any pair of users, each has a one-half probability of capturing the channel and being received correctly. We will also assume that, if three users collide, none of the three has sufficient strength to dominate the other two, and thus all three (or more) packets are lost.

Accounting for channel capture requires only a slight modification of the analysis for the slotted ALOHA case. If we have a slotted channel, and one user captures the channel if exactly two users collide, the success probability for user i can be expressed as:

$$s_i = g_i \prod_{\substack{j=1 \\ j \neq i}}^{j=n} (1 - g_j) + \frac{1}{2} \sum_{\substack{m=1 \\ m \neq i}}^{m=n} g_i g_j \prod_{\substack{k=1 \\ k \neq i \\ k \neq m}}^{k=n} (1 - g_k)$$

The first part of this expression is exactly the same as for the slotted ALOHA case. The second part of the expression, which sums up the results of many products, takes the probability that user i will transmit at the same time as one, and only one, other user, and considers each other user in turn. By multiplying this probability by $\frac{1}{2}$ allows for the fact that user i will be successful in the collision one-half of the time, and that half the time the other user will be successful.

As in the previous section, for n identical users, we can substitute:

$$s_i = S/n$$

and

$$g_i = G/n,$$

resulting in:

$$S/n = G/n(1 - G/n)^{n-1} + \tfrac{1}{2}(n - 1)G^2/n^2(1 - G/n)^{n-2}$$

or

$$S = G(1 - G/n)^{n-1} + \tfrac{1}{2}(n - 1)G^2/n(1 - G/n)^{n-2}$$

$$s_i = g_i \prod_{\substack{j=1\\ j\neq i}}^{j=n} (1 - g_j) + \frac{1}{2} \sum_{\substack{m=1\\ m\neq i}}^{m=n} g_i g_m \prod_{\substack{k\neq i\\ k\neq m}}^{k=n} (1 - g_k)$$

$$s_i = \frac{S}{n} \qquad g_i = \frac{G}{n}$$

$$\frac{S}{n} = \frac{G}{n}\left(1 - \frac{G}{n}\right)^{n-1} + \left(\frac{n-1}{2}\right)\frac{G^2}{n^2}\left(1 - \frac{G}{n}\right)^{n-2}$$

$$S = G\left(1 - \frac{G}{n}\right)^{n-1} + \frac{n-1}{2}\frac{G^2}{n}\left(1 - \frac{G}{n}\right)^{n-2}$$

$$S = G\left(1 - \frac{G}{n}\right)^{n-1}\left[1 + \frac{n-1}{2}\frac{G}{n}\frac{1}{\left(1 - \frac{G}{n}\right)}\right]$$

$$S = G\left(1 - \frac{G}{n}\right)^{n-1}\left[1 + \frac{\left(1 - \frac{1}{n}\right)G}{2\left(1 - \frac{G}{n}\right)}\right]$$

As n gets very large, $n \to \infty$

$$S = Ge^{-G}\left[1 + \frac{G}{2}\right]$$

Box 14-2

By grouping and rearranging the terms in this expression, as shown in Box 14-2, we can arrive at an expression that can be approximated, for a relatively large value of n (the number of users), by:

$$S = Ge^{-G}(1 + G/2)$$

This expression, together with the capacity curves for the unslotted and slotted ALOHA cases, is plotted in Figure 14-3. Not unexpectedly, the capture effect has permitted a roughly 50% increase in channel capacity over the slotted ALOHA without capture, to an overall capacity of about 57% of the gross channel bit rate. Furthermore, the value of capture probability can be easily changed by simply changing the denominator in the last term of the capacity expression. For instance, if in actual operation we find that, when two packets collide, one-third

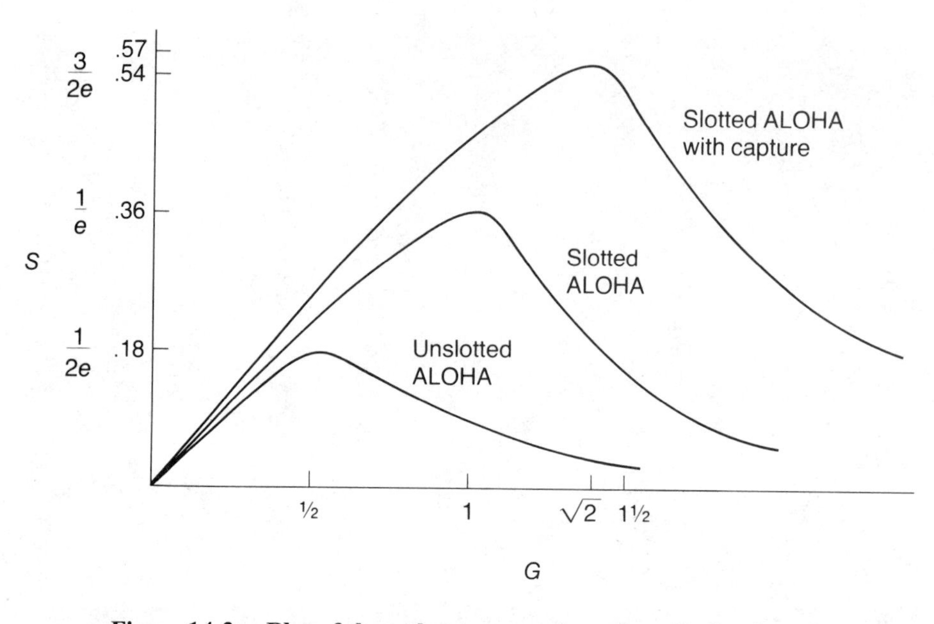

Figure 14-3. Plot of throughput versus channel traffic for slotted, unslotted, and capture ALOHA channels.

of the time user i gets through, one-third of the time user j gets through, and one-third of the time neither of the two users gets through, we can adjust the collision success probability to 1/3. The capacity expression thus becomes:

$$s = Ge^{-G}(1 + G/3)$$

which naturally results in a throughput reduction compared to the case where each of the users gets through with probability of $\frac{1}{2}$.

PACKET BROADCAST CHANNELS WITH CAPACITY RESERVATION

In analyzing the satellite broadcast ALOHA channel, in both the unslotted and the slotted cases, we made many idealizing assumptions in order to get results that could be readily expressed mathematically or graphically. We had to assume that the user community was very large (n approaching infinity) and that the packet originations obeyed the Poisson statistics. Furthermore, we had to assume that the users were functionally identical, behaved in the same statistical fashion, and shared the useful channel throughput equally. Rather than attempting to resolve the mathematics of these assumptions, we will briefly discuss some of the real-world situations that would affect the operation of an actual system. One response to real-world conditions is the use of capacity reservation techniques.

Unbalanced Users

An unbalanced user is one with much more traffic than the average user. Let us imagine a slotted ALOHA channel with no traffic in it and only one user in the system who is currently active. This one user has a great deal of traffic and begins to transmit in every time slot of the channel. Since he is the only active user, he never interferes with his own traffic and therefore achieves a throughput of 100% of the basic channel rate, even though the theory for the slotted ALOHA channel predicts a maximum throughput of 37%. Clearly, the assumption for many small users is not correct for the case where only one large user is operating in the channel.

Now let's assume there is one large user in a slotted ALOHA system who has an average demand of 80% of the basic channel rate. Even if this user transmits so often as to consume 80% of the channel capacity, in theory at least part of the remaining 20% is available to other users. We will assume that there are also a large number of small users, whose aggregate demand is 5% of the basic channel rate. The group of small users operate in the channel as if it were the same slotted ALOHA channel that we described earlier in the chapter.

The channel now sees a mixture of traffic demands, totalling 85% of the basic channel rate. Since 80% of that demand comes from a single user, who never interferes with himself, there is a good chance that the channel will be capable of satisfying the total demand. Again, this far exceeds the basic theoretical 37% slotted ALOHA capacity for small equal users. In fact, we can view this case almost as if we had two parallel channels. One channel is a dedicated channel of 80% of the basic channel rate, operating at nearly 100% capacity. The other channel is the remaining 20% of the basic capacity, carrying the remaining 5% of the total demand. The efficiency on the small channel is only 25% (5%/20%), well within the slotted ALOHA prediction. This is only a rough approximation of the view of the mixture of large and small demand users in a single slotted ALOHA channel, but it provides a useful perspective.

Without trying to compute exact boundaries, we can approximate a channel performance curve for a slotted ALOHA channel with mixed large and small users. Figure 14-4 shows the total channel throughput, S, as the mixture of throughput from a single large user, S_{large}, and many small users, S_{small}. When all of the traffic is from a single large user, a throughput of 1.0, or 100% of the basic channel rate, can be achieved. When the throughput is entirely due to small users, a throughput of 0.37, or 37% of the basic channel rate, is achieved.

But what about the general situation? What if the channel is used by many users, some of whom are large, some of whom are small, and some of whom fall in between. Worse yet, suppose that the users of the channel normally have very

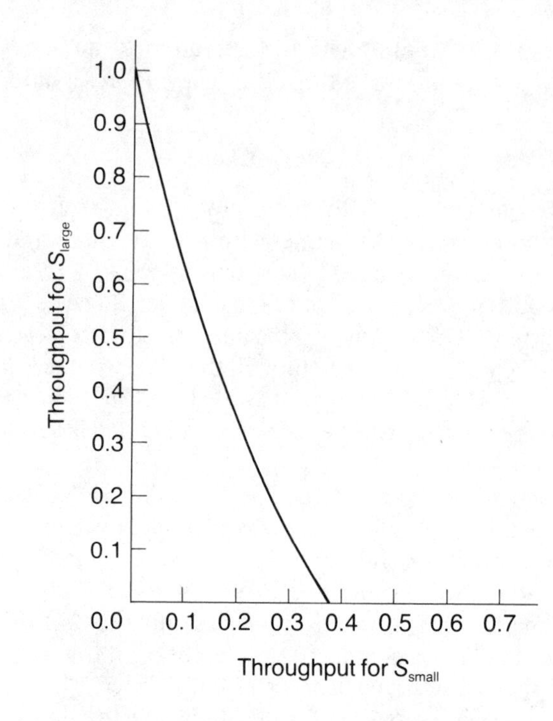

Figure 14-4. Throughput for a few large users versus many small users of a slotted ALOHA channel.

little to transmit, so their traffic generally fits into a single packet. But every so often they have much more information, and they would like to transmit five consecutive packets. With slight increases in the complexity of the channel protocols these cases can be accommodated very efficiently by use of reservation techniques.

Reservation Techniques

The first **reservation technique** we can look at is based on the assumption that, once a user starts to transmit, there is a good chance that he will have more than one packet's worth of data. To satisfy this assumption, a master frame structure is superimposed on the slotted ALOHA channel (Figure 14-5). The length of the master frame is R time slots (packet times), such that the total length of the frame exceeds the round-trip delay to the satellite. The beginning of each frame is indicated by a synchronizing preamble transmitted by the satellite or a master earth station. Each packet slot in the frame is numbered, 1 through R, and each terminal is required to keep track of which packet slots were used during the last frame. If a user desires to start transmitting, he must select a packet slot that was unused in the last frame.

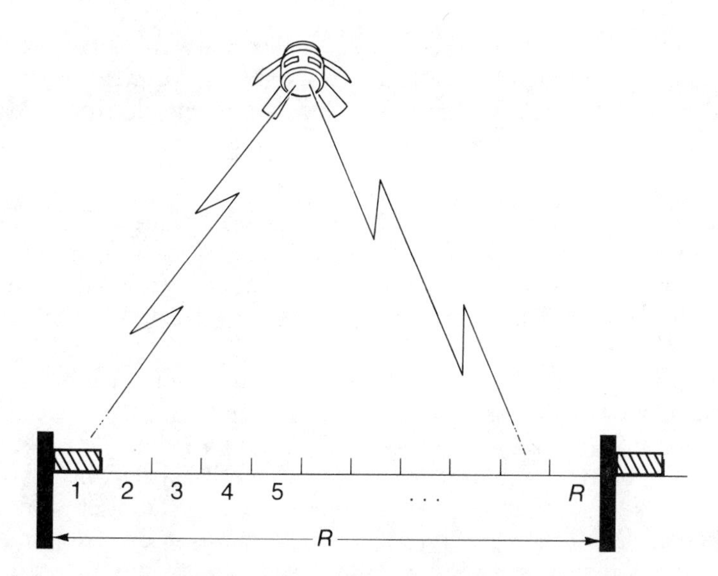

Figure 14-5. Frame structure superimposed on a slotted ALOHA channel: Total frame length, *R*, exceeds satellite round-trip delay.

The only way a collision can occur is if two users pick exactly the same vacant slot during the same quarter-second frame period and begin transmitting at the same time. Once a user successfully transmits a packet in a vacant time slot, he may continue to use that time slot in every successive frame, without further interference. When he has completed his transmission, he leaves the slot empty in the following frame. Hearing the slot empty, the remaining users can now select that frame to fill their subsequent demands.

As a specific example, let us return to our familiar 50,000-bits/second channel with 1000-bit packets, each lasting 20 milliseconds. If we let R be 12, in each frame there are 12 packet intervals, times 20 milliseconds each, totaling 240 milliseconds, which is essentially the round-trip delay to the satellite. If the channel is lightly loaded, most of the slots would be empty and the channel would be operating like the slotted ALOHA channel we studied earlier—that is, with a maximum capacity of 37% of 50,000 bits/second, or about 18,500 bits/second. But if one user becomes active, with a substantial amount of traffic, he will transmit full packets every frame. Since there are about four frames per second, he will be achieving a throughput of 4000 bits per second, without any collisions, while the remaining 11 packet slots are being used in the slotted ALOHA fashion.

As the channel becomes busier with high-volume users, we will eventually reach the condition where there are 12 high-volume users, each in one of the time slots and each achieving an average throughput of about 4000 bits per second. The remaining users would be forced to wait until one of the active users stopped transmitting. The channel in this condition has become a time-division, multiple-access channel, operating at essentially 100% capacity, but now serving only 12 users rather than the larger community that was intended. Hopefully, such a condition would not happen too often.

Borrowing Slots. A variation of the reservation technique avoids having the channel potentially locked up with large users, out of reach of the small users. Instead of R packet slots in the frame, we have N slots, where N is the total number of users in the network. This, of course, makes the overall length of the frame much longer.

Each user is assigned his own slot in the frame, and when he initiates his transmissions, he must transmit in his own time slot. However, if a user hears that some other user's slot was empty on the previous frame, he is permitted to assume control of the slot in order to increase the total capacity he has available. Once a given user assumes the use of another's slot, no other user is permitted to try to use that slot. If the rightful owner of the slot has information to transmit, however, he can regain control of his own slot by simply initiating transmission in it. This causes a collision, which is heard one round trip later by the borrower of the slots. The borrower is then required to cease transmission and return the slot to its rightful owner. The only capacity lost to collisions using this protocol are the intentional collisions used to signal the return of a packet slot to its rightful owner.

Subdividing the Channel. Probably the most efficient approach to dynamic packet reservation is to divide the slotted channel into a master frame consisting of two subframes. At the beginning of each frame is a short reservation subframe, during which the users transmit reservation requests on the numbered packet slots that follow. The reservations are heard by all other users, who must keep a record of all outstanding reservations. A reservation, which is a very short, standard-format message, consists of a slot number and the number of succeeding frames in which that slot is to be reserved. For example, user K has 6000 bits to transmit, and, according to his tables, slot number 7 has no reservations on it. So during the next reservation subframe user K would transmit the reservation message "K,7,6," meaning that user K is reserving slot 7 for the next six frames. Each other user hearing the reservation would log the reservation in his local table, and, presumably, would not try to use packet slot 7 anytime during the next six frame periods.

Collisions might occur in this form of packet broadcast channel only if two users attempt to transmit their very short reservation messages at exactly the same time during the reservation subframe. This reservation technique is extremely efficient for a community of users whose messages are consistently longer than a single packet. Messages that are only a single packet (or less) in length are forced to go through the time delay of making a reservation for just a single slot in a single frame. On the other hand, successful transmission of that packet on the first attempt, without any likelihood of collision, is essentially assured.

All of the reservation or capacity allocation techniques achieve higher efficiency than is possible under random operation. Naturally, they also impose some additional overhead, processing, and complexity on the operation of the channel, the terminal equipment, and the users. Under ideal conditions the reservation techniques reach the level of performance achievable with static time-division multiplexing, but they do it so that the resources are adaptable to changing demands of the user community.

HIGHLIGHT COMPARISON OF MULTIPLE ACCESS TECHNIQUES

Table 14-1 provides a generalized comparison of all multiple access techniques, including those we covered in Chapter 13. In an ideal case we could have random arrivals, of deterministic (constant) length, in a single-server system (represented by the queueing theory notation **M/D/1**) that has sufficient buffering to hold temporary overloads. Such an ideal case can consistently achieve nearly 100% occupancy of the channel, at the expense of sometimes considerable delay. Preassigned allocation techniques, such as time-division (**TDMA**) or frequency-division multiple access (**FDMA**), provide a static allocation of capacity, which can grossly waste capacity if some of the assigned users have little or nothing to transmit over a period of time.

Demand-access techniques, such as polling and the reservation methods in the slotted broadcast channel, are highly dynamic and serve a mix of high- and low-capacity users well. However, if we have a community of truly homogeneous, low-demand users, the demand access techniques employ too much complexity for the benefits gained. Finally, the collision-based techniques—ALOHA and slotted ALOHA—provide a relatively uncontrolled, technically simple mode of access to the broadcast channel.

Table 14-1. Multiple Access Technique Summary

Technique	*Examples*	*Characteristic*
Ideal	M/D/1 queue	Statistical multiplexing
Preassigned	TDMA FDMA	Static
Demand access	Polling Reservation	Dynamic
Collision	Pure ALOHA Slotted ALOHA	Uncontrolled

Table 14-2. Characteristics of Resource-Sharing Techniques

	Overhead	*Empty Slots*	*Collisions*
Static assignment	No	Yes	No
Dynamic reservation	Yes	No	No
Uncontrolled	No	No	Yes

The characteristics of these access techniques with regard to overhead, empty slots, and collisions are summarized in Table 14-2. The static assignment methods have no operational overhead and no collisions, but when users have nothing to transmit, the capacity is wasted as empty time or frequency slots. Dynamic reservation techniques have some capacity allocated to system overhead, but they have no capacity wasted because of either empty slots or collisions. The uncontrolled, or random, systems have neither overhead nor empty slots when the system demand is great, but they lose capacity to collisions.

SUMMARY

In this chapter we found that the theoretical capacity of the purely random ALOHA channel can be remarkably improved by additional discipline and control over the way users access the channel. Compared to the pure ALOHA case of about 18% of the basic channel bit rate, slotting the channel could reach 37% of the basic rate, and permitting one of the users to capture the receiver might achieve as much as 57% capacity. By adding additional intelligence and mixing users with different average utilization demands, we can impose dynamic reservation techniques that can get nearly 100% utilization of the channel under many conditions.

SUGGESTED READINGS

JACOBS, IRWIN MARK, BINDER, RICHARD, and HOVERSTEN, ESTIL V. "General Purpose Packet Satellite Networks." *Proceedings of the IEEE*, vol. 66, no. 11 (November 1978), pp. 1448–1467.

This paper surveys the range of techniques applicable to satellite packet broadcasting for use in a general purpose environment. It introduces and discusses a technique, called priority-oriented demand assignment (PODA), which schedules the use of satellite channel capacity according to the highly variable needs of the different user classes. The paper concludes with the results of experiments using these techniques in actual operation on a trans-Atlantic link as part of the ARPANET.

KLEINROCK, LEONARD, and LAM, SIMON S. "Packet Switching in a Slotted Satellite Channel." National Computer Conference, New York, June 1973. *AFIPS Conference Proceedings*, vol. 42. Montvale, N.J.: AFIPS Press, 1973, pp. 703–710.

This is one of the fundamental papers exploring the slotted channel both in a random mode of access and with various mixes of large- and small-demand users. The paper presents a comprehensive treatment of the various performance trade-offs, delay/capacity curves, and user loading curves.

ROBERTS, LAWRENCE G. "Dynamic Allocation of Satellite Capacity Through Packet Reservation." National Computer Conference, New York, June 1973. *AFIPS Conference Proceedings*, vol. 42. Montvale, N.J.: AFIPS Press, 1973, pp. 711–716.

This paper presents a complete treatment of the most flexible of the reservation techniques we discussed in this chapter, one using a reservation subframe at the beginning of each master frame. Quantitative and graphical results are derived and presented for a wide range of operating conditions. The paper includes numerous comparisons with the operation of the random channels.

15

Terrestrial Packet Radio Systems

THIS CHAPTER:

will look at the ways a packet broadcast system could operate in a limited geographic area, using terrestrial repeaters rather than satellites.

will show that, in a terrestrial system, users can improve throughput by listening to the channel before transmitting.

will look in some detail at the carrier sense multiple access system.

In the last few chapters we learned about random access packet communications techniques, with emphasis on satellite-based systems covering large geographic areas. There are, we discovered, many ways that such a distributed network of users could be interfaced to the system, such that a very large number of users could be simultaneously served by the limited capacity of the satellite channel. In this chapter we will bring our thinking "back to earth" and look at the concepts of packet broadcasting applied to an entirely terrestrial network.

TERRESTRIAL REPEATERS AND LOCAL NETWORKS

The General Situation

The general situation we will be looking at is shown in Figure 15-1: a large number of users, each with relatively low total transmission demand but a potential need to transmit at high data rates, within a circumscribed area. We will initially assume that the users are within line-of-sight distance from the central location. The central location may be a single computer all of the users are trying to interact with, or it may be a switch that acts as an entry point into the larger network covering a broader geographic region. The central station may even be a satellite earth station shared among all the users. If the channel is operating at frequencies in the VHF, UHF, or higher ranges, then the maximum distance between the

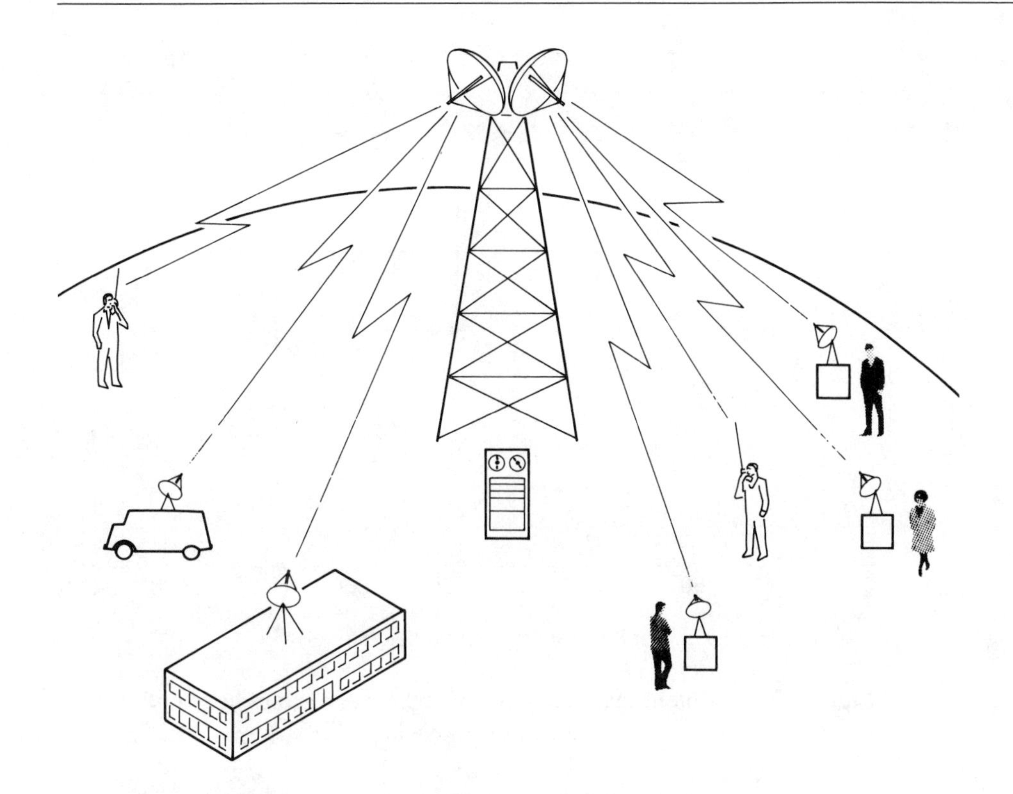

Figure 15-1. Operational environment of users communicating over a shared terrestrial radio channel.

transmitters and the receiver depends on the radio line-of-sight, accounting for the curvature of the earth. This distance is given by the approximate formula:

$$D_{\max} = 1.2(\sqrt{H_t} + \sqrt{H_r})$$

where $D_{\max}$ is given in miles, and H_t and H_r are the height of the transmitting and receiving antennas, respectively, in feet above the ground.

Range and Other Limitations

Figure 15-2 shows the maximum radio range as a function of the height of the central station antenna, assuming that the individual user antennas are approximately 20 feet above the ground (e.g., mounted on mobile vans, small office buildings, utility poles, etc.). This range is, of course, just an estimate, since it depends on local terrain, interfering obstacles, reflections from natural and man-made objects, and atmospheric conditions.

More important, however, when we are dealing with a packet broadcast system operating with terrestrial repeaters which can amplify the signals and extend

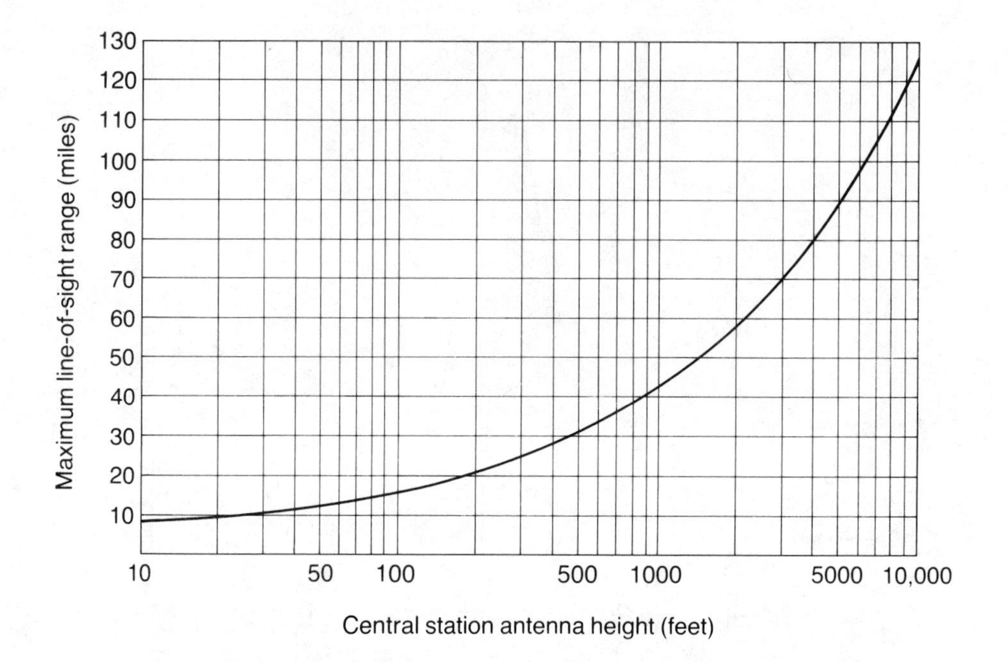

Figure 15-2. Plot of maximum line-of-sight range versus height of central station antenna.

the communications range, is the fact that the line-of-sight distance is not the only limitation to the successful delivery of packets to a central location. The major problem is the decrease in received signal power from a remote transmitting station, which is proportional to the square of the distance between the transmitter and the receiver. This means that if user B is 4 miles from the receiver and user A is 2 miles from the receiver, then user A's signal will be four times stronger than user B's, as detected by the receiver (assuming, of course, that they are using similar equipment and the same initial transmitter power). Therefore, when two users transmit at the same time, using packet broadcasting, the user who is closer to the central station is much more likely to capture the receiver and be received successfully. The expected delay is thus very dependent on the distance between the user and the central station. The impact of this phenomenon is that there is a theoretical range beyond which a terrestrial packet broadcasting user cannot expect to use the system successfully.

Gaining Information from the Channel

Another very significant aspect of terrestrial packet broadcast systems is the ability to gain useful information simply by listening to the common channel. Listening to the current information on a satellite channel was of no use in determining whether or not the channel was available since the information was at least a quarter of a second old. In the terrestrial system the circumstances are quite different. Though the typical packet may last 20 milliseconds (i.e., a 1000-bit packet on a 50,000-bits/second channel), the propagation time between the transmitters and receivers is on the order of 0.1 millisecond (i.e., 20 miles distance at a propagation velocity of 186,000 miles per second). In other words, since the packet is about 200 times longer than the typical delay encountered in its arrival at the central location, it is very probable that hearing a currently inactive channel means the channel really is quiet, and a new packet has a good chance of being received without interference.

This notion introduces a number of possible modes of packet broadcast operation, known collectively as *carrier sense multiple access* (**CSMA**). As the name implies, users sense the state of the channel by listening for the carriers associated with other users' transmissions.

CARRIER SENSE MULTIPLE ACCESS—LISTEN BEFORE YOU SEND

The ability of users to listen to the channel before transmitting eliminates most of the problem of overlapped packets. The CSMA techniques have been the object of a great deal of theoretical study, with many different algorithms and possible modes of operation evolving. In addition, the combination of the CSMA principles and the concepts of packet reservation can create very powerful and flexible systems, capable of operating very close to 100% of capacity over a broad range of conditions.

Operational Principles

Let us look first at the operational principles of the carrier sense techniques. When we talk about sensing the channel, we mean that there is some way for a user's terminal device to determine if some other user's terminal is currently transmitting. Since, in general, individual users can communicate with the central station but not directly with each other, it may be necessary for the central station to transmit a special beacon signal, on some other frequency, to indicate when the users' input channel is occupied. Even when users' terminals are able to sense the operational state of the channel, some problems remain.

The CSMA delay is very short compared to the round-trip satellite delay we encountered in our first analysis of the ALOHA technique. However, the propagation delay is not negligible, particularly when the channel rate is relatively high or the packets are relatively short. For example, if a channel rate of 1,500,000 bits per second is used, a 1000-bit packet will have a packet length of about 0.67 millisecond. The propagation time to a user about 60 miles from the central station would be about one-half this value. This means that, if a user started transmitting from a distance of 60 miles, half of his packet would already be transmitted before the central station could even begin to detect that the transmission had begun. Thus, even after listening to the channel, a user closer to the central station might still conclude that the channel was vacant.

In order to account for this problem, let us consider a parameter, often designated *a*, which is simply the ratio of the maximum propagation delay to the packet length. In the example we just gave, the value of *a* would be 0.5; that is, the propagation delay to the most distant users in the system would be about one-half of the packet length. For terrestrial systems, *a* generally has a small value, most likely less than 1. For satellite systems, *a* is a larger number—typically 10, 50, 100, or greater—because of the minimum 250-millisecond round-trip satellite propagation delay.

User Persistence

Another key characteristic of the CSMA is user persistence, or simply **persistence**—user behavior when the user senses that the channel is busy. The relevant operating algorithms are classed as either persistent or nonpersistent.

The nonpersistent algorithm works as follows: Upon receiving a transmit command from the user, a user terminal device senses the channel. If the channel is sensed vacant, the terminal transmits. There is a good probability, especially if *a* is very small, that the packet will succeed in getting to the central station. However, if the terminal senses the channel to be occupied, the nonpersistent terminal sets a random timer and, after the timer elapses, senses the channel again. Sensing the channel is repeated until the channel is sensed to be vacant, and the terminal then transmits the packet.

The persistent algorithm works quite differently. The terminal senses the channel and, if it finds it vacant, of course transmits the packet. However, if the persistent terminal senses the channel and finds it occupied, it continues to sense the channel. The instant the terminal senses the channel becoming vacant, it begins to transmit. This can achieve very high channel occupancy since the channel is not permitted to go idle before the next persistent user begins transmitting. A problem occurs when two or more persistent users have packets to transmit. As soon as the channel becomes idle, all persistent users who have been waiting begin to transmit immediately, causing immediate collisions and reduced traffic throughput.

The persistent user algorithm can be improved by the addition of one parameter. Each user senses the channel persistently. However, when the channel becomes idle, the users do not always transmit. In fact, the users transmit with some probability—*p* (where *p* lies between zero and unity). For example, let us assume two users and *p* equal to one-half. When the channel becomes idle, each user essentially flips a coin. If the coin comes up heads, one user—say, user A—transmits; if it comes up tails, he does not. If user A does not transmit, he waits a time sufficient to account for the maximum propagation delay, then reinitiates normal carrier sense operation. By this time user A would begin to sense any other user's transmission. In other words, one-fourth of the time user A would transmit, and user B would wait; one-fourth of the time user B would transmit, and user A would wait; one-fourth of the time both users would transmit; and one-fourth of the time both users would wait.

The probability of a collision is reduced from unity for the fully persistent case to one-fourth in the so-called $\frac{1}{2}$-persistent case. Persistent algorithms are generally defined by the nomenclature of p-persistent, where p is the transmission probability. The original persistent algorithm—that is, where the users all transmit immediately upon hearing the channel vacant—would thus be 1-persistent since users transmit with the probability of unity on hearing the channel become vacant.

Mathematical Analysis

As might be expected, the mathematical analysis of the persistent and non-persistent carrier sense multiple access techniques is quite complex. Although the analytical principles follow many of the same approaches we have already used, the interplay of the many probability parameters must be incorporated.

In Figure 15-3 we see the throughput curves for a 1-persistent channel for various values of a ranging from $a = 0.0$ to $a = 1.0$. Notice that a value of $a = 0$ means that the propagation delay is truly negligible compared to the packet length (a theoretical fiction), and therefore the users have "perfect" information about the channel when they listen to it. This assumption naturally leads to the highest possible throughput for the 1-persistent channel.

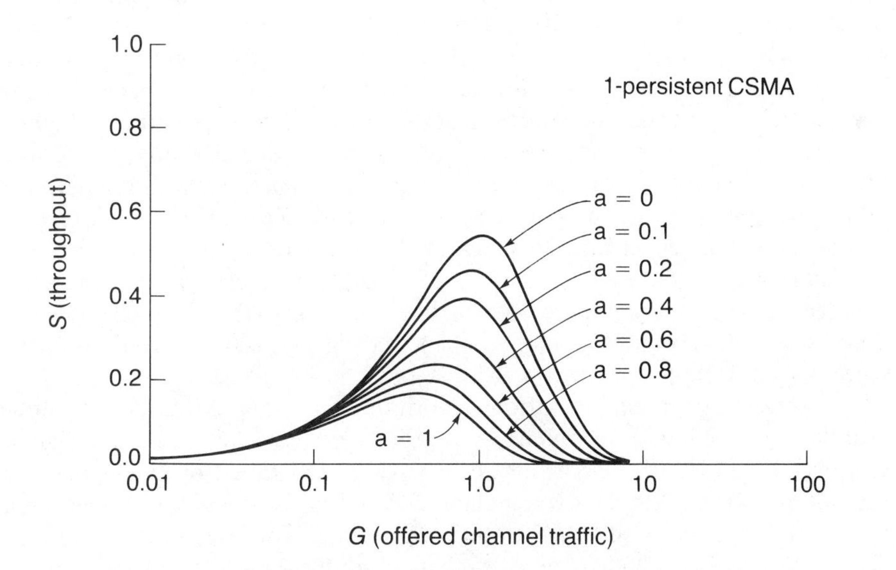

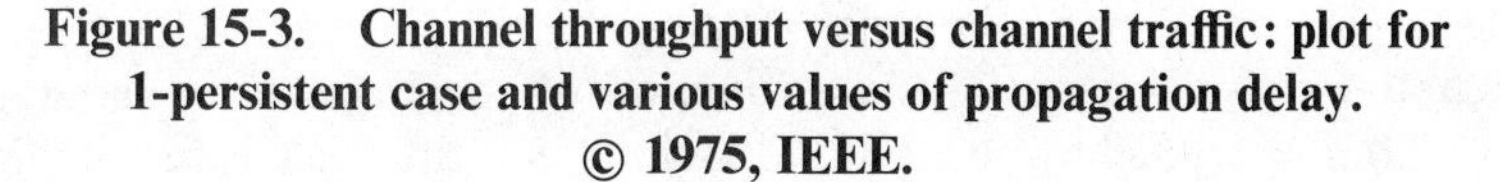

Figure 15-3. Channel throughput versus channel traffic: plot for 1-persistent case and various values of propagation delay. © 1975, IEEE.

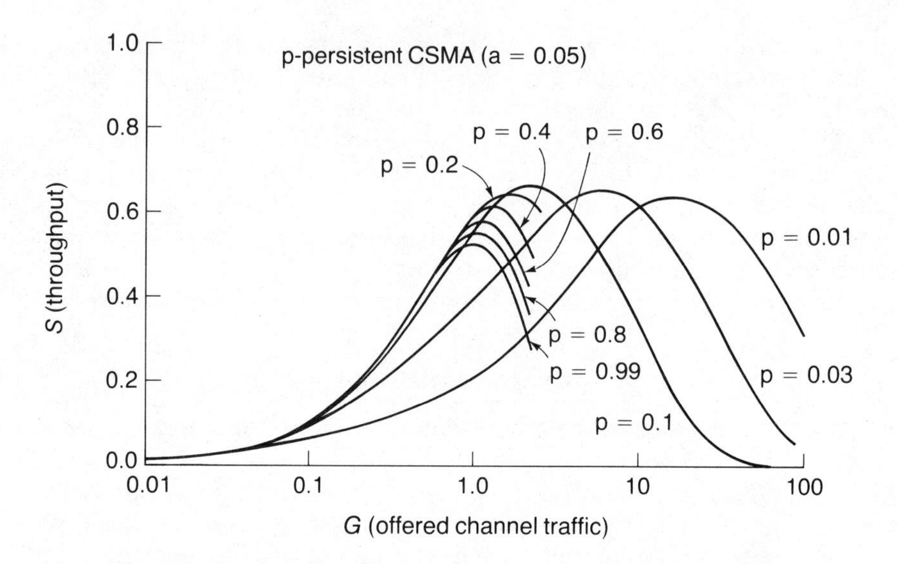

Figure 15-4. Channel throughput versus channel traffic: plot for propagation delay of 5% of packet length and various values of transmit probability when channel becomes clear. © 1975, IEEE.

In Figure 15-4 the relationship of the channel throughput and traffic is shown for a p-persistent algorithm with the value of p varying from 0.01 to 0.99. In each case the value of a was held constant at $a = 0.05$. Note the interesting fact that, at a value of p between 0.01 and 0.20, the maximum channel capacity changes very little. By contrast, the traffic corresponding to the maximum throughput does shift from higher to lower values as p is increased. Recall that the ratio of traffic to throughput is a measure of average delay since it is a direct reflection of the average number of times a given packet has to be repeated before it is received successfully. This effect clearly indicates the existence of an optimal value of p—that is, the one that achieves the greatest possible channel capacity at the lowest average value of delay. Figure 15-4 further shows that the channel capacity is decreased, without significant improvement in the delay, as the value of p is further increased above the value of $p = 0.2$.

Figure 15-5 summarizes various algorithms, including ALOHA and slotted ALOHA, all computed at a value of $a = 0.01$. The significant increase in channel capacity for the various algorithms can be readily observed on this figure. It is interesting to note that the overall channel capacity for the nonpersistent algorithms is as great as for any of the other algorithms. However, since the value of throughput occurs at higher values of channel traffic, the average delay is longer.

The impact of propagation delay is shown in Figure 15-6. This figure plots the maximum channel capacity, as a function of the value of a, ranging from $a = 0.001$ to $a = 1.0$. The value plotted is, in effect, the peak value obtained on

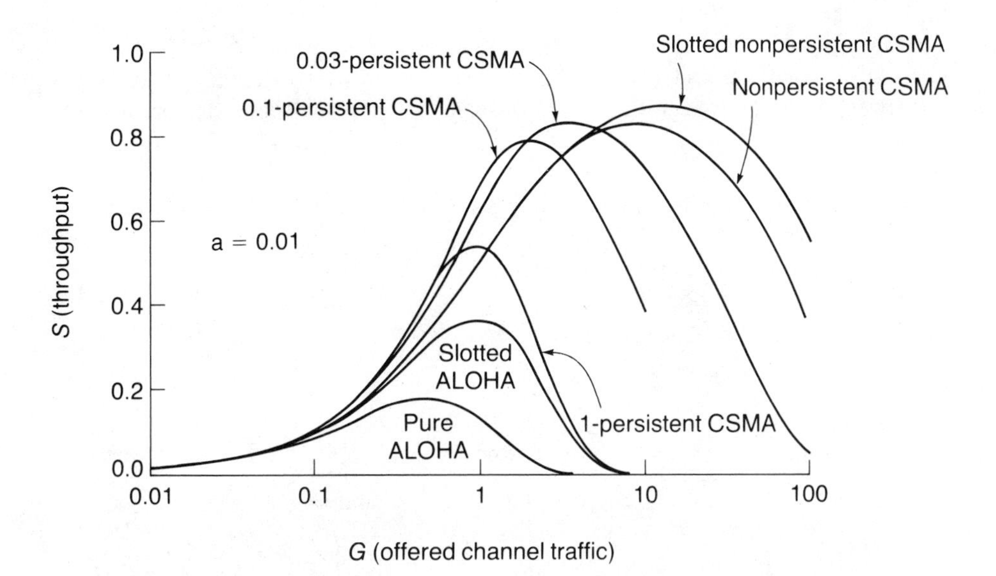

Figure 15-5. Channel throughput versus channel traffic: Delay of channel is assumed to be 1% of the length of a packet. © 1975, IEEE.

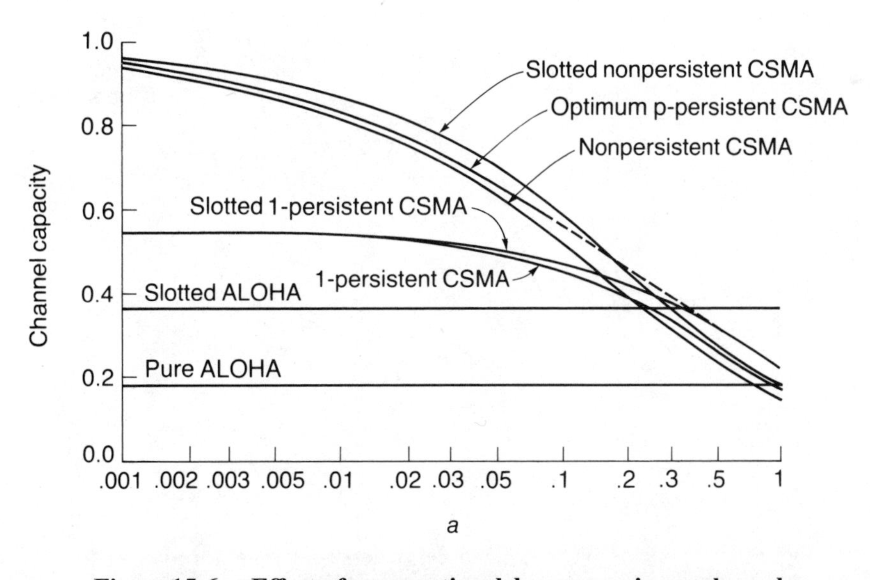

Figure 15-6. Effect of propagation delay on maximum channel capacity. © 1975, IEEE.

the kind of curves shown on the previous figures. The point here is that, as the value of a reaches the vicinity of 0.2, the carrier sense process fails to yield any improvement compared to the basic nonsensing ALOHA techniques because the information sensed in the channel is really too "old." This is important when we consider a carrier sense mode of operation with fairly high-capacity channels since, if the basic bit rate is high, the packet durations will be short, leading to large values of a.

Finally, the information in Figures 15-3 through 15-6 can be displayed in terms of delay versus throughput, as shown in Figure 15-7, which shows the throughput/delay tradeoff for various algorithms, computed at a value of $a = 0.01$. For any

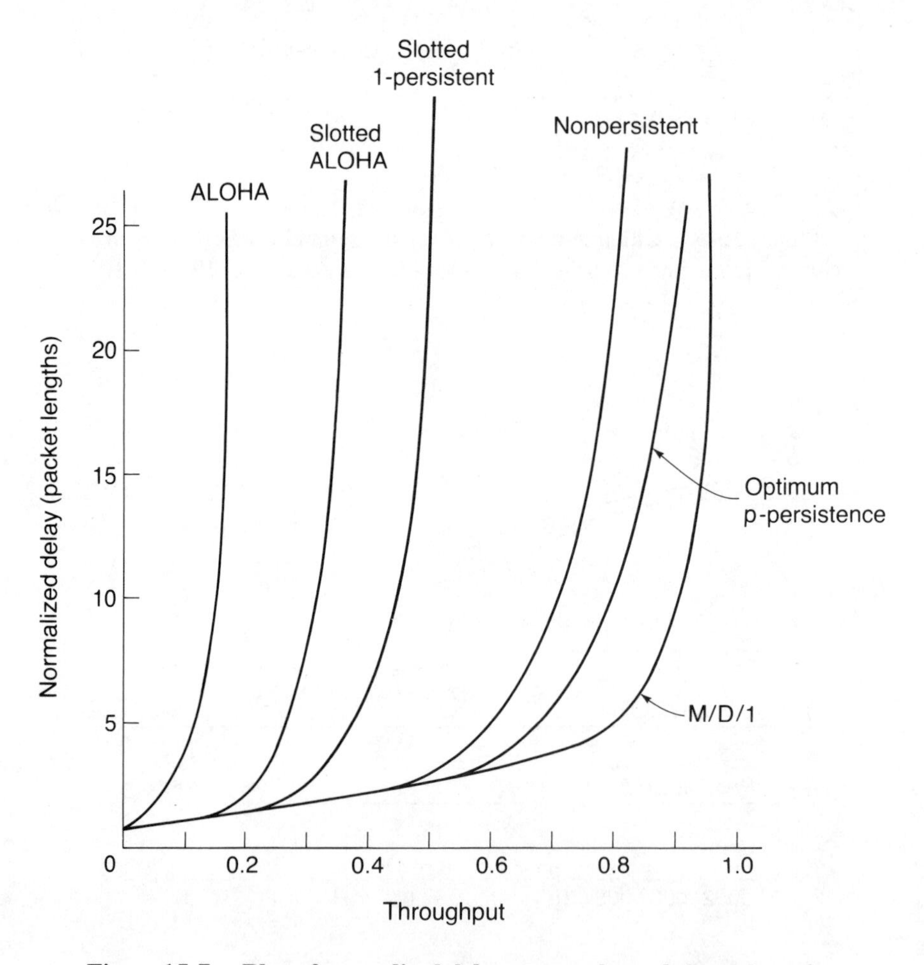

Figure 15-7. Plot of normalized delay versus channel throughput for various network access techniques.

given value of delay (in this figure normalized to the packet length), the overall channel throughput can be substantially increased by the choice of the proper channel operational algorithm. Recall, however, that this is computed for a value of $a = 0.01$, which would correspond roughly to a 30-mile system operating at 50,000 bits per second with 1000-bit packets. As the value of a increases, the advantages shown by the carrier sense systems in the figure begin to fade away rapidly, as the information sensed by the users becomes increasingly stale.

SUMMARY

We have now been able to extend many of the concepts of packet broadcasting and packet switched networks without switches to a localized terrestrial environment. The basic idea of all members of a network freely transmitting on a common radio frequency channel is carried over directly from the satellite broadcast concepts we developed earlier. Because of the much shorter distances in the terrestrial environment, users can use information in the channel, in real time, to affect and modify their behavior. As a result, much higher channel capacity is possible than with satellite transmission. However, a possible limitation is that the signals of users located far from the central station would generally be dominated by closer signals, and would have difficulty being received correctly without many retransmissions.

There are a number of different algorithms by which users can sense the current state of the channel and decide whether or not to transmit. We introduced the techniques of both persistent and nonpersistent carrier sense multiple access (CSMA) and compared CSMA operation to the basic ALOHA and slotted ALOHA techniques. If the propagation delay is short relative to the packet length (approximately 20% or less), the information users hear by monitoring the channel is sufficiently "up to date" to permit a useful decision and significantly enhance channel performance. However, if the information is too old, the ALOHA techniques provide service as good or better than the CSMA techniques.

Only an overview of terrestrial packet broadcasting technology could be covered in this chapter. The Suggested Readings provide a detailed coverage, including an extensive experimental system developed under contract to the Advanced Research Projects Agency (ARPA) of the U.S. government.

SUGGESTED READINGS

ABRAMSON, NORMAN. "The Throughput of Packet Broadcasting Channels." *IEEE Transactions on Communications*, vol. COM-25, no. 1 (January 1977), pp. 117–128.

A consolidated treatment of the theory of packet broadcasting, this paper includes the author's practical experiences in applying the theory in operational networks at the University of Hawaii. The derivation of the spatial capacity of the terrestrial networks and the limitations experienced in practical applications are discussed.

KAHN, ROBERT E. "The Organization of Computer Resources into a Packet Radio Network." *IEEE Transactions on Communications*, vol. COM-25, no. 1 (January 1977), pp. 169–178.

This paper describes a general packet radio network and plans for an experimental implementation that extends the operation beyond line-of-sight by use of packet repeaters. The paper introduces possible solutions to problems of security and privacy, spectrum utilization and allocation, and overall architectural organization of the network resources.

KAHN, ROBERT E., GRONEMEYER, STEVEN A., BURCHFIEL, JERRY, and KUNZELMAN, RONALD C. "Advances in Packet Radio Technology." *Proceedings of the IEEE*, vol. 66, no. 11 (November 1978), pp. 1468–1496.

This is a comprehensive summary of the practical aspects of packet radio technology. Included are discussions of resource organization, control of the network, and the application of a spread spectrum transmission technology to achieve high-transmission bandwidths with low incidence of interference. Of greatest interest are the descriptions of experimental operational equipment that was deployed and tested in the San Francisco Bay Area of California.

KLEINROCK, LEONARD, and TOBAGI, FOUAD A. "Packet Switching in Radio Channels: Part I—Carrier Sense Multiple Access Modes and Their Throughput-Delay Characteristics." *IEEE Transactions on Communications*, vol. COM-23, no. 12 (December 1975), pp. 1400–1416.

This paper, which was the source of many of the figures used in this chapter, reports on the operation and performance characteristics of carrier sense multiple access. The practical results, which are discussed in terms of the relationship of propagation delay and packet length, provide a solid basis on which to build the technology of packet radio systems.

KLEINROCK, LEONARD, and TOBAGI, FOUAD A. "Packet Switching in Radio Channels: Part II—The Hidden Terminal Problem in Carrier Sense Multiple Access and the Busy-Tone Solution." *IEEE Transactions on Communications*, vol. COM-23, no. 12 (December 1975), pp. 1417–1433.

In this paper CSMA is extended to the more realistic situation where users are unable to directly hear the transmissions of other users. It is demonstrated that, if the central station provides a separately transmitted busy tone to indicate the presence of a user in the common channel, the theoretical performance advantages of the CSMA technique can be retained.

TOBAGI, FOUAD A. "Analysis of a Two-Hop Centralized Packet Radio Network—Parts I and II." *IEEE Transactions on Communications*, vol. COM-28, no. 2 (February 1980), pp. 196–216.

The two-part article extends the theory of both slotted ALOHA and CSMA to the situation where user transmissions go through multiple relays before reaching the central station. The throughput-delay relationships are derived, and the complex interdependencies of protocol, topology, and storage capacity of the repeaters are discussed. The results show that the introduction of multiple relays tend to reduce overall capacity.

16

Combination of Satellite and Terrestrial Connectivity— The General Optimization Problem

THIS CHAPTER:

will examine the problem of designing networks that involve a large amount of satellite capacity.

will consider the optimum number of satellite earth stations relative to the density of users and the terrestrial connection costs.

will show the application of the results to practical situations early in the network design process.

In the last few chapters we have seen that, when satellites are combined with any one of many different access protocols, packet network functions can be accomplished without discrete packet switching nodes. When satellites form the essence of a distributed packet network, all users must have immediate local access to a satellite earth station, or else some other means must be used to connect terrestrial users to the satellite and to each other.

A first approach is, in essence, to employ the satellite as a "cable in the sky," tying the terrestrial nodal facilities together. The next step would logically be to use the satellite facilities in a broadcast mode to provide all of the interswitch trunking among the network nodes. However, only when the satellites can provide the connections completely back to the individual users will the total benefit of the networkwide broadcast and distributed access control be fully realized.

In this chapter we will introduce a model of the typical user, who needs a terrestrial connection to get from his location to the nearest satellite earth station from which he derives his overall network services. From the general model we will be able to estimate the optimum number of earth stations, which will permit us to estimate the overall network complexity and the mix of satellite and terrestrial services that is required to meet the users' needs.

SERVING USERS WITHOUT DEDICATED EARTH STATIONS—THE GENERAL PROBLEM

We will begin with the basic assumption that the number of users, and the number of user locations, is sufficiently large that the geographical placement can be approximated by random user distribution over the area served by the system. This assumption is analogous to the random traffic generation rates, leading to the Poisson traffic arrivals, that we have been using in capacity-delay relationships, except that the randomness applies to distribution in space rather than in time. Although the users are randomly distributed, we will assume initially that the satellite earth stations are arranged in a regular, geometric pattern (called a **tessellated pattern**). Later we will generalize the situation to randomly distributed earth stations and compare the results to the tessellated case.

Consequently, we will consider a fixed geographic area, in which network users are uniformly but randomly distributed. We desire to optimize the number of earth stations to provide access to the satellite for all users at the least cost. At most, there could be one earth station associated with each user, but in general an earth station acts as a concentration point and serves a number of users in its vicinity.

In order to conceptually simplify the geometry, we will assume that the service area is a square, each side of which is S miles. The results will be derived in terms of user densities. This means that the actual shape of the service area is not particularly important, providing that the area is large enough to support a

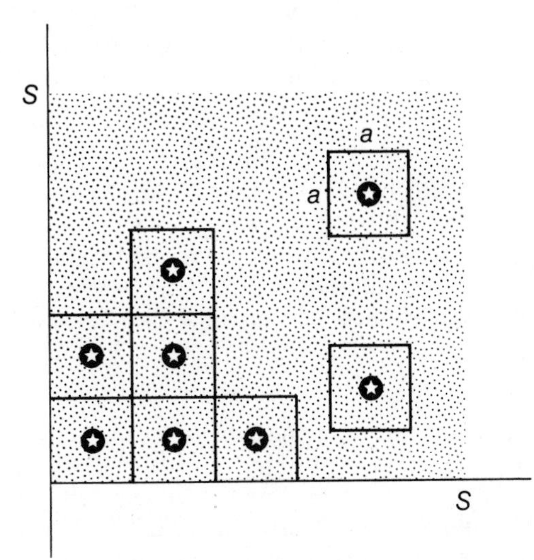

Figure 16-1. Satellite network geometry for random users and tessellated servers. © 1976, ICCC.

fairly large number of earth stations, and the average distance between earth stations is a small fraction (1/4 or less) of the smallest dimension of the service area. The assumed geometry based on the square service area is shown in Figure 16-1.

The overall service area of S^2 square miles is assumed to be divided into q service areas. Each small service area, surrounding one of the earth stations, is a small square, with sides of dimension a. In order to completely cover the service area, we note that:

$$qa^2 = S^2$$

or

$$a = S/\sqrt{q}$$

The parameters we will be using are:

n = number of user locations

q = number of earth station locations

S = dimension of total service area

a = dimension of a side of the area serviced by single earth station

ρ = density of users = n/S^2

C_q = average cost of an earth station

C_1 = cost per mile of line needed to tie each user to the nearest earth station

$\bar{d}$ = average distance from each user to the nearest earth station

The overall cost of the network comprises four major components: the earth station cost, the space segment cost, the terrestrial line costs, and the terrestrial fixed plant costs. Let us look at each in turn.

Earth Station Cost

The earth station cost, represented by C_q, is the average installed cost of the satellite earth stations. Since an economic analysis is involved, we will make the earth station cost represent the life-cycle cost of the earth station—including its acquisition cost, installation cost, cost of operation, maintenance, and repair over a 10-year life cycle.

Space Segment Cost

The space segment cost represents the overall cost of the satellite, including launch and continuing control costs. Since no single-purpose network is likely to use the full capacity of an entire satellite, this cost would be the pro-rated share of total capacity allocated to the particular user network of interest. In addition, since the amount of satellite capacity used is dependent on the amount of user traffic and not on the number of earth stations, we will assume that the space segment cost is constant, independent of the number of earth stations that ultimately optimizes the network. The underlying implication is that all user traffic eventually reaches the satellite, via either a local or a remote earth station.

Terrestrial Line Costs

The terrestrial line costs are the mileage costs associated with the connection from each user to his nearest earth station. For n users, each an average of $\bar{d}$ miles from the nearest earth station, the total line costs are just: $n\bar{d}C_1$ where C_1 is the cost per mile of the terrestrial lines. To be commensurate with the earth station costs, the terrestrial line costs would have to be estimated for a comparable life-cycle period—say, 10 years—based on lease tariff charges, which are generally given in monthly costs.

Terrestrial Fixed Plant Costs

The terrestrial fixed plant costs are independent of distance and traffic—for instance, attachment charges, modem costs, port costs, or any other costs that are related to the number of users but not to the cost of the earth station itself. Since all users will eventually be connected to an earth station, these costs are essentially independent of the number of earth stations.

The problem we are attempting to solve, therefore, is simply to minimize total system cost, C, where:

$$C = \text{(earth station cost)} + \text{(terrestrial line costs)} + \text{(space segment cost)} + \text{(terrestrial fixed plant costs)}$$

$$C = qC_q + nC_1\bar{d} + \text{(space segment cost)} + \text{(terrestrial fixed costs)}$$

In the minimization the independent variable is the number of earth stations, q, and our intention is to find the number of earth stations that minimizes the overall cost. Since the space segment and terrestrial fixed plant costs are dependent only on the number of users and the total amount of traffic flowing through the satellite, and not the number of earth stations, they need not be considered in the minimization process. The problem thus reduces to: minimize C, with respect to q, where:

$$C = qC_q + nC_1\bar{d}$$

The key to the minimization process is the recognition that the average distance between the users' locations and the nearest satellite earth station, $\bar{d}$, is directly related to the number of earth stations, q. If q is very large—that is, there are many earth stations—the average distance to the nearest one will be small. If, on the other hand, q is small, and there are few earth stations, then the average distance to the nearest one from each user will be larger and more cost will reside in the terrestrial lines connecting the users to the nearest earth stations.

DERIVING THE OPTIMUM VALUE OF THE NUMBER OF EARTH STATIONS

Random Distribution of Users

The general service area of the network, as was shown in Figure 16-1, consists of a random scattering of users, divided into small square service areas, with a satellite earth station located in the center of each small service area. Inside the small service area there are a sprinkling of user locations, some close to the earth station (the center of the square) and some much further away.

The average distance from the earth station to the users is determined by a mathematical integration process, which in effect computes the distance to every possible user point within the square and takes the average of those distances. Evaluating this average distance through geometric integration is, with the aid of a good table of integrals, not overly difficult. Readers who are interested in the details of this derivation should consult Abramson and Rosner (1976). The result is simply:

$$\bar{d} = 0.381a$$

which says that the average distance from the center of a square to every other possible point within the square is 38% of the side of the square. We can take this value and go back to the cost formulation, where we find that:

$$C = qC_q + nC_1\bar{d} = qC_q + nC_1(0.381a)$$

or, since $a = S/\sqrt{q}$,

$$C = qC_q + 0.381nC_1S/\sqrt{q}$$

This gives us the expression fully relating the overall cost of the satellite-based network as a function of the number of earth stations, which we can proceed to optimize by taking the derivative of the expression and setting it equal to zero, with the result that:

$$q_{\text{opt}} = \left[\frac{0.381nC_1S}{2C_q}\right]^{2/3}$$

By remembering that the user density, ρ, is given by $\rho = n/S^2$, we can generalize this optimization expression by substituting for S, with the result that:

$$q_{\text{opt}} = 0.332n\left[\frac{C_1}{C_q\rho^{1/2}}\right]^{2/3}$$

This equation is the optimum number of earth stations relative to the number of users, the user density, the cost of the lines connecting the users to the earth stations, and the cost of the earth stations themselves. From this expression we find that the optimum number of earth stations increases if there are more users and if the cost of the lines connecting users to earth stations is high. The optimum number of earth stations is lowered by high earth station costs and high user densities, which allow many users to be served by fewer earth stations.

Random Distribution of Earth Stations

If the earth stations are not neatly placed in a regular geometric pattern, a different mathematical approach is required to estimate the optimum number of earth stations. The opposite extreme would be a random distribution of satellite earth stations.

The analysis of this situation requires a different viewpoint from the previous, tessellated case. Within the total service area there are q possible randomly scattered earth stations that can serve the user, but we desire to compute the average distance to the nearest station that can serve each user. There is, in effect, a double averaging over two random processes. For each user there is a computation of the average distance to the nearest earth station. The overall average has to be computed over the total user population within the service area. The results (see Abramson and Rosner, 1976, for details) for the random earth station case are:

$$q_{\text{opt}} = 0.396n \left[\frac{C_1}{C_q \rho^{1/2}} \right]^{2/3}$$

This result differs from the previous result only in the change of scale factor: The optimum number of earth stations for the random case is about 20% higher than when the earth stations are arranged in a regular, geometric pattern. This effect is primarily due to the fact that when the earth stations, as well as the users, are randomly distributed, there is a reasonable probability that the random clusters of earth stations may be away from probabilistic clusters of users, leaving some earth stations much less loaded than others. In the case where the users are random but the earth stations are distributed regularly, there will, on the average, be a much more balanced usage of the earth stations, leading to a lower optimum number.

For convenience, both of these results are summarized in Box 16-1. The application of these formulations can best be demonstrated by means of examples. Remember, the intention and application of this approach to the satellite/terrestrial network leads only to an estimate of the number of earth stations, on an averaging basis, that are needed to optimize the network configuration. The actual location of the earth stations would depend on the particular, rather than random, topology of the actual user situation. In addition, the general solutions do not include the effects of every possible factor that could be important. For example, the notion of quantity discounts for large numbers of satellite earth stations procured at the same time was not included, nor was the relationship of the overall capacity used within the satellite depending on the amount of terrestrial connectivity and terrestrial switching that might be present. The explicit effect of

$$q_{opt} = 0.396n \left[\frac{C_l}{\sqrt{\rho C_q}} \right]^{2/3} \quad \text{(random earth stations)}$$

$$q_{opt} = 0.332n \left[\frac{C_l}{\sqrt{\rho C_q}} \right]^{2/3} \quad \text{(tessellated earth stations)}$$

q_{opt} = the optimum number of satellite earth stations

n = the number of network users

C_l = the cost of the terrestrial connection from the users to the nearest earth station (expressed on a per-mile basis)

C_q = the cost of the satellite earth station (note that C_l and C_q must be in comparable measures)

ρ = the user density = n/a where a is the total area being served by the network

Box 16-1

reducing or eliminating most terrestrial switching facilities by application of one of the random access packet broadcasting techniques is not treated. However, we have illustrated a macroanalysis technique that gives us the ability to treat large communications networks in an aggregated or random network fashion without having to deal with fine-grain structural detail in the early phases of network design.

APPLICATIONS AND EXAMPLES OF THE RANDOM NETWORK ANALYSIS

The application of these formulations can best be illustrated by several examples. In the first example we will look at a network of many small users, distributed throughout an area the size of the continental United States.

The applicable parameters are:

n = number of users = 30,000

S^2 = area of the continental United States = 3,000,000 sq. miles.

ρ = user density = 30,000/3,000,000 = 0.01

C_1 = line cost = \$2.00 per mile per month. The monthly line cost can be equivalanced to an approximate ten-year, present-worth figure by multiplying the annual cost by a ten-year, 10% present-worth factor of about 6.5. Thus the ten-year present-worth of a \$2.00-per-month expenditure is approximately $2 \times 12 \times 6.5 = \$156$.

C_1 = \$156 per mile for 10 years.

C_q = \$250,000 = ten-year installed and operated cost of the satellite earth station.

By substituting these parameters in the equation for the optimum number of earth stations, using the regular, geometric pattern we find:

$$q_{\text{opt}} = 0.332 \times 30{,}000\left[\frac{156}{(250{,}000) \times \sqrt{(0.01)}}\right]^{2/3}$$

or

$$q_{\text{opt}} = 9960(6.24 \times 10^{-3})^{2/3}$$

$$q_{\text{opt}} = 9960(3.37 \times 10^{-2}) = 336 \text{ earth stations}$$

Using the formulation for random, rather than tessellated, distribution of earth stations results in an optimum number of 401 stations. Consequently, a network of 30,000 users—which, while large, is not uncommon for many shared network applications, such as point-of-sale networks, transactional banking terminals, and the like—results in a rather large number of earth stations. In addition, the ability to achieve a ten-year cost of a modest earth station of about \$250,000 is not at all difficult using present state-of-the-art technology.

As a second example, let us consider a situation where the earth stations are intended to serve clusters of users rather than individual users. This case not only illustrates the basic point that it is possible to treat user clusters in the same way we have treated individual users, but suggests a formulation that can be compared directly to a detailed satellite optimization study (see Rosner, 1975).

We will assume 2500 separate geographic locations around the continental United States, at which potential network users are clustered. After a local traffic analysis on the "average" location, it is determined that each location needs ten lines to connect the installation to the nearest satellite earth station. The applicable parameters are now:

n = number of users = 2500

S^2 = area of the continental United States = 3,000,000 sq. miles

ρ = user density = 2500/3,000,000 = 0.000833 users/sq. mile

C_1 = line cost = \$2.00 per line per mile per month times an average of ten lines per user location = \$1560 per mile for the ten-year comparison period.

C_q = \$1,500,000 for the ten-year period, for a relatively large earth station capable of serving many users.

By substituting these parameters into the equation for the optimum number of earth stations, using the regular geometric pattern, we find:

$$q_{\text{opt}} = 0.332 \times 2500\left[\frac{1560}{(1{,}500{,}000) \times \sqrt{0.000833}}\right]^{2/3}$$

or $$q_{\text{opt}} = 830(36.1 \times 10^{-3})^{2/3}$$
$$q_{\text{opt}} = 830(11.0 \times 10^{-2}) = 91 \text{ earth stations}$$

Using the formulation for random, rather than tessellated, distribution of earth stations results in an optimun number of earth stations of 109 earth stations. The number of earth stations is considerably smaller for this case because the number of individual user locations is smaller, and the assumed cost of the earth station is much higher than in the first example. In fact, the optimum number of earth stations is so highly dependent on the cost of the earth stations that it is generally useful to plot the optimum number of earth stations as a function of the assumed life-cycle cost of the earth stations.

This is done, for our second example, in Figure 16-2. Here the optimum number of earth stations is plotted over a range of earth station costs between zero and about $4 million. Plots are shown for both random and tessellated arrangements of satellite earth stations. In addition, a plot representing the results

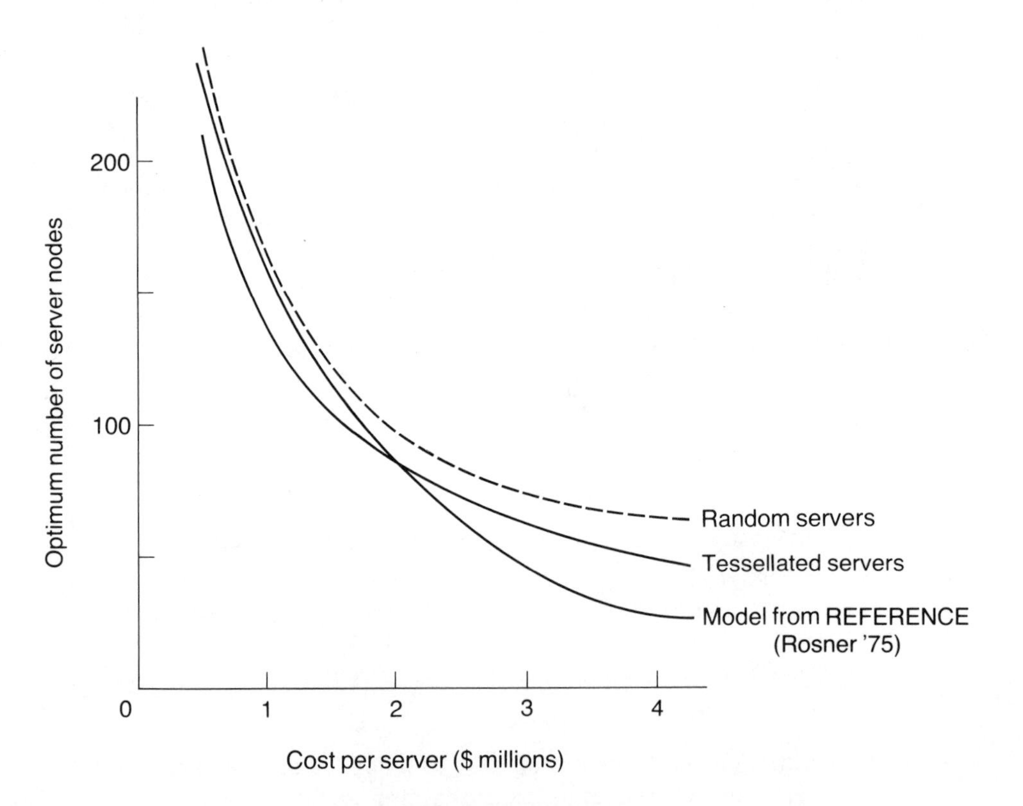

Figure 16-2. Plot of optimum number of server nodes versus cost per server node. © 1976, ICCC.

taken from a much more detailed analysis of the model for the second example is shown. The agreement is amazingly good, considering the ease with which the general equation we have developed in this chapter can be applied. The agreement is quite good for relatively large numbers of earth stations. However, when the earth station cost is greater than $2.5 million, resulting in optimum values of the number of earth stations of 50 or less, the averaging assumptions inherent in our macroanalysis are poor, and we cannot expect the methods of this chapter to yield very accurate results.

SUMMARY

In this chapter we have addressed the problem of estimating the number of satellite earth stations in a randomized network of many users spread over a broad geographic area. The technique would be particularly useful in the early phases of a distributed network to determine the feasible range over which packet satellite operation might make sense. The design of such a network is sensitive to earth station cost, user density, and the cost of the terrestrial interconnections between the users and their nearest earth stations.

Two different expressions were shown, relating the optimum number of earth stations in a mixed satellite/terrestrial network to the number and density of users and the satellite earth station and terrestrial line costs. Several examples were given where networks employing as many as 400 earth stations would result in optimum configurations.

This chapter completes the general topic of satellite and packet broadcasting networks and the achievement of the attributes of a packet switched network without the need for discrete nodal packet switching elements. The ability to imbed microprocessors into terminal devices and do the distributed protocol monitoring and control is rapidly becoming available at relatively small incremental cost for the terminal devices. As a result the practicality of many of the techniques discussed in this part of the book will likely find increasingly broad application in the near future. In addition, commercial network implementations, which make satellite communications accessible directly from the users' premises, are rapidly being developed and deployed. The combination of packet switching and the potentially high efficiencies achievable even with bursty or occasional users of the various broadcasting techniques makes the application of packet broadcasting for both satellite and terrestrial networks highly desirable.

SUGGESTED READINGS

ABRAMSON, NORMAN, and ROSNER, ROY D. "Optimum Densities of Small Earth Stations for a Satellite Data Network." *Proceedings of the International Conference on Computer Communications*, ICCC '76. Toronto, July 1976. Pp. 123–127.

This paper provides a detailed mathematical analysis of the results that were discussed in this chapter. The formulation of the optimizing equation and the derivation of the optimizing condition for the network where both the users and earth stations are randomly distributed are included.

CLARK, DAVID D., POGRAN, KENNETH T., and REED, DAVID P. "An Introduction to Local Area Networks." *Proceedings of the IEEE*, vol. 66, no. 11 (November 1978), pp. 1497–1517.

This article deals with the specialized problem of local area networks, which would be analogous to connecting many local users to their nearest satellite terminal. Substantial emphasis is given to the hardware and protocols that can exploit the possibility of using very high bandwidth communications media in the local area.

HUYNH, DIEU, KOBAYASHI, HISASHI, and KUO, FRANKLIN. "Optimal Design of Mixed-Media Packet Switching Networks: Routing and Capacity Assignment." *IEEE Transactions on Communications*, vol. COM-25, no. 1 (January 1977), pp. 158–169.

This paper presents a different approach to the optimization of packet networks where satellite and terrestrial connectivity are combined. The basic system operation employs packet broadcasting techniques in combination with limited-capacity terrestrial connectivity. MASTER, a protocol where packets that encounter collisions are retransmitted over the terrestrial network, demonstrates significantly improved delay performance and good stability characteristics with overload at very nominal increases in total system cost.

MCGREGOR, PATRICK V., and SHEN, DIANA. "Network Design: An Algorithm for the Access Facility Location Problem." *IEEE Transactions on Communications*, vol. COM-25, no. 1 (January 1977), pp. 61–73.

This paper proposes an algorithmic technique for locating access facilities or concentration points (or shared satellite earth stations) to obtain economical connection of users to the shared facilities. In addition, the paper provides more than 40 references to the literature of localized network design.

ROSNER, ROY D. "Optimization of the Number of Ground Stations in a Domestic Satellite System." *Proceedings of the EASCON Conference, EASCON '75 Record.* Washington, D.C., September–October 1975. Pp. 64A–64F.

This paper shows the results of a detailed topological study of the applicability of satellite communications to large distributed networks. It introduces the concept and practicality of placing satellite ground stations at the premises of large users or within concentrations of such users. This reference provides a baseline against which to measure the utility of the approximation technique we have developed in this chapter.

PART FIVE

Data Networks, Packet Switching, and the Common Carriers

With the grounding in the basics of packet switching that we have developed over the first four parts of this book, we are now ready to take a thoughtful look at the full variety of worldwide telecommunications services and technologies.

The situation can be quite confusing. Packet switched networks are available in many countries; in fact, some countries even have competitive packet switched services. In addition, it is possible for users to synthesize their own packet switched networks by using relatively low-cost processing and control elements and leasing the much more capital-intensive intercity or international transmission facilities from the common carriers. Consequently, it is important to understand the services, facilities, and networks that are offered under a variety of charging arrangements from the common carriers and telecommunications suppliers.

In this part we will look at the general telecommunications carrier marketplace. We will provide some detailed information about several of the "value-added" packet switching carriers operating in the United States, with interfaces to packet carriers in other nations. We will then talk about tariff structures and attempt some projections that can be useful in economic analyses. This part will conclude by looking at some prospects for future carrier-provided services that implement new, advanced technology.

17

The Carrier Marketplace

THIS CHAPTER:

will describe the types of services offered by common carriers of telecommunications, including packet switched services.

will present a brief survey of key carriers providing domestic services in the United States and international services.

In order to judge the present and future viability of packet switching in the marketplace, it is necessary to get an overall picture of the services that are available from the **common carriers**—those organizations, usually franchised by a governmental body, that provide communications services to the general public. The communications marketplace is extremely complex throughout the world. It is particularly so in the United States because entry into the marketplace is relatively easy, and specialized suppliers can compete with the "phone company" in many voice and nonvoice services.

COMMON CARRIER SERVICES AND FACILITIES

The Economics of Common Carrier Services

Competition in communications services within the United States has developed quite rapidly since the landmark **Carterfone decision** of 1968 and the approval of Microwave Communications Incorporated (**MCI**) construction of interstate microwave transmission facilities in 1969. These two rulings by the U.S. Federal Communications Commission (**FCC**) declared, in effect, that it was not in the public interest to restrict all telecommunications to the regulated monopolies of the established (voice) phone companies. While competition was dawning in the United States, no such changes were taking place in most countries outside of the U.S., where telecommunications are provided by government-owned monopolies or are under extremely restricted competition. In fact, the telecommunications administrations are generally managed together with the postal service, where, in most cases, the telecommunications charges are specifically designed to subsidize the postal rates.

At the time, most experts expected that the decisions would lead to greatest competition in the area of terminals and end instruments. The construction and operation of long-haul transmission facilities is extremely capital intensive, with very high start-up costs and strong economies of scale. Despite this fact, which heavily favors the large, established carriers, many suppliers have moved into the telecommunications transmission marketplace and are successfully competing with the established phone companies.

This success is due partly to some protective regulation and fully allocated pricing demanded by the FCC, and partly to the ability of the new suppliers to apply the most modern satellite and computer-based techniques to their services. Furthermore, the new entrants can enter the most lucrative service areas first—those where potential customers are most densely concentrated and thus can be served with large-capacity, high-efficiency facilities. In addition, overall common carrier regulation continues to be reduced, with the effect of further easing the entry of new carriers into the market.

Classification of Carrier Facilities

Carrier facilities can be broadly classified according to speed, system arrangement, and rate structure. In Figure 17-1 these characteristics are interrelated as a logical decision tree of possible service combinations. Note that packet switching is not included in this basic classification; at best, it is a subcategory of switched services.

Speed. The speed (or bandwidth) characteristic recognizes the use of a voice channel as the fundamental building block of telecommunications services. Low speed refers to bandwidths typically 300 bits/second or less, many of which can fit into a single voice-grade channel. Medium speed (or voice-grade) describes the capacity of a single voice-equivalent channel. Such a channel is required for voice-to-voice conversation, but, with present technology, it is also capable of providing 9600 bits/second of data transmission. High speed (or wideband) applies to all capacity requirements exceeding a single voice-equivalent channel, all of the way up to the hundreds of megabits of data needed for video and sensor data. By and large, normally tariffed common carrier services go up to 56,000 bits/second, with higher rates generally needing special agreements and arrangements, often available only via satellite facilities.

System Arrangement. The system arrangement characteristic divides telecommunications services into switched and private-line (nonswitched) services. As we have seen, in a switched service or network there are fewer lines than users, on the expectation that not all the users will need to communicate at the same time. Users gain not only economy but also the flexibility of being able to communicate with a large number of other users.

Private-line systems use dedicated, point-to-point lines between the communications endpoints. The line is always available, but it is fixed in connectivity so that only the users connected to that line are able to use the point-to-point

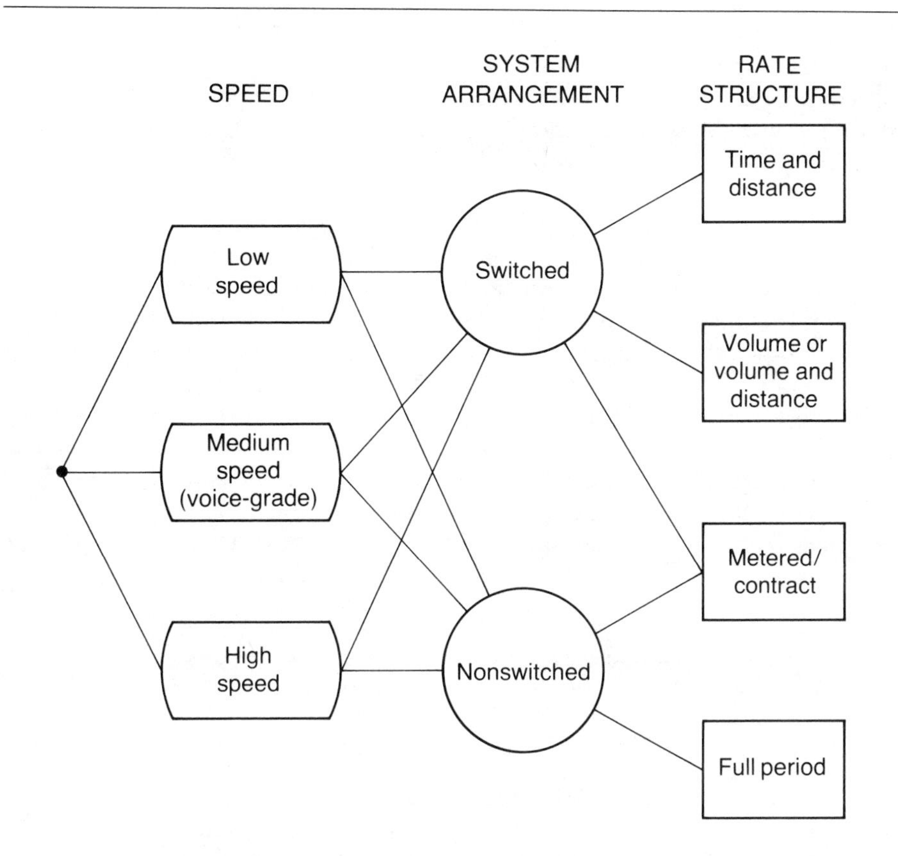

Figure 17-1. Interrelation of common carrier services.

capacity. When the dedicated users are idle, the capacity is wasted. It is possible, however, to lease point-to-point circuits on a dedicated basis and to use customer-owned switches, processors, or controllers to effectively establish a switched network at the customer level. This is the essence of what have come to be known as value-added networks (**VANs**). As far as the leasing carrier is concerned, the leased circuit is permanently terminated at the customer endpoints, even if those endpoints happen to be the customer's switches.

Rate Structure. The variety of rate structures makes it very difficult to compare the services of the many common carriers. Transmission rate, distance, data quantity, connection time, total communications load, time of day, geographical locations, and a host of other factors all must be taken into account. For simplicity Figure 17-1 indicates four broad rate structure classifications.

Full-period service, applicable only to private-line system arrangement, permits 24-hour-per-day, 365-day-per-year usage of the line. Private-line services are also provided on a metered, or monthly contract, minimum-usage basis. The fact that usage is metered implies that, although the user thinks he has a dedicated private line, the carrier is somehow able to use that line (or portions of it) for other customers and is thus willing to offer a lower rate than for full-period service. Metered and monthly minimum contract services are also available using the switched networks, the most common example being the Wide Area Telephone Service (**WATS**) for nationwide long-distance ("toll-free") calling.

Many rate structures are sensitive to time, distance, and possibly time of day. The best example is the direct-distance dialing (**DDD**) service of the phone company, where the cost of the call depends on its length (in minutes), the distance between the end offices completing the call, and the time of day and day of the week that the call takes place.

Finally, services that are tailored primarily to data communications have rate structures based primarily on volume or a combination of volume and distance. This structure is especially applicable to packet switching services, where usage charges are based on the number of packets actually transmitted. Note that volume is not synonymous with time; volume of data involves a combination of usage time, data rate, and idle periods. The value-added and satellite services have been tending toward rate structures of this type.

COMMON CARRIERS IN THE UNITED STATES

Suppliers of POTS

The common carrier market in the United States consists of more than 1500 telephone and telecommunications companies. Most of these companies are very small, independent phone companies, supplying POTS service (plain old telephone service) to a town, region, or community.

Approximately 82% of the telephones in the United States are served directly by the American Telephone and Telegraph Company (AT&T) through one of its 21 operating companies. AT&T also owns the Western Electric Manufacturing Company and Bell Telephone Laboratories, which do the research, development, and product manufacturing for the AT&T operating companies. AT&T supplies most of the nation's interstate long-haul communications through the AT&T Long Lines Department.

About 15% of the nation's phones are served by the General Telephone and Electronics Corporation (GT&E), also through local and regional operating companies. GT&E owns Lenkurt Manufacturing Company (transmission equipment), Automatic Electric Company (switching equipment), and Sylvania Electronics. GT&E Research Laboratories complete the corporate picture.

The remaining phones in the United States are served by about 1500 other local and regional companies, some as small as manual offices serving only 50 subscribers. The key point here is that, within its areas of jurisdiction or franchise, each of these companies acts as a regulated monopoly supplier of POTS service. In addition, all of these companies use compatible technical standards, procedures, protocols, and parameters to insure total interoperability and interconnectivity of telephone equipment anywhere in the total system.

Specialized Services

By contrast, competition flourishes in the so-called specialized services market. Specialized services is a catch-all phrase used to describe all communications services (voice and data) that do not require the total interconnectability with the rest of the world's telephones associated with POTS service. Principal offerings originally stemmed from data communications services and were designed for computer-based applications, but the specialized market has rapidly expanded to include voice, video, graphics, facsimile, teletype, message, and other communications functions.

Many specialized carriers, local and regional as well as national, operate within the United States. Tables 17-1 through 17-4 present a very brief sketch of

Table 17-1. Packet Switching/Data Communications Carriers

GTE TELENET COMMUNICATIONS CORPORATION

Nationwide value-added, switched service network
First commercial packet switcher
Distance independent services—volume-dependent costs
Gateway services to non-U.S. networks
Electronic mail services
Virtual private networks
Tariffed in 90 U.S. cities (February 1980)

TYMNET, INCORPORATED

Nationwide value-added, switched service network
Evolved from Tymshare time-sharing ADP services
Distance independent services—volume-dependent costs
Direct and gateway services to non-U.S. locations
Electronic mail services
Virtual private networks
Service to over 200 metropolitan areas (February 1980)

GRAPHNET COMMUNICATIONS

Nationwide value-added, switched service network
Provides graphics and facsimile services
Limited gateway services to non-U.S. locations
Provides data format processing and conversion
Service to approximately 50 cities (January 1979)

services offered by the nationwide specialized carriers. It should be noted that, in virtually all cases, AT&T—either directly or through its operating companies—can offer a competitive service. The only exceptions, at the present time, are direct point-to-point satellite transmission and packetized value-added services. Fur-

Table 17-2. Domestic Satellite Carriers

AMERICAN SATELLITE CORPORATION

Joint venture of Fairchild Industries and Continental Telephone
First U.S. domestic satellite carrier
Uses capacity leased from various satellite spacecraft
First to use customer-premises earth stations
Service to more than 50 locations, both dedicated and shared, by 1979

COMSAT GENERAL CORPORATION

U. S. domestic subsidiary of international COMSAT Corporation
Presently a "carrier's carrier"—capacity leased to GT&E, AT&T
Owns and operates three COMSTAR satellites
Provides direct services via one-third partnership in SBS

RCA AMERICAN COMMUNICATIONS CORPORATION

Owns and operates two satellites—SATCOM I & II
Major distributor of cable TV services to customer-premises (cable TV operator–owned) earth stations
Private-line services via shared and customer earth stations
Wideband (56 Kb/s) data point-to-point services

WESTERN UNION CORPORATION

Owns and operates three satellites—WESTAR I, II, and III
Customer-premises earth stations or microwave interconnect
Private-line voiceband and wideband digital services
Leased transponder as well as leased channel services

SATELLITE BUSINESS SYSTEMS (SBS)

Partnership of IBM, Comsat General, and Aetna Insurance
Operations commenced in early 1981
Integrated digital voice, data, facsimile service
Customer-premises and shared earth stations
Service to 20 cities (1981), 80 cities (1982)

Table 17-3. Domestic Terrestrial Carriers

MICROWAVE COMMUNICATIONS, INCORPORATED

First authorized specialized interstate common carrier
Nationwide private-line service serving more than 70 cities by 1980
Switched voiceband services—EXECUNET
Extensive private terrestrial microwave network
Primarily uses phone company services for local access

SOUTHERN PACIFIC COMMUNICATIONS CORPORATION

Microwave carrier using railroad right-of-way
Extended to nationwide service with acquisition of DATRAN assets
Nationwide private-line services to more than 70 cities by 1980
Switched voiceband services—SPRINT
Extensive private terrestrial microwave network
Uses phone company services for local access

ITT DOMESTIC TRANSMISSION SYSTEMS, INCORPORATED

IT&T entry into domestic private-line and switched voice services
Evolved from IT&T acquisition of United States Transmission Services
Switched voiceband services—CITY-CALL
Domestic value-added services—data, facsimile, and message
FAXPAK and electronic mail services
Gateway services to non-U.S. locations

thermore, a number of technically advanced services are currently under development by various organizations, including the Advanced Communications Service (ACS) being developed by AT&T (see Chapter 20).

Another attribute of the specialized common carriers is that they can limit their services to only those areas where the market is large enough to support the entry of new carriers. Because not all carriers operate in all cities and states, the applicability of many of these services depends on the carrier's area of coverage. It is common, therefore, for users to lease a point-to-point circuit from AT&T to carry the connection from the user's location to the nearest service point of a particular carrier. Even if the specialized carrier is tariffed directly to the city where the customer has his facilities, the connection from the customer's premises to the specialized carrier's offices has to be supplied by the local phone company (generally, either AT&T or GT&E).

Table 17-4. Overseas Gateway Carriers

ITT WORLD COMMUNICATIONS, INCORPORATED

Major gateway carrier between domestic services of many countries
Provides telex, telegram, leased channel, and message services
Universal Data Transfer Service provides packetized gateway services between Telenet, Tymnet, and foreign networks

RCA GLOBAL COMMUNICATIONS CORPORATION

Major international carrier among 200 locations in U.S. and overseas
Provides telex, telegram, leased channel, and message services

WESTERN UNION INTERNATIONAL, INCORPORATED

Major gateway carrier between domestic services of many countries
Provides telex, telegram, leased channel, and message services
Handles satellite maritime, television, and voice/data services
WUI Database Service provides packetized gateway data facilities between U.S. and about 20 foreign countries
Provides high-speed data services via satellite

TRT TELECOMMUNICATIONS CORPORATION

International carrier between United States, Caribbean, Latin America
Provides telex, telegram, and message services
Produces and distributes STORTEX—high-speed telex system

COMMUNICATIONS SATELLITE CORPORATION (COMSAT)

Furnishes satellite services to common carriers providing satellite service between U.S. and foreign countries
Owns 23% interest in International Telecommunications Satellite Organization (INTELSAT)
Operates major research laboratory devoted to satellite technology

SUMMARY

The multitude of specialized common carriers, each with its own features, service areas, and attributes, leads to a confusing picture of the best way to meet telecommunications needs. In addition, in most countries outside the United States telecommunications are provided by government-owned monopolies or are under much more restricted competition. Finally, the rapidly evolving technology, regulatory environment, and competition create very rapid changes in the pricing and tariffing of the communications services.

This chapter introduced the variety of available telecommunications facilities, classifying them according to speed, system arrangement, and rate structure. We then looked at the common carriers in the United States. POTS is supplied by regulated monopoly companies, primarily AT&T. By contrast, specialized services are provided by many different companies offering a great variety of applications.

SUGGESTED READINGS

CASWELL, STEPHEN A. "Coming to Grips with Planned Carrier Services." *Data Communications*, vol. 8, no. 11 (November 1979), pp. 48–54.

This paper introduces the major new planned offerings in the common carrier market from AT&T, Xerox, and Satellite Business Systems, and places them in context with the presently available services. It shows how these new services might be applied and how present services can meet many of the same requirements. Excellent graphical presentations are included.

SHAW, LOUISE C. "Data Communications Carriers." *Datamation*, vol. 26, no. 8 (August 1980), pp. 107–112.

This article highlights the companies that provide data communications services, primarily in the United States or at gateways between the United States and various other countries. In addition to a brief description of the services each company provides, the article provides key financial data.

TAYLOR, CAROL A., and WILLIAMS, GERALD. "Considering the Alternatives to AT&T." *Data Communications*, vol. 9, no. 3 (March 1980), pp. 47–62.

This article provides a comparative description of Tymnet, Telenet, American Satellite, and Western Union Satellite services. The overall technology utilized by each of the carriers is briefly described, along with information on the service areas, and basic tariffs and charges for each service.

18

Value-added Networks (VANs) and Packet Switching

THIS CHAPTER:

will describe the value-added network and the role of packet switching in its development.

will look at the Telenet packet network—its structure, operation, user interface, and service costs.

will look at the Tymnet network—its "packetlike" structure, operation, user interface, and service costs.

will briefly describe packet switching services in countries other than the United States.

Of all of the communications common carriers mentioned in Chapter 17 only a few employ the packet switching technology and principles discussed throughout the first four parts of this book. Some of the carriers use a "packetlike" technology, and others treat the data blocks that flow according to some of the standard protocols as if they were packets. Up to this time, no commercial carrier services have attempted to exploit the flexibility and features attained by combining packet switching and satellite broadcast communication.

Through the commercial implementation of packet switching in the United States the concept of a value-added communications network has evolved. VANs and packet switching have become complementary approaches that allow a particular technology to serve a unique segment of the communications marketplace.

PUBLIC NETWORKS AND THE ROLE OF VANs

The concept of value-added networks developed in the early 1970s out of two specific needs of data communications users. The first need was to efficiently utilize the broad-bandwidth, high-capacity communications facilities that were available from the common carriers at a low unit cost. The second was to adapt primarily analog communications facilities to the unique characteristics of rapidly increasing numbers of digital data communications users.

Economic Considerations

Public network **tariffs** available in the late 1960s and early 1970s, particularly for full-period, point-to-point services, carried impressive economies of scale and bulk discounts for high-bandwidth and high-capacity circuits. For example, under AT&T's **TELPAK** tariff alone, capacities equivalent to 60 or 240 voice channels were available at unit costs two to four times less than costs on a single-channel basis. Restrictions in the tariffs precluded the resale and redistribution of such bulk-discounted capacities for profit, but the tariffs did permit the cooperative leasing and sharing of such facilities among user groups. The public resale of communications capacity is still an active technical and policy issue in the common carrier environment. However, utilizing high-bandwidth facilities as the basis for more comprehensive and more elaborate communications services appeared to most observers to be legal.

Adaptation to Digital Communications

At the same time, all carrier communications were being handled on a purely analog basis. Even where digital channels were beginning to be used, they were being applied on a voiceband and analog equivalent basis. However, transmission impairments that were acceptable to voice users—such as the accumulation of noise along an extended route or an occasional electrical transient—have serious deleterious effects on data communications users. Single-channel remedies for these ill effects were both inefficient and expensive. Enhanced processing techniques on a facilitywide and shared basis gave analog and broad-bandwidth facilities significant potential for reducing their shortcomings.

The ARPANET's successful experiment in resource sharing and enhanced operational reliability of wideband services applied to data communications provided the technological foundation for value-added networks based on packet switching. Communications lines operating at 50,000 bits per second were individually too expensive and too unreliable to operate between a single pair of computers. However, when they were linked among interface message processors, which shared the capacity and provided multiple connections and adaptive routing, large-scale computer internetworking became feasible and cost-effective.

Government Regulations. The prohibition of division and resale of communications capacity acquired in bulk from the established carriers appeared to preclude the commercial use of packet switching to provide this shared service. Early commercial entrants into this field—Packet Communications, Incorporated,

ITT Domestic Transmission Systems, and Telenet Communications Corporation, without using the term *value-added network*, asked the U.S. Federal Communications Commission to rule specifically on the legality of such offerings.

Although Packet Communications, Incorporated, disappeared as an entrant, and the ITT system was slow to develop, favorable FCC rulings in 1973 and 1974 allowed Telenet to develop as a viable data communications common carrier. The overriding argument in support of this approach was the need for a second-tier carrier to add services and features to the bulk facilities leased from the first-tier carrier in order to add value to the communications services seen by the end user. Notice that we are dealing exclusively with a communications service, as opposed to a remote time-sharing or teleprocessing service. Numerous vendors, such as Tymshare, General Electric Time Sharing, and Infonet, had been providing data communications services to end users, but only as a necessary adjunct of their own remote computing services. While intelligent processors are very much a part of value-added services, in essence they provide the means for supplying the service, rather than supplying the service itself.

Defining the Concept of VANs

The concept of adding services to those supplied by a primary carrier led to the formal definition of a value-added carrier as one who "adds value" to the service or facilities of one carrier to meet the specific needs of a retail end user. The resulting service must be distinct from the basic offerings of the underlying carrier. In the case of a packet switched data carrier, error control, enhanced connection reliability, dynamic routing and failure protection, and logical multiplexing are all examples of service features that add value to the basic intercity transmission medium.

Packet switching is by no means the only basis for a VAN. For example, IT&T's **FAXPAK** service adds value to the communications medium by doing format conversions between different facsimile terminals. A pure communications function is performed—that is, transmitting an image from one terminal to another—but considerable processing of the transmitted data may be required in the process.

Questions quickly arise as to the boundaries between value-added services, pure communications services, and data processing. Message mailbox and electronic mail systems appear to meet the broad definition of a value-added service. But if messages are stored in semipermanent files, or if message formats are stored in the network and can be locally modified by the users, the service begins to look at least partially like a data processing, rather than a direct communications, function. The importance of such distinctions is that they become the boundary of jurisdictional responsibility of the Federal Communications Commission, as well as rulings and regulations limiting the operation of the franchised common carriers to communications and not data processing activities. At this point it is not clear how far VANs will eventually go in providing unique and highly marketable communications, teleprocessing, and information-based services.

THE TELENET PACKET SWITCHED NETWORK

The **GTE Telenet** commercial packet switched network was developed as a commercial venture of many of the same principals who developed the ARPANET. The capital-intensive nature of the network led to its acquisition by General Telephone and Electronics in 1979. With a capital and marketing base provided by the second-largest communications carrier in the United States, GTE Telenet will be expanding network coverage and services such as electronic mail and private networks, in addition to providing public packet switched data services.

Operation. Telenet's network operation and internal protocols evolved from the ARPANET experience, with additional capabilities and redundancies built into each of the switching nodes. The network is a virtual circuit–based packet switching protocol, meeting the requirements of the CCITT X.25 protocol at the user interface. In addition, Telenet provides customized user interfaces to meet the needs of individual users. It also provides emulation (simulation) interfaces, which interpose the packet network on a transparent basis in the communications line by emulating the computer responses to the terminals and the terminal responses to the computer. The latter situation is illustrated in Figure 18-1, where the network functions as a multistation controller and as a terminal multidrop polling control station.

User Access. User access to the network is through one of three classes of Telenet Central Offices. Class I offices, such as those in Boston, Dallas, and San Francisco, support user access speeds up to 56,000 bits per second. Class II offices, such as those in Detroit, Miami, and Spokane, provide connection speeds of up to 9600 bits per second. Class III offices support rates up to 1200 bits per second; such offices are located in Cincinnati, Philadelphia, and Tucson, as well as many other cities across the country. User access can be made to public dial-in ports (connections), private dial-in ports, or fixed ports dedicated on a full-time basis to a single user. Users can implement X.25-compatible software in their host computers, or they can utilize Telenet-provided interface processors to provide network transparent service. Terminal clusters can be accessed to the network very efficiently by use of Telenet access controllers placed at the customer's premises.

Tariffs. Telenet charges to the users are determined by a network access charge and a traffic-based charge. The access charge for dedicated ports is a fixed monthly amount based on the bit rate of the access port. For public dial-in ports the charge

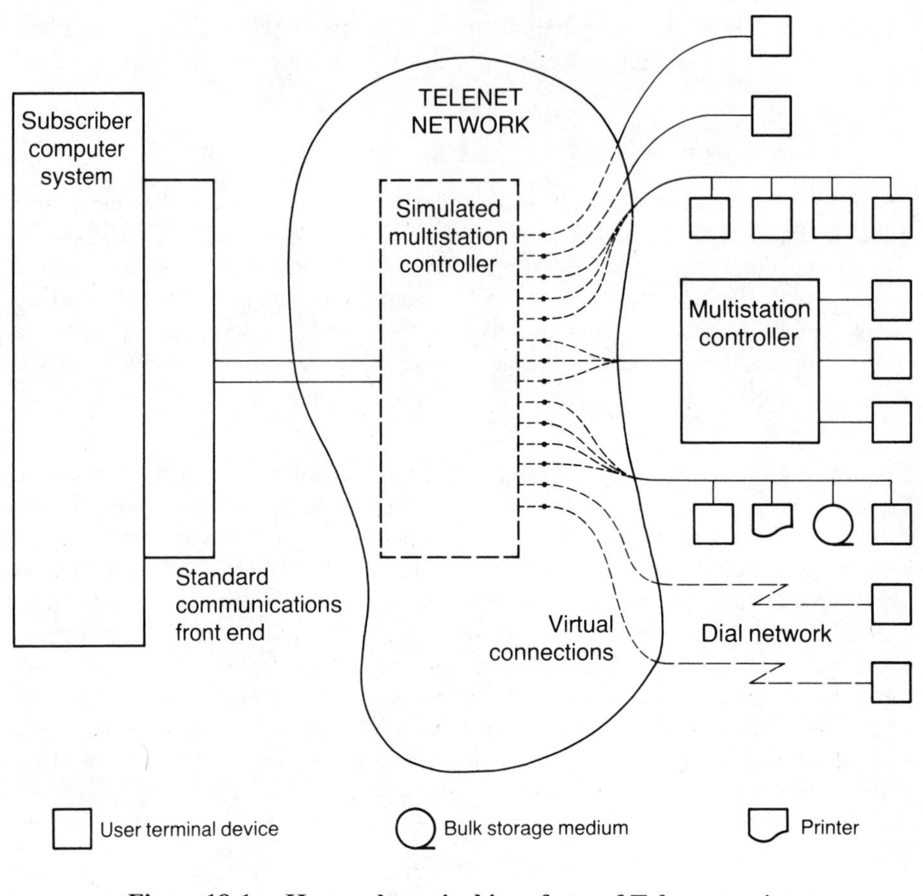

Figure 18-1. Host and terminal interfaces of Telenet carrier packet network. (Courtesy GTE-Telenet, Inc.)

is based on both port speed and the amount of usage at an hourly service rate. The traffic-based charge is calculated on the number of data packets actually transmitted, with a packet containing anywhere from 1 to 128 data characters. Table 18-1 illustrates the Telenet rate schedule extracted from the Telenet tariffs filed with the FCC. Additional charges are made for customer-premises network interface equipment (Telenet processors) and optional service features, such as additional privacy identification or detailed usage reporting. In addition, various discounts based on monthly billings, reaching as much as 50% for monthly billings over $18,000, are applicable. It is important to remember that this rate schedule is included only as an example. Actual rate structures and specific charges of course will change with time and conditions.

Table 18.1

GTE Telenet Public Network Rate Schedule

Abstract of GTE Telenet Tariff FCC No. 1 effective November 1, 1980

I. Network Access Charge

Dedicated Access Facilities

The rates below include a leased channel port at a Telenet Central Office (TCO), the access line between the customer's location and the TCO, and the associated modems or digital interface units.

Port Speed	Installation Charge[2]	Monthly Charge**
110-300 bps	$460	$ 320
1200 bps	575	360
1800 bps	635	475
2400 bps	690	600
4800 bps	805	800
9600 bps	920	1100
56,000 bps	1035	2100

Public Dial-in Service

	Port Speed	Hourly Charge
Local Dial[1]	110-300 bps	$ 3.90*
	1200 bps	3.90*
In-WATS	110-300 bps	17.00
	1200 bps	17.00

Note 1: A **Nightline** rate of $.75 per hour, including up to two thousand packets of user data, applies between 6 p.m. and 7 a.m. on weekdays (local time at the TCO where the call originates); all day Saturday and Sunday; and on certain holidays. There is a six minute minimum charge per call and a $7,500 overall minimum charge per month for connection time.

Private Dial-in Service

Port Speed	Installation Charge[2]	Monthly Charge*
110-300 bps	$370	$190
300 bps (Bell 407C modem)	390	240
1200 bps	390	260
TWX	390	250

Private Dial-out Service

Port Speed	Installation Charge[2]	Monthly Charge*
110-300 bps	$485	$360
TWX	485	360

Table 18.1 GTE Telenet Public Network Rate Schedule (*continued*)

Private Packet Exchange Service (PPX)

A special access arrangement enabling a customer to buy a group of ports and associated facilities at any Telenet Central Office for the exclusive use of his organization. Overflow calls to PPX dial-in ports are automatically switched to Telenet public dial-in ports and charged at the hourly rate.

	Installation Charge	Monthly Charge
Packet Exchange Control Arrangement	—	$400
Local Switching Option	$400	400
Dial-in ports/each		
110-300 bps	140[3]	70[4]
1200 bps	160[3]	110[4]
Dedicated access facilities/each (Intraexchange only)		
110-300 bps	160[3]	110[4]
1200 bps	230[3]	155[4]
Leased access ports/each		
2400-9600 bps	85[3]	85[4]
56,000 bps	230[3]	200

Note 2: When multiple private dial ports at the same TCO location, or when multiple dedicated access facilities terminating at the same customer address, are ordered at the same time for subsequent concurrent installation, a $200 discount applies for each nonrecurring installation charge beyond the first one.

Note 3: A $200 charge applies (in addition to installation charge) to the initial order or to any change of service per PPX.

Note 4: A minimum monthly charge for four ports per category applies.

II. Traffic Charge

Regular Service*

$.50 per thousand packets. Each packet contains up to 128 characters of user data.

Hotline Data Service*

An optional service arrangement providing for a fixed monthly traffic charge in lieu of packet charges for all traffic between two specific network stations.
Monthly Charge:
$60.00—110-300 bps ports
$90.00—1200 bps ports

THE TYMNET NATIONWIDE DATA NETWORK

Development

The public data network services provided by **Tymnet** grew from the private intercomputer network started in 1970 by Tymshare, Incorporated. The purpose of this network was to link Tymshare's geographically dispersed computers and to permit their time-sharing customers low-cost access to any one of those computers. Computers belonging to companies other than Tymshare were brought into the network on a limited basis under the shared usage provisions of the applicable communications tariffs. By 1976 about 60 non-Tymshare computers were included. With the FCC's 1975 approval of the Telenet operation as a precedent, Tymnet, Inc., was formed as a subsidiary of Tymshare. In December 1976 the FCC approved Tymnet's application to operate as a data common carrier, and Tymnet began providing tariffed services in April 1977.

Functions and Applications

For many user functions and applications Tymnet and Telenet are directly competitive; however, there are significant differences in the philosophy and technology of the two networks. Tymnet services are primarily terminal oriented, with the protocols and data structures optimally matched to user terminal devices available from many vendors. Network connections are available up to 9600 bits per second, as opposed to the maximum rates of 56,000 bits per second available from Telenet. Because of these lower rate limits, most of Tymnet's computer interconnects are designed to support remote terminal operations rather than bulk computer-to-computer transfers.

Operations. Telenet employs packet switching techniques based on the principles we have developed throughout this book. By contrast, all routes through the Tymnet network are centrally controlled and directed, and the user information is controlled through the network on a character by character, rather than a packet, basis. User connection to the network is initiated by a request that goes to a centralized network controller. Multiple redundant controllers are used for reliability, but at any given time a single controller is in full command of the network operation. Having full knowledge of the connectivity and traffic status of the network, the controller can assign a route for the new connection, which will remain fixed for as long as that connection exists.

Table entries made in each of the nodes along the route reference particular switch buffers to each connection. User data is carried in the form of characters or 8-bit **bytes**, in logical records. Each logical record needs to carry as overhead only the logical record number, which acts as the cross-reference to the buffer and table entries, identifying the connection with which this data is associated. At each node individual logical records are grouped together, forming an internodal

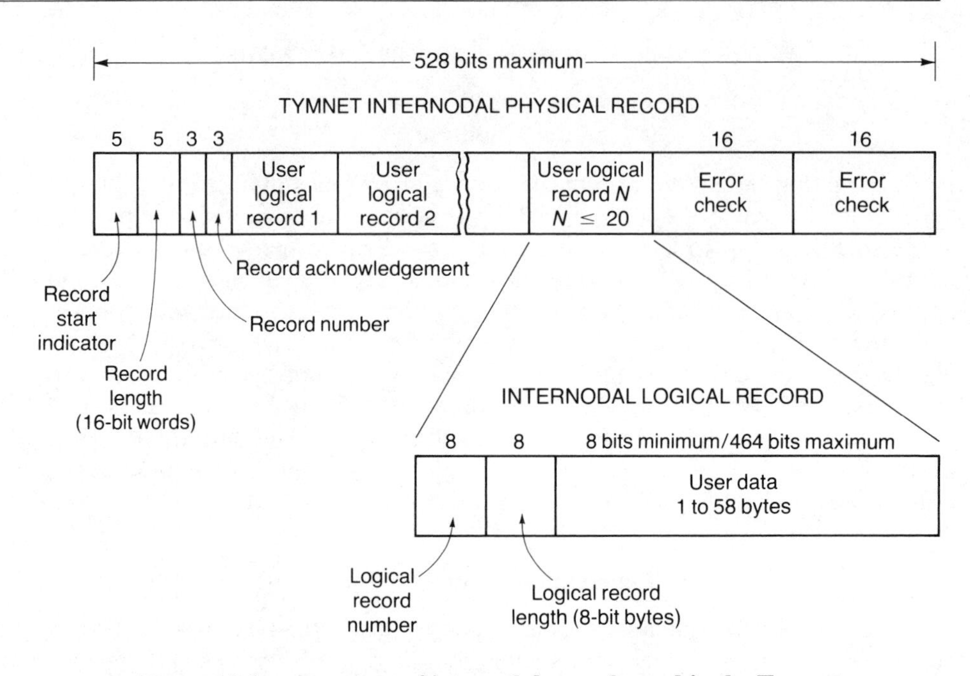

Figure 18-2. Structure of internodal records used in the Tymnet data network.

block or internodal physical record, for transmission between the nodes. An internodal physical record may be as long as 528 bits, which include the overhead, acknowledgement, and error-check information. The structure of Tymnet's logical and physical internodal records is shown in Figure 18-2.

For a single high-speed user, operating under relatively light network loading, a single internodal physical record may contain 64 bits of overhead and up to 464 bits of user data from a single user connection. At the other extreme—with very low-speed users—a single physical record may contain up to 20 logical records, from each of 20 different user connections, each logical record containing 16 bits of overhead plus a single 8-bit character of user information. Internodal physical records are assembled and disassembled at each node of the network. If we were to trace the transmission of a single block of user data through an end-to-end connection, we would likely see that data grouped together with other user data at each node and on each link.

To the external observer the physical internodal records look like packets, but the overall network operation is considerably different from that which handles complete packets as distinct entities on an end-to-end basis. The major advantage of the Tymnet operation is that, for low-speed, character-by-character users, good line efficiencies can be achieved without incurring the delays associated with waiting for a single user to fill a packet.

Services and Tariffs. The Tymnet network provides the full range of value-added features and user access adaptations. As with Telenet, both public and private dial-in ports, as well as synchronous and asynchronous dedicated host ports, are available. In addition to Tymnet-unique synchronous interfaces, the network will also support the X.25 packet switching standard interface. Services are provided throughout the United States, and **gateway** services are provided to data carriers in more than 30 other countries. Direct access is provided at more than 170 metropolitan areas, and within the United States toll-free inward WATS service is provided from any possible user location.

Again as with Telenet, the cost of service is composed of an access (port) charge, usage (volume) charge, and a charge for any network interface equipment that may be provided to the customer. The usage charge is computed on the basis of an actual character count of the transmitted data. The rate of $0.03 per thousand characters compared to Telenet's $0.50 per thousand packets indicates a substantial difference in network application.

For example, a data quantity of 125,000 characters (nominally, 1 million bits) sent in the Tymnet network would have a usage charge of $3.75. The same data sent through Telenet in the most efficient way, as nominally 1000 full packets, would incur a usage charge of only $0.50. However, the same data sent inefficiently by the user, at an average of, say, 5 characters per packet, would incur a usage charge of $12.50. In addition, the nature of the user application may make it difficult to efficiently use the 128-character capacity of a full packet. Consequently, the comparative cost of Tymnet and Telenet services is highly application oriented.

WORLDWIDE PACKET SWITCHING SERVICES

Development and Extent

The development, deployment, and operation of packet switched networks is by no means unique to the United States. Conceptual development of the ARPANET in the late 1960s was closely paralleled by similar research at Great Britain's National Physical Laboratory (NPL) led by Donald Davies. These experiences encouraged the practical implementation of commercial services in Canada (DATAPAC), Great Britain (Experimental Packet Switching System, EPSS, later to become Public Packet Switching System, PSS), France (CYCLADES and later TRANSPAC), Spain (Spanish PTT), and other countries. Quasi-commercial networks or cooperative networks, such as the European Informatics Network (EIN), spread both the technology and the applications of high-speed data communications exchange systems and packet switching throughout the continent.

By the end of the 1970s packet switched or packetlike networks were being developed or at least planned by the communications administrations of most of the European countries, many Latin American countries, Japan, Australia, and other nations. In addition, arrangements had been made in conjunction with the overseas gateway carriers to extend the data services of Telenet, Tymnet, DATAPAC, TRANSPAC, and other networks into many countries without immediate capabilities for national packet switched networks. International cooperation through the CCITT has rapidly led to agreement on a standard—designated X.75—which allocates functions and responsibilities for the interfacing of different data communications networks.

Numerous vendors operating on a worldwide basis, such as Siemens (German), SESA (French), Tran Telecommunications (U.S.), Northern Telecommunications (Canada), and others have developed product lines of communications processors that permit users to establish self-contained private packet or hybrid packet/circuit switched networks within their own organizations. This approach is particularly valuable for organizations that employ equipment of different computer vendors or different product generations.

Services

International data communications services are evolving for more than point-to-point interchange of digital data. Electronic mail, message retrieval, and videotext information systems have been introduced on a service or trial basis in many countries.

Applications-oriented services and highly efficient data communications exchange services are particularly important in areas outside the United States, or on international links, because geographical, competitive, and political factors tend to make the fundamental communications channels quite expensive. Packet switching services and other techniques to make nonvoice user telecommunications more efficient and less expensive are rapidly being implemented throughout the world.

SUMMARY

This chapter has looked at the application of packet switching technology to operational nationwide and worldwide data communications networks. We found that packet switching was directly linked to the evolution of value-added communications networks. The development of VANs and increasing regulatory encouragement of telecommunications competition together have resulted in highly cost-effective communications services to support distributed data and information processing systems.

We considered the Telenet and the Tymnet networks, indicating both their parallels with and departures from the principles of packet switching. These two networks presently carry the substantial portion of the publicly switched data communications traffic in the United States, but the future looks bright for a number of new systems, which we will explore in Chapter 20.

Recent developments of packet switched networks have made many communications services available throughout the world. In addition, it is increasingly possible for organizations to implement private, distributed packet switched systems.

SUGGESTED READINGS

CRINER, JAMES C. "What Is Value Added Network Service?" *Telecommunications*, vol. 11. no. 10 (October 1977), pp. 45–52.

Criner, a staff member of the U.S. Federal Communications Commission, traces the legal and regulatory evolution of value-added communications carriers in the United States. The paper shows the similarities and differences between value-added services, capacity resale, capacity brokerage, and composite data service vendors, all of which have had an impact on the development of public switched data communications services.

HARCHARIK, ROBERT. "The International Spread of Packet-Switching Networks." *Telecommunications*, vol. 13, no. 9 (September 1979), pp. 103–104.

National packet networks are growing in France, Great Britain, Japan, and other countries, and Tymnet is playing a significant role, through the international record carriers, in providing interconnections among those systems and the United States' systems. The paper describes the modular implementation of the network gateways and presents information on the range of service costs in the international environment.

HUBER, J. F. "Packet Switching by Siemens." *Telecommunications*, vol. 14, no. 7 (July 1980), pp. 47–52.

This article describes the application of the Siemens EDX-P packet switching system to the INFOSWITCH system of the Canadian National/Canadian Pacific (CN/CP) public data system. The hardware and software and the interfaces of the system are described. Of particular interest is the fact that the Siemens product line is only one example of commercial communications processors that can be employed by organizations to implement private or closed/shared packet switched systems.

MATHISON, STUART L. "Commercial, Legal, and International Aspects of Packet Communications." *Proceedings of the IEEE*, vol. 66, no. 11 (November 1978), pp. 1527–1539.

This paper traces and reviews the policy issues relating to the structure and regulation of national data and packet networks and the interconnection of national networks into an international packet switching system. The author highlights the legal and regulatory difficulties associated with distinguishing between nonregulated data processing services and regulated telecommunications services. The paper shows what policy and regulatory issues will have to be substantially solved before the various packet communications systems of the world will be unified in the same sense that the present telephone and telex systems are unified.

RINDE, J. "Tymnet I: An Alternative to Packet Switching Technology." *Proceedings of the Third International Conference on Computer Communications,* ICCC '76. Toronto, July 1976, pp. 268–275.

This paper describes the internal operations and protocols of the Tymnet network and traces the flow of physical and logical records through the network. Comparisons are drawn between the technology applied in Tymnet and the packet switching technology of the ARPANET.

TRIVEDI, ASHOK. "Emerging Services in International Public Packet-Switching Networks." *Communications News,* vol. 17, no. 7 (July 1980), pp. 26–27.

This article highlights the spread of packet switched networks and advanced information systems, such as videotext and teletext services, throughout the world. The role of IT&T is emphasized, and the ability to interconnect the overseas national PTT packet networks to both Tymnet and Telenet using IT&T's Universal Data Transfer Services is described.

19
A Tariff for Every Occasion

THIS CHAPTER:

will look at examples of existing tariffs for communications services.

will describe a projection technique for estimating what the tariffs for digital services might be over the next 25 years.

In the last chapter we found that the cost of commercial packet switched services had, in general, three components: (1) an access charge, which was associated with the use of the input/output port of the switch; (2) a usage charge, which was associated with the quantity of information actually transmitted and was generally distance independent; and (3) an equipment lease charge, which was associated with any special interface equipment that the carrier provided on the customer's premises. Possible additional charges might have to be paid to the local telephone company for access connections to the packet network.

In addition, we alluded to the possibility that a user could establish a closed, private packet switching network by acquiring the appropriate communications processors and leasing the needed high-capacity lines from a major common carrier. The discussion of tariff structures in this chapter will be useful in understanding the different alternatives for such an implementation.

A SAMPLING OF COMMON CARRIER COMMUNICATIONS TARIFFS

The tariffs presented here are for the purpose of illustration. While they were accurate at some point in recent history, they are subject to changes and modifications, so it would be impossible to insure that they are presently accurate. Individual carriers will provide a copy of their latest tariffs on request, together with any projected or pending changes to those charges. What changes more slowly is the structure of the tariff with regard to the service speeds, system arrangements, and overall rate structures. This sampling of actual tariffs also illustrates the difficulty of comparing similar, but not identical, services. The total cost to the user becomes very application dependent, based on total volume; average message length; distribution of usage over the day, week, and month; combinations of user access speeds; and so forth.

The Basic Tariff for Specialized Carrier Services

The basic tariff for specialized carrier services is the Series 260 Tariff maintained by AT&T for private-line, point-to-point service. An extract from that tariff, known as Multischedule Private Line (**MPL**), is shown in Table 19-1. The service provided is purely voiceband. In order to use such a line for digital data services, a modem (modulator/demodulator) or data set is required; this can be either leased from the phone company or purchased by the user. The tariff shows the rate per mile over different distance ranges and for various densities. The phone company estimates the density of both customers and facilities. When the density of both is high, the incremental cost of service is low, as reflected in lower tariffs. Part of the tariff is a list of cities and service areas that designates those service points considered to be high density; all others are considered to be low density.

Table 19-1. Monthly Cost Per Mile, Point-to-Point Voiceband Full-Period Circuit (as of January 1980)

	DENSITY		
Mileage	*High to High*	*High to Low*	*Low to Low*
First mile	$51.00	$52.00	$53.00
Next 15 miles (2–15)	1.80/mi.	3.30/mi.	4.40/mi.
Next 10 miles (16–25)	1.50	3.10	3.80
Next 15 miles (26–40)	1.12	2.00	2.80
Next 20 miles (41–60)	1.12	1.35	2.10
Next 20 miles (61–80)	1.00	1.35	1.60
Next 20 miles (81–100)	1.00	1.35	1.35
Next 100 miles (101–200)	.50	.50	.68
Next 800 miles (201–1000)	.40	.40	.40
Each add'l mile (1001 and up)	.40	.40	.40
Plus additional monthly charges for station terminal			

Note: These tariffs were subject to overall rate increases totalling approximately 35% by mid-1981.

Table 19-2. Monthly Circuit Cost, AT&T Dataphone Digital Service (DDS) and MCI Digital Private Line (as of January 1980)

Rate (KB/S)	*AT&T-DDS*	*MCI-Digital PL*
2.4	$218.30 + M	$160.80 + .87/mi
4.8	369.20 + M	160.80 + .87/mi
7.2	N/A	205.80 + .87/mi
9.6	611.86 + M	205.80 + .87/mi
56.0	1546.00 + M	N/A
1344.0	3400.00 + M	N/A

The DDS mileage charge (M) is computed from the following table:

Mileage	CHANNEL RATE		
	2.4–9.6	*56.0*	*1344.0*
First 15 miles (1–15)	$1.80/mi.	$9.00/mi.	$64.00/mi.
Next 10 miles (16–25)	1.50	7.50	64.00
Next 75 miles (26–100)	1.12	5.60	64.00
Next 100 miles (101–200)	.66	3.30	64.00
Next 300 miles (201–500)	.66	3.30	50.00
Next 500 miles (501–1000)	.66	3.30	40.00
Additional miles (1001 and up)	.40	2.00	40.00

Note: The AT&T tariffs shown here were subject to rate increases totalling approximately 35% by mid-1981.

Tariffs for Digital Services

Table 19-2 shows two tariffs for fundamentally digital services, the AT&T Dataphone Digital Service (**DDS**) and the MCI-provided digital Private-Line Service. These rates, like the Series 260 Tariff, are distance sensitive but are bit-rate sensitive as well. They provide fundamentally digital services, directly usable by data communications facilities. In actuality, these services are, for the most part, provided over analog facilities, with the data terminal devices provided by the carrier as an integral part of the service. More significantly, the performance specifications of these services are in terms that have particular utility to a data communications user. For example, performance for such channels would be

Table 19-3. Examples of Metered (Switched) Digital Services (as of January 1980)

Rate (Kb/sec)	*Datadial (SPCC)*	*Telenet Class 1 TCO Packet*
2.4	\$170 + \$2.00/mile* + 25¢/minute	\$600 + 50¢/1000 packets**
4.8	\$270 + \$2.00/mile* + 30¢/minute	\$800 + 50¢/1000 packets**
9.6	\$320 + \$2.00/mile* + 35¢/minute	\$1100 + 50¢/1000 packets**
56.0	N/A	\$2100 + 50¢/1000 packets**

* Fixed monthly mileage charge applied to access line from customers premises to SPCC service office only.

** A packet contains up to 128 consecutive data characters.

Note: GTE-Telenet was in the process of filing rate revisions as this went to press. Fixed charges are probably reduced, but usage charge of \$1.00/1000 packets is likely.

guaranteed on the basis of achieved bit error rate or "error-free seconds," rather than amplitude, phase, and noise-related parameters associated with voiceband channels.

Table 19-3 shows two representative switched-metered digital services provided by Southern Pacific Communications Corporation (SPCC) and GTE-Telenet. The tariffs shown assume, in effect, high-density service areas, where these two companies would have major switching offices. Note that the SPCC tariff is time sensitive, whereas the Telenet tariff is data quantity sensitive. Distance sensitivity enters only for the access line charges between the customer's premises and the carrier's serving offices at each end. Note again the difficulty in comparing two competitive services. For example, at 4800 bits per second it is possible to transmit, at most, 288,000 bits per minute. With the SPCC service a minute of usage entails \$0.30 as an incremental cost. If the Telenet service is used efficiently, filling each of the packets with 1000 bits of data, a minute's usage could generate at most 288 packets, with an incremental cost of less than \$0.15. On the other hand, if the packets are filled inefficiently, with an average of 100 bits per packet, the cost of a minute's transmission would be \$1.50.

At the other extreme, AT&T can provide limited services among a number of the larger metropolitan areas in the United States with switched, 56,000-bit-per-second digital tariffs. This tariff, the Dataphone Switched Digital Service (**DSDS**), is time and distance sensitive (Table 19-4). The incremental cost of one minute of service under this tariff is about \$0.69 at a distance of 1000 miles; as many as 3,360,000 bits could be transmitted during this time. Even grouped efficiently into full 1000 packets, this service would generate an incremental charge of about \$1.70 on Telenet. On the other hand, the DSDS service charges for the full connection period even if data is not sent continuously during that time. It would not be

Table 19-4. AT&T Dataphone Switched Digital Service (DSDS)—56,000 b/s (as of January 1980)

Distance	*Cost per Minute*
Up to 50 miles	$ .39
51–150 miles	.45
151–300 miles	.48
301–600 miles	.57
601–1200 miles	.69
1201–2000 miles	.90
Over 2000 miles	1.20

Add a fixed monthly charge of $275 or ($325 + $6/mi) depending on proximity to serving office.

unreasonable, during a 1-minute connection, to average an overall 25% channel utilization rate. With Telenet service this would generate about 850 efficiently filled packets, and a charge of about $0.42. Here again, depending on the particular application, there is a very large divergence between the two competitive services.

Tariffs Involving Satellite Facilities

Table 19-5 presents some representative tariffs for voice-grade private-line channels derived from satellite facilities. The tariffs shown are on an earth terminal–to–earth terminal basis; access line charges from the user endpoint to the satellite earth terminals at each end must be added. Note that, for users in the proximity of the earth stations, the costs can be substantially less than for terrestrially derived channels. There are potentially substantial discounts applicable to users of groups of 12, 24, 60, or more channels. For many users the installation of a customer-premises earth station can substantially reduce cost per channel as well as improve service quality and reliability.

Table 19-6 shows some representative international tariffs—a 75-bit-per-second teletype/data channel and a voice-grade channel authorized alternate voice and data service. Note that, on a per-mile basis, these tariffs are between five and fifteen times more expensive than the equivalent services within the United States. This is due partly to the expense and difficulty of providing transoceanic services, which until recently were mainly by extremely expensive undersea cables. Technical

Table 19-5. Some Representative Satellite Tariffs (American Satellite Corp., RCA, and Western Union) (as of June 1980)

		Cost per Month
Washington D.C. to	Atlanta	$540.00
	Los Angeles	1175.00
	Houston	1075.00
	Chicago	875.00
New York City to	San Francisco	$1175.00
Dallas to	Los Angeles	$775.00
	New York City	775.00
	San Francisco	875.00
	Pittsburgh	925.00

Local access charges must be added.

Additional Cost Factors		
Terminal charge (each end)		25.00
Installation charge (one-time, per end)		90.00
Quantity discounts:	12 channels	20%
	24 channels	30%
	60 channels	35%

advances in both cable construction and undersea cable laying has kept the cost of international undersea cable roughly comparable to satellite communications, and most countries have tried to maintain a balance between cable and satellite for their international telecommunications connectivity. Other reasons for the relatively high international tariffs are the complex political and regulatory arena in which these tariffs have to be negotiated, and the complexity of providing all the administrative and technical interfaces among both domestic and international carriers.

In any case, the sheer magnitude of the international tariffs encourage operating such channels at the maximum possible efficiency. It is also a factor in the difficulty of individual organizations' affording their own, dedicated facilities for overseas connectivity. These two factors have heavily influenced the rapid evolution and development of packet switched services and the international

Table 19-6. Representative International Tariffs (as of June 1979)

	75 Bits/Second	*Alternate Voice/Data*
Eastern Gateway to:		
Bahamas	$3000/mo.	$5000/mo.
Brazil	5223	13,057
Germany	3607	10,260
Iran	6338	15,846
Italy	3950	11,886
United Kingdom	3261	8683
Western Gateway to:		
Australia	$5232/mo.	$16,345/mo.
Hawaii	225	3770
Hong Kong	7050	17,624
Singapore	5734	18,442
Japan	7072	21,303

agreements to CCITT X.25 and X.75 protocols as an efficient usage-based methodology for international services.

A CURRENT AND FUTURE TARIFF COST MODEL

Despite the difficulty of doing so, it is often necessary to project costs for initial network feasibility studies and for network design algorithms. Uniform models can help designers allow for the tremendous uncertainties in the economic, technical, and competitive environment of leased telecommunications and establish a range of confidence in the results of a design or economic analysis.

Assumptions of the Model

In order to project tariff costs for digital transmission services into the future, we will use known datapoints within existing tariffs and extrapolate over the range of approximately 4 to 64 kilobits per second. This will involve an averaging of purely digital tariffs with private-line voiceband tariffs in combination with data modems. An estimated tariff curve can then be extrapolated for both high- and low-density service areas, and for the mileage (distance-dependent) costs and the

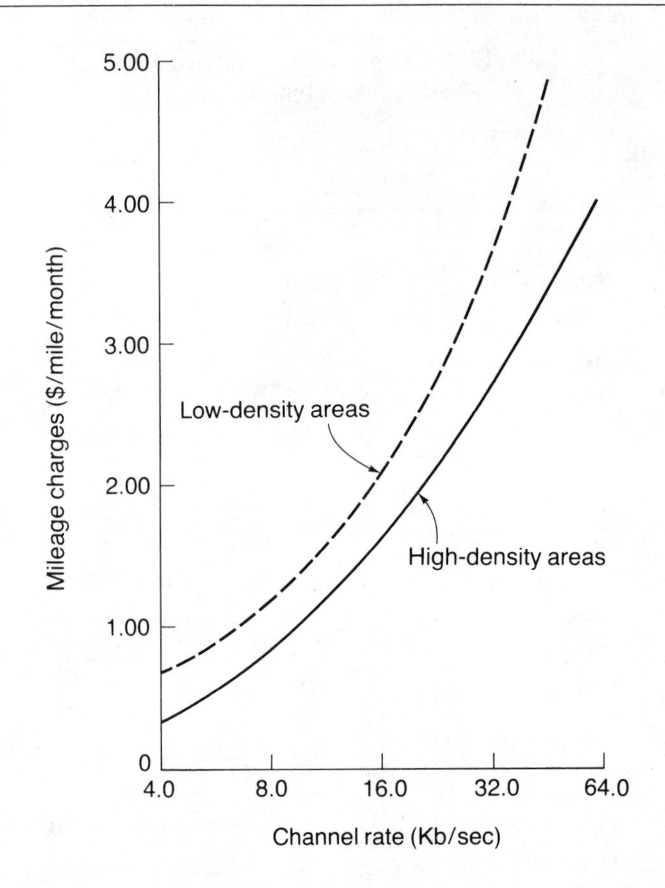

Figure 19-1. Plot of estimated mileage charges as a function of channel bit rate.

termination (connection-dependent) costs. The resulting extrapolated tariff curves are shown in Figure 19-1 (high- and low-density mileage costs) and Figure 19-2 (high- and low-density termination costs).

Since any analysis of networks, packet switched or otherwise, only makes economic sense if it is extended over the life cycle of the network, it is necessary to project these tariffs into the future. The approach we will take is based on the technology employed to derive the digital channels. To simplify the analysis, we will disregard the impact of regulatory and competitive pressures and limit ourselves to the situation within the United States.

Digitizing Transmission Facilities

Our projection is based on the fact that the common carriers are converting and replacing much of the analog transmission plant with digital transmission

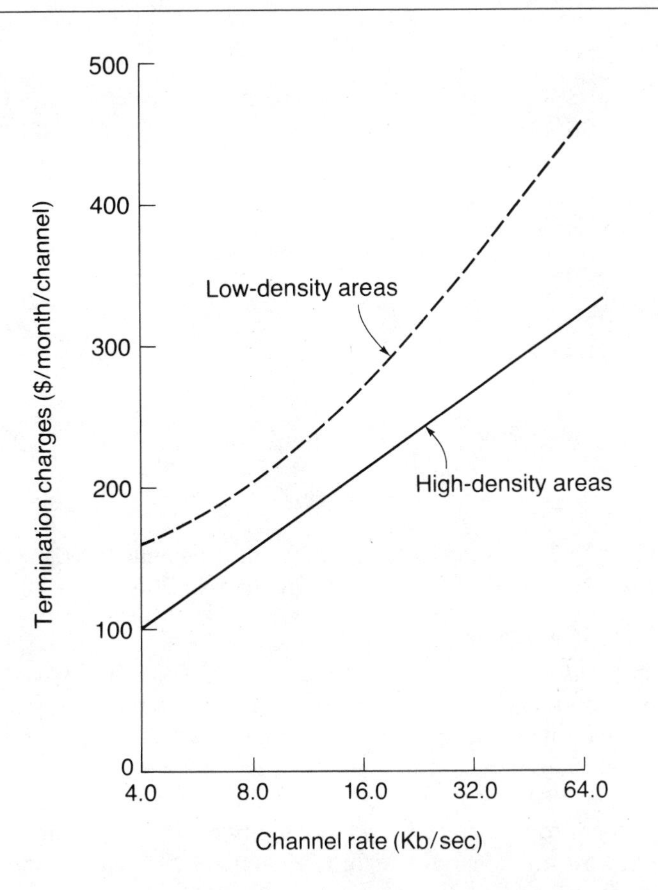

Figure 19-2. Plot of estimated termination costs per voice channel as a function of channel bit rate.

facilities known as pulse code modulation (**PCM**). This is being done because the technology is making it cheaper and more reliable to convert analog voice conversations to 64,000-bits-per-second digital streams, and to transmit these digital signals over the long-haul communications plant. Figure 19-3 estimates the percentage of common carrier plant that will be digitized, using PCM-based transmission equipment, by the year 2000. The digitization of the common carrier plant began in the early 1960s with certain short-haul applications, particularly in high-density areas for distances up to about 50 miles. Between 5 and 10% of the total plant was digital by 1970, and about 25% by 1980. Application of digital technology will continue to increase as both the demand for digital services and the technology to meet that demand increase. The plant may never become 100% digital. However, the implementation of a digital/electronic telephone or integrated telephone/computer terminal, leading to eventual digitization of the local loops to the customers' premises, makes an entirely digital plant seem possible.

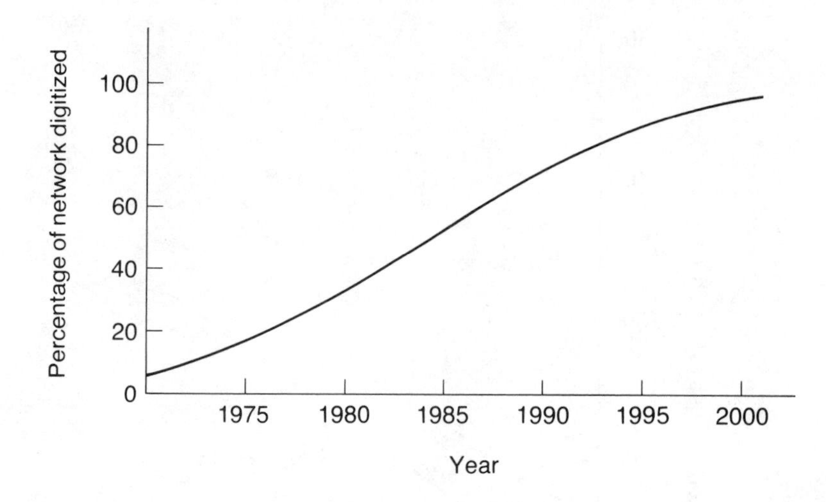

Figure 19-3. Estimated percentage of the common carrier plant that will be digitized by the year 2000.

The key point in estimating plant digitization is the fact that, despite exponential growth of digital and nonvoice services, the overwhelming percentage of the total information flowing in the public networks will continue to be voice-based conversations and information. In order to carry a single 56,000-bits-per-second digital signal, in the conventional analog telephone plant a 12-channel group of voice channels has to be displaced and a group bandwidth data modem used. In a digital system, on the other hand, only a single PCM voice channel need be displaced, since that single voice channel is actually being transmitted as a 64,000-bits-per-second digital data stream. Thus not only is a 56,000-bits-per-second data signal possible in a single PCM voice channel, but also an extra 8000 bits per second of capacity is available for extra overhead information, signaling, or other user/network functions.

Estimating Future Cost of Digital Channels

We can now estimate the future cost of digital channels on the basis of the technology used. With analog systems the "technical" cost of a 56,000-bits-per-second digital channel is about 12 times the cost of a voice channel, whereas in a digital PCM system the cost is essentially equal to a voice channel. Therefore, in theory, if a system were half digital and half analog, the average cost of a 56,000-bits-per-second digital channel would be six times the cost of a voice channel since half the time we would have to displace 12 voice channels, but half the time we would have to displace only one.

Figure 19-4 shows the changing cost ratio of a 56,000-bits-per-second digital channel compared to a voice channel. The shape of the curve is actually the inverse

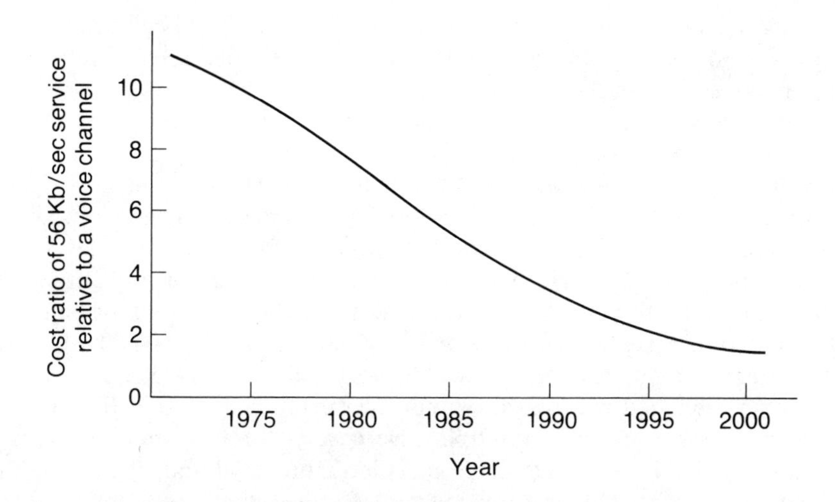

Figure 19-4. Cost ratio of a 56-Kb/sec digital channel to a voice channel.

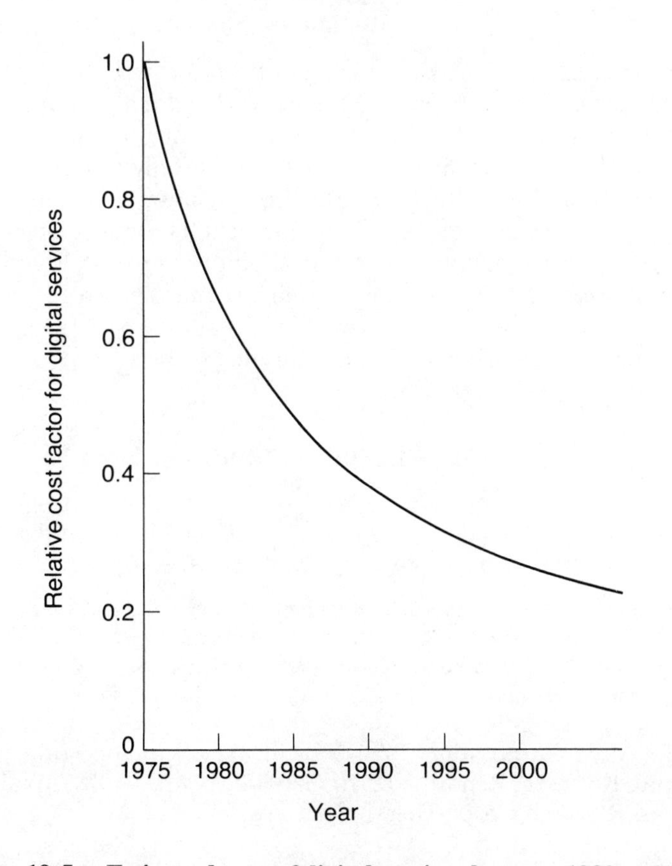

Figure 19-5. Estimated cost of digital services between 1980 and 2000 compared to 1975 cost.

of the digitization growth rate curve shown in Figure 19-3. It begins in the essentially all-analog plant of the late 1960s with a cost ratio of 12 to 1, and reaches a ratio of about 2 to 1 in the late 1990s.

Technically, a 56,000-bits-per-second data channel will displace only one voice channel in an all digital system. However, we can hypothesize that there will always be a higher cost associated with the digital channel because of more rigorous performance standards and more stringent user demands. Digital PCM voice channels can accommodate a considerable number of errors (perhaps as many as one error per 1000 bits) before significant degradation of voice quality occurs. However, for data services, where every bit must either be correct or be retransmitted until it is correct, error rates of 1 in 100,000 or 1 in 1,000,000 are needed.

This analysis can be combined with historical data on voice channel costs to produce an estimated relative cost trend for digital transmission (Figure 19-5). This trend curve provides a multiplicative cost factor that should be applied against the digital service cost curves given in Figures 19-1 and 19-2. These three figures taken together give us a starting point for the estimation of the future cost of digital common carrier, full-period, point-to-point circuits in the United States.

SUMMARY

This chapter has presented a sampling of the tariff structures and tariffs that are available for packet switching services as well as for the transmission components that, when combined with switching nodes, can establish a private packet switched network. Again, we must realize that these are only representative samples of existing tariffs, and the actual tariffs change rapidly with the introduction of new services, new technology, and new competitors. Individual carriers will provide their latest charges upon request. In addition, a number of consulting and publication services keep track of all tariff filings and can provide current tariff information. This chapter concluded with a technology-based projection technique, by which we can make at least an educated guess at the future cost trend of digital data private-line services.

SUGGESTED READINGS

Center for Communications Management, Inc. "Executive Telecommunications Planning Guide." Updated periodically. Available from the Center for Communications Management, Inc., Ramsey, N.J., 07446.

Center for Communications Management is one of the consultant publication services that provide extracts of current communications tariffs for ready use by communications planners. The publisher provides replacement pages when tariff changes are made.

Datapro Research Corporation. "All about Data Communications Facilities." Datapro Research Report No. 70-G-100-01a. May 1978. Available from Datapro Research Corp., Delran, N.J. 08875.

This report compares the data communications carriers and their services. It presents various tariffs at the time of publication, ways to optimize the use of services, and expected changes in the services. Information about the various termination arrangements and their charges is also provided.

COVIELLO, GINO J., and ROSNER, ROY D. "Cost Considerations for a Large Data Network." *Proceedings of the Second International Conference on Computer Communications*, ICCC '74. Stockholm, August 1974. Pp. 289–294.

This paper indicates and discusses a number of sensitivities that impact the cost of computer networks, including topology, channel rates, switching versus connectivity costs. It projects future costs compared to present costs. This was the first public presentation of the cost projection model discussed in this chapter. The paper was reprinted in Computer Networks: A Tutorial, by Marshall Abrams, Robert Blanc, and Ira Cotton, Long Beach, Calif.: Computer Society Publications Office, IEEE, 1975.

20

Some New Common Carrier Prospects

THIS CHAPTER:

will look at some new service offerings proposed by common carriers that either provide or supplement packet switched services.

will highlight the Advanced Communications Service (ACS) proposed by the Bell System.

will provide an overview of the system being implemented by Satellite Business Systems.

will present an overview of a system concept originally proposed by Xerox Corporation as the Xerox Telecommunications Network (XTEN).

SOME PROPOSED NEW TELECOMMUNICATIONS SERVICES

Now that packet switched and other value-added networks are established in the common carrier marketplace, it is not surprising that the large primary carriers are looking for an appropriate competitive response. Further, the success of relatively small companies in the value-added network business has stimulated larger corporations to look at communications services as a natural extension of their basic information-based businesses. General Telephone and Electronics, for example, acquired Telenet Communications Corporation as an owned subsidiary.

Following this trend, the American Telephone and Telegraph Company (AT&T) is developing a nationwide data communications service with many novel features—the Advanced Communications Service (**ACS**). At the same time, changes are likely in both regulatory and legislative restrictions on AT&T's activities. As a result of these developments AT&T can become a major force in the business of nonvoice information movement. Although AT&T is not likely to offer data-processing services per se directly to the public, the company's greater latitude in applying advanced processing and computational techniques will make information services more available and more responsive to the end users. ACS—which includes, but is not limited to, packet switched services—is a first step in this direction.

A very different approach to the communications and information services market has been undertaken by Satellite Business Systems (**SBS**), a partnership of Comsat General Corporation, International Business Machines (IBM) Corporation, and Aetna Life & Casualty Insurance. The original concept of the SBS network was to provide an advanced communications utility to large, multilocation organizations, with primary emphasis on voice and wideband services. While data communications, on both a private-line and a switched basis, would be accommodated, the SBS concept was much broader than either a VAN or a packet switching network.

Xerox Corporation also proposed an advanced approach to the total communications and information systems market. This service, known as **XTEN,** had many features in common with the SBS network. However, it used a much more flexible mix of terrestrial and satellite services and provided complete end-to-end service. The large capital outlays required to initiate the proposed XTEN service have apparently caused Xerox to withdraw, at least temporarily, from the market and concentrate their resources on the Ethernet product line for local communications and information distribution. Nevertheless, because of XTEN's technical practicality and useful features, it is likely that communications services will eventually evolve that use many of its proposed techniques.

In the next few sections we will look at each of these proposed communications services, with particular emphasis on the role that packet switching may play as part of those services or in competition with them. We will conclude this chapter by comparing various features of each of the proposed services.

AT&T'S ADVANCED COMMUNICATIONS SERVICE (ACS)

In response to increasing competition in the data communications and special services field, American Telephone and Telegraph Company announced plans, late in 1978, to provide an Advanced Communications Service (ACS) to meet a wide range of data communications customer requirements. ACS will provide a nationwide digital data service, accessible from any location in the United States. Presumably it would also be gateway compatible with overseas data services. Although ACS was not explicitly announced as a packet switched service, many details of the user features and the network design imply that the fundamental operation of ACS will be implemented using packet switching. However, since a number of other network modes and features will be available, the network implementation may not be limited to packet switching.

Modes of Operation

There are three distinct modes of operation that can be selected by any user. These modes are call mode, message mode, and transaction mode.

Call Mode

This is a circuit switched type of operation, with data transmitted through the network in an essentially transparent format. No conversion or alteration of data rates, formats, codes, or protocols will take place. The network presumes user compatibility at the input and the output of the network.

Message Mode

This is the most service-oriented mode of operation, where user information is handled on the basis of complete messages. The network will provide locally generated support features, such as message storage and retrieval, message formatting, format checking and validation, multiple addressing, and message broadcasting.

This mode of operation is essentially an electronic mailbox service. An additional benefit is that input or output can be in machine format, with the translation to user-readable format provided by the network. For example, a salesman may enter his daily sales data via a local terminal, on which the network provides a locally generated order form image. The salesman fills in the blanks, with each entry being locally validated. When the order form is ready for transmission, a copy is locally journaled for later retrieval if necessary. Only variable data is actually transmitted through the network and delivered in a previously defined, machine-readable format.

While the message mode is clearly a message delivery service, the advanced processing techniques provide increased utility to the users as well as very efficient movement of data across the network.

Transaction Mode

This mode provides the features most typical of a packet switched service. Individual user transactions, for the most part assumed to be relatively short, are handled on a terminal-to-host, host-to-host, or terminal-to-terminal basis. The network provides for speed, format, code, and protocol conversion. In addition, the network can provide normally host-controlled features such as polling control of remote terminals, user authorization and validation, error recovery, journaling, device handling, and broadcast/ retransmission of information.

Topological Structure

The overall topological structure of the network is illustrated in Figure 20-1. For the most part, ACS nodes will be colocated with major AT&T digital system hub offices, particularly Number 4 Electronic Switching Systems, where there is easy access to digital transmission capacity at rates of 56,000 bits per second or higher. Interswitch trunks among the ACS nodes will generally be 56,000 bits per second, although higher bit rates could be used to accommodate high traffic conditions. If necessary, an ACS node could be located away from a major digital hub office and served using a 50- or 56-Kb/sec channel from group-band transmission

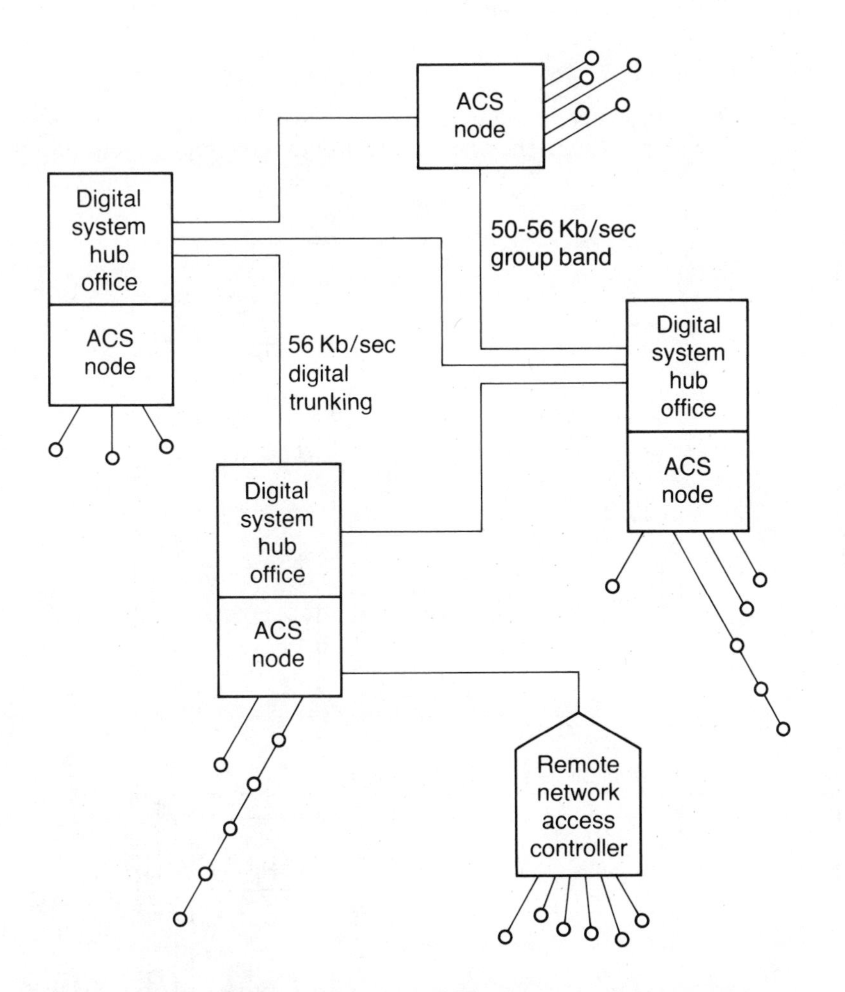

Figure 20-1. Basic topology of AT&T Advanced Communications Service (ACS).

or analog facilities. Finally, a remote access controller, which is primarily an intelligent multiplexor, will permit economical extension of ACS services to areas with too little demand to justify the construction of a full ACS node.

Design Elements of ACS Nodes

The functional design of the ACS nodes is shown in Figure 20-2. The major elements are the data switch, the message handler, and the network access controller. In addition, each node has central office access equipment and a data set frame. Each of the elements will be duplicated for reliability at each node and will be implemented with dedicated minicomputers or microprocessors.

Data Switch

The data switch is responsible for the switching and routing of user data through the network. It provides the logical connections

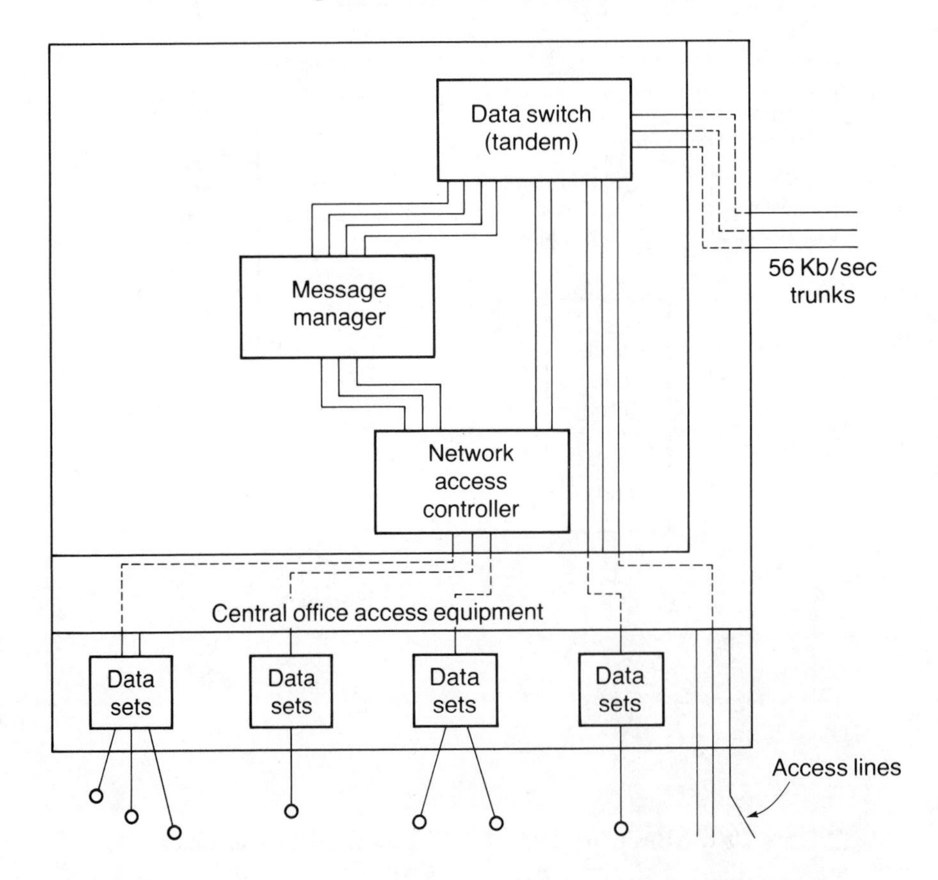

Figure 20-2. Functional design of a typical ACS switching node.

through the network, controls traffic flow, and provides for the tandem movement of traffic passing through the node. If ACS supports the X.25 CCITT protocol (and it is very likely to do so), the protocol will be implemented in the data switch.

Message Manager

The message manager formats, processes, and stores the non-real-time messages and transactions of the users. Preformatted forms and messages are stored for later completion and transmission, and processed data is held for later retrieval.

Network Access Controller

The network access controller provides the protocol and physical interface between nonintelligent terminals and the data switch, and performs terminal emulation protocols for commonly available host computer interfaces. Host computers programmed to support the network interface protocols will be able to bypass the network access controller and interact directly with the data switch.

Central Office Access Equipment

The access equipment provides the physical connecting frames, monitoring equipment, cabling, and local equipment-switching facilities.

Data Set Frame

The data set frame provides the physical equipment for terminating the physical access and transmission lines, particularly those that require data modems for digital access to analog transmission facilities.

Services

As initially proposed, ACS will support a wide range of protocols, including private-line emulation as well as public switched service. Private-line and data-phone digital service, provided on a point-to-point full-time basis, will operate at rates below 1.8 Kb/sec asynchronous, and rates of 2.4, 4.8, 9.6, and 56 Kb/sec synchronous. Public switched network services will initially be supported at 1.2 Kb/sec asynchronous and 2.4 or 4.8 Kb/sec synchronous.

The network will be capable of a great many host functions, thereby off-loading the mainframe computers of many communications processing functions. The network interface will perform both synchronous and asynchronous polling, with network-based queueing, thus freeing host computer front-end processing and storage power. The network will provide device handling for a variety of terminals, along with speed, protocol, format, and code conversion. Broadcast, multiaddressing, and message conferencing will be accommodated, as will format and validation routines for control of network access. Closed user groups can be accommodated by use of the validation routines. The message management functions will allow for journaling and storage of messages and transactions. Positive acknowledgement of delivered traffic and recovery from network and user errors will be provided.

Charging and accounting for ACS services will depend on the total resources used. The user charge will be based in part on the total amount of data transmitted—in bits, characters, or data blocks—and the elapsed connect time. Network resource units, comprising specialized services such as message formatting, will be another charging basis. Finally, stored journaled messages or formatted information will be charged on the basis of total amount of storage used.

The capabilities of ACS will undoubtedly evolve with changing demands for service and the experience AT&T gains from its initial offerings. Early announcements of ACS set an ambitious schedule for an extremely comprehensive set of services. Since then slippages in schedule and changes in intended services and capabilities have occurred. However, AT&T's depth of total resources, as well as the potential extension of many ACS services to the huge home telephone and communications market, insures the eventual widespread availability of ACS.

THE NATIONWIDE NETWORK OF SATELLITE BUSINESS SYSTEMS (SBS)

Development

The initial motivation for the communications concepts developed by Satellite Business Systems (SBS) is probably the same as that which led to the original development of packet switching and the ARPANET. That is, long-haul communications facilities were simply incapable of providing the quality and capacity to tie medium- and large-scale computer facilities together in an economical fashion. Evolving computer hardware, software, and protocols permitted resource sharing, load sharing, distributed data-base management, and distributed networks of computers, but intersite communications were a severe limitation, with links over a few miles in length generally limited to 4800 bits per second or less. The SBS network evolved from what IBM originally envisioned as the ultimate computer peripheral device: a rooftop satellite antenna that tied large computer facilities together through a common satellite channel.

IBM's original proposals to establish a satellite-based data carrier met with strong opposition from both computer and communications suppliers. The other computer manufacturers feared that the huge resources of IBM would establish a communications capability tailored to IBM equipment, yielding the company a tremendous competitive advantage in the overall marketplace. Similarly, communications carriers feared that the resources of IBM could rapidly make such a system a dominant carrier by adapting it to many communications users besides the IBM computer users. It was a "no-win" situation since limitations on the system concept and deployment that satisfied one community heightened the concerns of the other.

In order to accommodate the concerns of both the computer and the communications industry, and still be consistent with its policy of stimulating competition among the common carriers of the United States, the Federal Communications Commission essentially approved IBM's satellite-based system. However, it required that a three-way partnership be formed, in which neither IBM nor the satellite carrier would have majority ownership. It further required that the organization so formed operate in an "arm's-length" fashion, giving no preferential treatment to the parent organizations.

This ruling had two effects. The organization, finally established in December 1975, included, besides IBM, Comsat General Corporation as the satellite partner, and Aetna Life & Casualty Insurance Company. Each company owned an equal one-third share of SBS. The second effect was the broadening of the SBS concept to be much more than a data communications carrier among computer facilities. In fact, it led to a system design based more on the need of human users than on the requirements of their computers.

Characteristics

What has evolved is a flexible, satellite-based approach to the total communications needs of large businesses, government agencies, public-service organizations, industry groups, and other entities having large volumes of communications traffic among widely dispersed locations. A highly efficient compressed digital speech capability is provided. Digital transmission capabilities accommodate the full range of computer, data, facsimile, video, and other wideband communications requirements.

The SBS design is based on at least two synchronous satellites, in orbit over the United States, operating at the previously unused frequencies of 12 and 14 gigahertz. (All previous commercial satellites operated at frequencies of 4 and 6 gigahertz.) Access to the satellites would be via relatively small earth stations, with antenna diameters of either 16 or 23 feet, placed directly at the end user's location. This approach completely bypasses the potentially limiting problem of local connectivity and distribution between the system end user and the nearest satellite earth station. Electronic processing and interface equipment at each earth station would digitally integrate voice, data, facsimile, and other services into a single, dynamically managed, time-division multiple access bit stream for transmission over the satellite to the appropriate destination earth station.

The overall network structure has many features in common with the packet satellite broadcast techniques we discussed in Chapters 13 through 15. However, the SBS network is a centrally controlled, switched network, in contrast to the highly distributed control that characterized most of the packet broadcast techniques we looked at previously. A conceptual diagram of the SBS system is shown in Figure 20-3.

Though the SBS system was originally developed to serve large customers with their own on-premises earth stations, recent SBS plans provide for shared earth stations, located in major metropolitan areas, designed to serve smaller

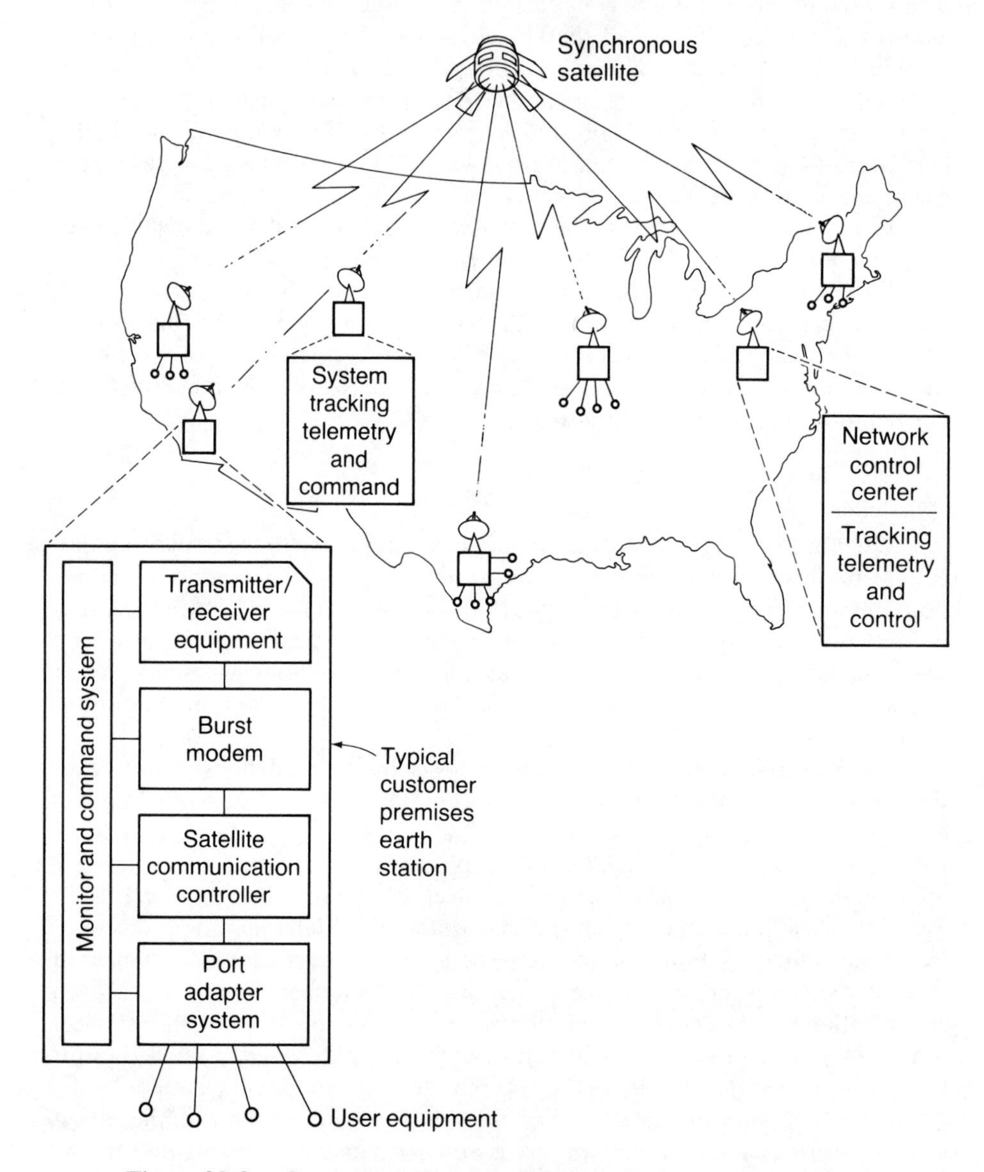

Figure 20-3. Conceptual diagram of the Satellite Business Systems (SBS) network.

customers. Twenty metropolitan areas are planned for shared services by early 1982, expanding to nearly 100 total areas by early 1983.

It is important to recognize that in either configuration—using either dedicated customer-premises earth stations or shared earth stations—the SBS system is designed to provide "network services" to the end user. It is not configured to be cost-effective simply to meet a collection of point-to-point circuit requirements. Minimum billings, for the smallest customer configurations, will probably be on the order of $25,000 per month to service at least three locations with the equivalent of 20 or more full-time voice channels (or equivalent data capacity). However, where the customer demands and traffic patterns require the flexibility and capacity of a large, distributed communications network, SBS services are likely to be technologically highly desirable, and economically extremely competitive.

THE XEROX TELECOMMUNICATIONS NETWORK (XTEN)

Development

In November 1978 Xerox Corporation announced their intention to develop and build a nationwide telecommunications network. Initially the system appeared to have many similar features to the proposed SBS system. Closer inspection revealed considerable differences in the technical approaches, despite the fact that a major portion of the Xerox capability was also based on satellite communications. However, the Xerox proposal, using the name XTEN, utilized terrestrial microwave distribution and wired-city local communications to achieve extremely high-efficiency, shared utilization of the satellite earth stations. As a result, the XTEN approach was even more capital intensive than the SBS system. However, it appeared to have a broader potential market among small and medium-size businesses and organizations. In fact, the XTEN system would be able to provide many of the same kinds of dedicated-network features that SBS would offer and, at the same time, many of the public network features for smaller users that are provided by the value-added networks.

Characteristics

The overall XTEN approach is illustrated in Figure 20-4. Synchronous communications satellites would link together shared earth stations located in and around major metropolitan areas. The high-density business sections of the metropolitan areas would be linked to the earth stations via dedicated microwave links from local city nodes centrally located in the vicinity of the users. Individual user locations would be linked to the local city nodes using short-range dedicated microwave, or possibly optical/infrared transmission. Finally, individual user devices within a single building, office complex, industrial park, or campus would be connected to each other and to the network using a local high bandwidth packetlike connection, called **Ethernet.**

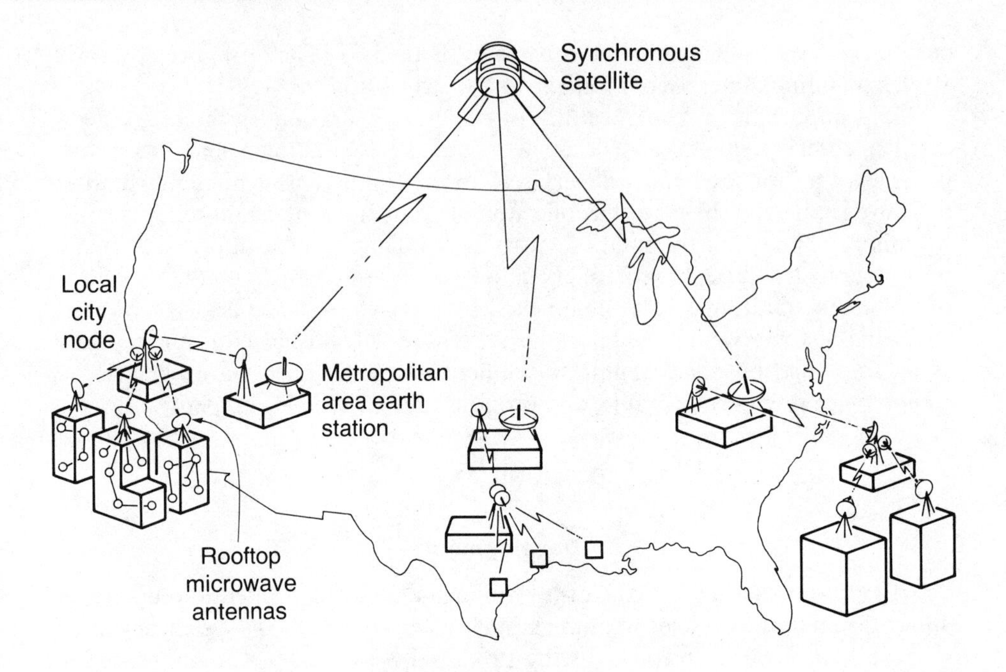

Figure 20-4. Conceptual diagram of the proposed Xerox telecommunications network (XTEN).

Ethernet is a packet-based communications technique utilizing the carrier sense multiple access discussed in Chapter 15. However, instead of using a common radio channel, the carrier sense operation utilizes a high-capacity coaxial cable operating at a data rate of 10 million bits per second. Information moving from one device to another within the same complex—say, from a computer to a local terminal or a line printer—would simply travel on the local cable. Data destined for a remote location would be picked off the cable by a local interface controller and broadcast to the local city node, from which it would enter the XTEN network.

Technologically, XTEN is one of the most flexible and versatile approaches to telecommunications possible. Major problems of "downtown" congestion of cabling, local distribution, and physical space for equipment are minimized by the Ethernet operation in conjunction with microwave distribution and interconnection with satellite earth stations. At the same time, the expensive earth station can be made cost-effective by the concentration of many users at a single earth station.

The Future of XTEN

Despite XTEN's advantages many difficulties have cast considerable doubt on the full implementation of the network as initially conceived. Problems include the cost of constructing such a network with enough connectivity for major market penetration; the many practical problems of microwave frequency allocations, electromagnetic compatibility, and interference possibilities; and establishment of many line-of-sight communications right-of-ways.

Nevertheless, the approach has such merit that it is highly likely that a network with XTEN characteristics will eventually evolve, either through the efforts of Xerox or some other major corporation, or through other communications networks, which can combine long-haul satellite communication with efficient local distribution systems. In the meantime, Xerox Corporation appears to be concentrating on the further development and marketing of the Ethernet techniques in order to facilitate office automation, wired-city electronic mail, teleconferencing, and highly efficient local distribution of information.

ACS, SBS, AND XTEN COMPARED

Table 20-1 summarizes the characteristics of the three revolutionary communications networks we have been discussing. Let us look at their similarities and differences in detail.

Basic Architecture

Both ACS and SBS use a distributed architecture, with no nodal facility being particularly more important than any other. In SBS, however, there is a centralized control facility essential to proper network operation. The XTEN network is hierarchical, with traffic being concentrated through a succession of shared facilities.

ACS is primarily a terrestrial-based network, although AT&T indicates that satellite connectivity may be used if it appears to be

Table 20-1. Comparison of Prospective Telecommunications Services

Feature	*AT&T Advanced Communications Service (ACS)*	*Satellite Business System (SBS)*	*Xerox Telecommunications Network (XTEN)*
Basic architecture	Distributed terrestrial	Distributed satellite	Hierarchical satellite
Target customer base	All data	Large voice and data	Data, graphics, facsimile
Getting-started costs	Low	Very high	Moderate
Applicability to low-density areas	Good	Poor	Fair
Suitability for voice	None	High	High

cost-effective. SBS and XTEN, of course, are fundamentally satellite-based networks.

Target Customer Base

ACS is targetted to all data communications customers, both large and small users. Eventually, ACS may become the POTS service of the home computer user. SBS, on the other hand, is seeking the large corporate or organizational user and, in so doing, is combining voice, data, and wideband services.

The XTEN approach lies somewhere in between. It attempts to provide primarily nonvoice services for both medium and large organizations.

Getting-Started Costs

Getting-started costs are the costs associated with a new user's joining the network. For ACS these costs are quite low—probably little more than adding a new telephone to the common user voice network.

Getting-started costs in SBS are very high, as they entail the installation of a new satellite earth station at a customer's premises, although sharing of some earth stations will reduce this effect somewhat.

XTEN would have moderate getting-started costs, requiring the installation of an Ethernet local system and a short-range microwave connection to the nearest XTEN local city node.

Applicability to Low-Density Areas

ACS service can be extended to low-density customer areas very well. In fact, the Remote Network Access Controller is designed specifically for this purpose. SBS cannot easily be extended to low-density areas. If there are not enough users or sufficient traffic density to economically justify an earth station, network extension is not readily possible.

XTEN would have fair extendability to low-density areas, depending on distance from the nearest XTEN earth station or local city node, terrain, and microwave link construction costs.

Suitability for Voice

ACS is specifically designed as a nonvoice service. SBS has evolved to be primarily a voice or voice-equivalent service, although it efficiently combines all modes of transmission into a digital format. While XTEN does not explicitly include voice, its operation and bandwidths make it highly suitable to carry voice services.

SUMMARY

In this chapter we have introduced and discussed three major new prospects in the communications marketplace that would either enhance packet switched network operation or provide nationwide alternatives to private or public packet switched networks. We looked in some detail at the ACS, SBS, and XTEN approaches. Although each of these networks has its own characteristics and advantages, many of the techniques of packet switching or packet broadcasting are inherent in all these prospective services. The packetlike operation may very well be totally transparent to the end users, however.

It is increasingly clear that, particularly with the continued decrease in packet switching hardware and software costs, an organization can establish a nationwide or worldwide private packet switched network. At the same time, a wide range of competitive services are available that can either emulate a private network or offer the flexibility of public network service and access.

SUGGESTED READINGS

CASWELL, STEPHEN A. "Coming to Grips with Planned Carrier Services." *Data Communications*, vol. 8, no. 11 (November 1979), pp 48–54.

This paper introduces the proposed services of ACS, SBS, and XTEN and compares them with services offered by other common carriers. Emphasis is placed on the potential impact these services will have because of the high transmission bandwidths that will be available. These impacts will extend to electronic mail, intelligent copiers, teleconferencing, and integration and interchange of information servicing the "automated office."

DAVIS, GEORGE R. "AT&T Answers 15 Questions About Its Planned Service." *Data Communications*, vol. 8, no. 2 (February 1979), pp. 41–60.

Davis, the managing editor of Data Communications, presents key information about AT&T's ACS proposal in a question-and-answer form. The article also provides additional details on the protocols, features, and facilities of ACS, and highlights the various kinds of interfaces that will be initially supported by the service. Most of the information was extracted from documentation AT&T provided to the FCC late in 1978.

PART SIX

Integrated Networks and the Future of Packet Switching

Packet switching is just one step in the evolution—sometimes the revolution—of communications technology. It has been an extremely important step, however. It has clearly demonstrated the divergence in cost between raw transmission capacity and sophisticated communications processors and switches. Moreover, it has been a catalyst in many new international agreements concerning information interchange and the primary driver in the initiation of value-added networks.

In this last part of this book we will look at the future of packet switching. Having stressed its advantages over circuit switching in Part One, we will now show that it is possible to combine these two switching techniques within one network in order to gain the advantages of both.

The rapid digitization of many transmission facilities and the reductions in cost of digitized voice communications has led to the integration of voice, video, data, graphics, and other services in a single network—for instance, the Satellite Business Systems network. In this part we will discuss the evidence that an integrated services network based primarily on packet switching may ultimately provide the most cost-effective general communications services.

Finally, we will characterize the key trends that will impact on the future growth and application of packet switching. We will try to provide some general guidelines for the communications user and supplier to follow in dealing with the rapidly changing and expanding technology of communications in general and packet switching in particular.

21

Hybrid Techniques: Combining Circuit and Packet Switching

THIS CHAPTER:

will look at the characteristics of communications that make it desirable to combine circuit and packet switching.

will look at various techniques and approaches to achieve integration within network switches.

will discuss the master frame approach to integrated switching, which has been applied to both experimental and practical networks.

In order to develop a thorough understanding of packet switching, we have compared it with classical circuit switching, particularly the common user voice telephone system. We found that, for short, bursty communications, packet switching offered advantages in terms of delay, processing, and system overhead. Even for a network that mixes long and short messages, packet switching offers substantial advantages in overall resource utilization and, therefore, overall network cost. For long messages, however, there is a point at which circuit switching is more efficient. As a result, it is reasonable to consider technologies that combine the advantages of both switching techniques within a single network and, if possible, within a single switch.

CLASSIFICATION OF INFORMATION AND SWITCHING TECHNIQUES

If all communications traffic flowing through a network were homogeneous, with possibly random, but uniformly distributed, message lengths and arrival rates, a single network could be designed to ideally match those characteristics.

However, communications characteristics are quite diverse. In addition, a wide variety of communications services and facilities—such as facsimile terminals; electronic mail services; still, limited-motion, and full-motion video services; communicating word processors; and many business and consumer information services—are developing very rapidly. These, together with the possibility of using the same network for voice communications, are leading to the need for networks that can efficiently mix traffic with a wide variety of data rates and traffic statistics.

Classes of Communications Traffic

It is useful to group communications traffic into several classes with similar characteristics. We will use a classification based primarily on the degree of continuity in the information flow. Table 21-1 lists the characteristics and gives examples of three general traffic classes—continuous, bursty, and interruptible.

Continuous. **Continuous traffic** is characterized by a continuous flow of information, over a fixed communications path, with real-time, end-to-end connectivity. In general, continuous traffic has long holding times and operates between compatible end users. Because of the relatively long holding times, some delay at the beginning of the communications, and a small constant delay (such as over a single satellite hop) can be tolerated. Though high-quality connections are desired, no error correction is possible because of the real-time connectivity. If the network resources for continuous traffic are limited, new requests for service are temporarily blocked from entry into the network. Typical examples of continuous traffic are voice, video, and facsimile transmission.

Table 21-1. Approach to Information Classification

Continuous	*Bursty*	*Interruptible*
Continuous-real time	Discrete messages	Long data streams
Connection delay	Near real-time	Not real-time
Fixed delay	Delay variable	Long delays vs. economy
No error control	Error controlled	Error controlled
Long holding time	Short total lengths	Indefinite lengths
Blockable	Nonblocking	Nonblocking
Compatible users	Arbitrary users	Arbitrary users
VOICE, VIDEO, FACSIMILE	INTERACTIVE, QUERY/RESPONSE	BULK DATA, FILES, DATA BASE

Bursty. **Bursty traffic** is composed of discrete messages, transactions, or portions of communications flows that can be handled as complete entities. Overall network performance must be nearly real time, but small transit delays and variability of the delay can be tolerated. Overall length of the individual messages is relatively small, but high transmission rates are desirable during the actual transmission time. The transmission is generally error controlled to prevent inadvertent alteration of the user data, and nonblocking service is highly desirable. Bursty traffic often involves dissimilar terminals, with the network required to make the conversions needed to effect user compatibility. Interactive computer operations, query/response, and distributed data-base operation are typical examples of bursty traffic sources.

Interruptible. **Interruptible traffic** is generally derived from long data streams that do not require real-time movement through the network. Long delays are tolerable, especially in support of overall economy of transmission. Error control is normally required, as is nonblocking service. Messages are long, with indefinite total lengths, and may frequently transit the network between dissimilar users. Examples of interruptible traffic are bulk digital data, large data files or remote program loads, or overnight electronic mail services. The key characteristic of interruptible traffic is that, although the individual message lengths are generally large, the transmission of such data need not occur in a single, continuous stream; it can be "**interrupted**" when necessary for the handling of more time-sensitive traffic.

In a certain sense, voice traffic can be considered interruptible. If we look at a typical voice conversation on an expanded time scale, we see significant "interruptions" in the communication on one-half of the circuit during the period that one of the parties is listening to the speaker. The capacity of the channel in the direction from listener to speaker is normally wasted. We will discuss this point further in Chapter 22, when we look at the integration of voice and data in common switched networks.

Matching Characteristics with Switching Techniques

Our classification of communications traffic helps in matching traffic with the different network techniques. Interruptible traffic, however, does not clearly match the capabilities of either packet or circuit switching. If interruptible traffic is passed through a circuit switched network, it is difficult to capitalize on its interruptibility. If, on the other hand, it is passed through a packet switched network, the high overhead associated with the long message lengths tends to waste capacity.

If we could integrate circuit and packet switching characteristics into a single switch, then the traffic could be handled in the most efficient way, depending on its own characteristics and the requirements of all the other traffic competing for network resources. Transmission efficiency would be maximized by pooling total capacity and making it available to whichever service had the highest current demand. An integrated network could also provide for many different user functions as well as for the interoperability of different classes of user terminals. For example, recent developments have led to the fielding of voice message services, where unidirectional voice messages are stored in the network for future delivery. In an integrated network the digitized voice message could be transmitted on an interruptible basis via packet switching, and delivered to the destination user in the form of a continuous, circuit switched message.

APPROACHES TO INTEGRATING CIRCUIT AND PACKET SWITCHING

Integration through Shared Transmission

There are a number of different ways circuit and packet switching can be combined within a single network. Let us first consider several that use shared transmission capacity.

Common Trunking. The most obvious approach is through common trunking, where both circuit and packet switches have equal access to common transmission facilities via multiplexing or concentrating equipment (Figure 21-1). However,

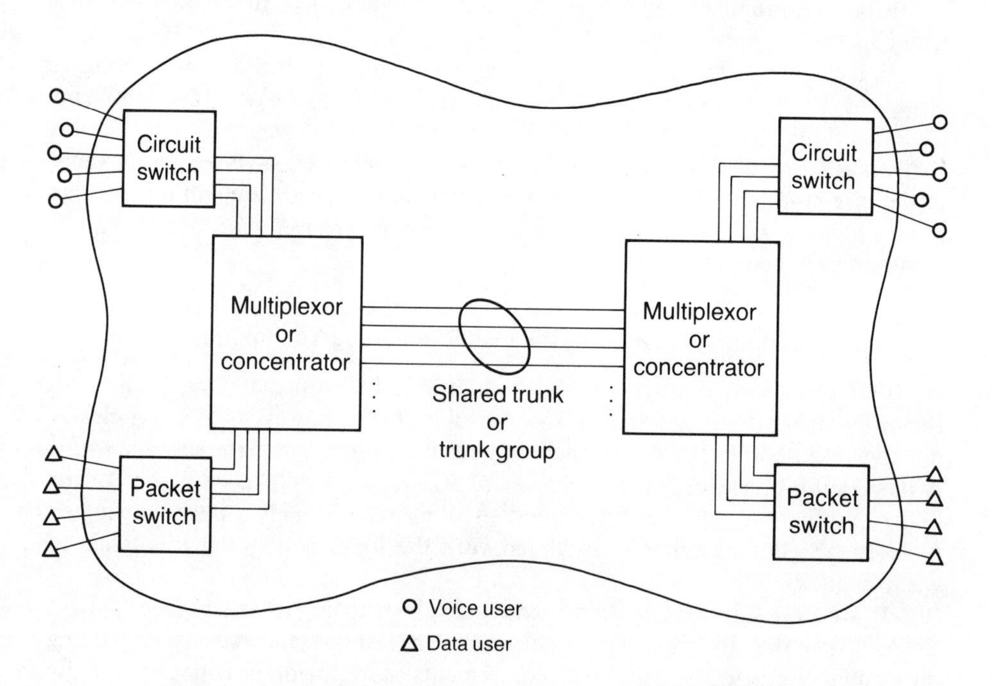

Figure 21-1. Network integration achieved through common trunking.

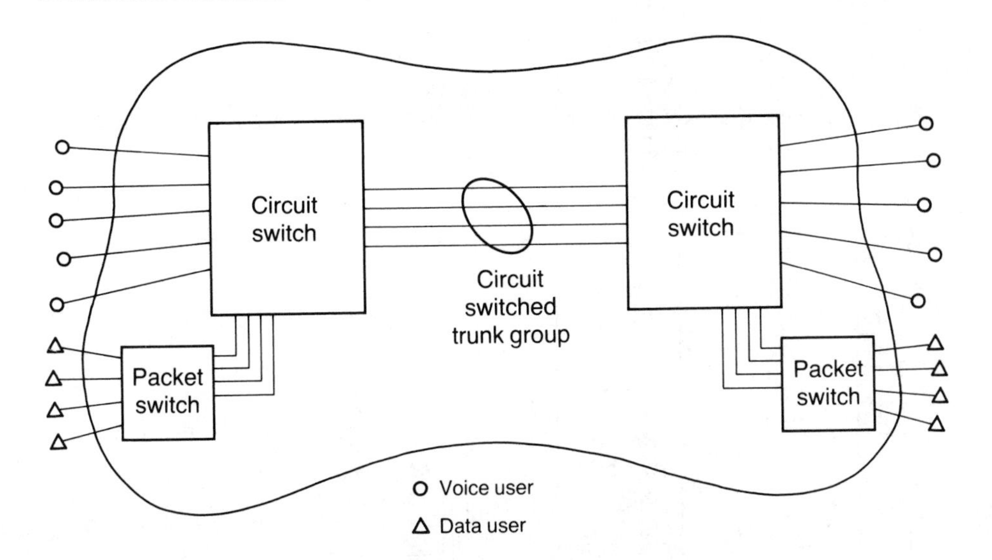

Figure 21-2. Network integration achieved by imbedding packet switches within a circuit switched network.

integration at this level does not really provide for much improvement in transmission efficiency, transmission utilization, or the exchange of traffic between different user communities or terminal types.

Imbedded Network. A more useful form of integration might be an imbedded network (Figure 21-2). Here, the packet switched network is "imbedded" in a circuit switched network, with each switch having its own community of users. The packet switched network shares the transmission facilities of the overall network by acting as if the packet switches were user terminals within the circuit switched network. Each packet switch would have a nominal connectivity pattern with other packet switches, derived with "held-up" connections through the circuit switched network. When the traffic between any pair of packet switches began to increase, the packet switch could request that the circuit switch provide additional connectivity or capacity by the addition of a physical circuit through the circuit switched network. Furthermore, the packet switches could complete the delivery of traffic to terminals homed off the circuit switches by establishing temporary connections to such users through the circuit switches.

Combined Matrices. Another approach, shown in Figure 21-3, is to combine two switching matrices, in effect, under the control of a single processor. One of the matrices would be a circuit switching matrix, while the other would be a system of buffers used to hold the information that is processed as packet switched data. Requests for service would come into the single processor, which would decide on the best way to meet the service demand and route the traffic through the most

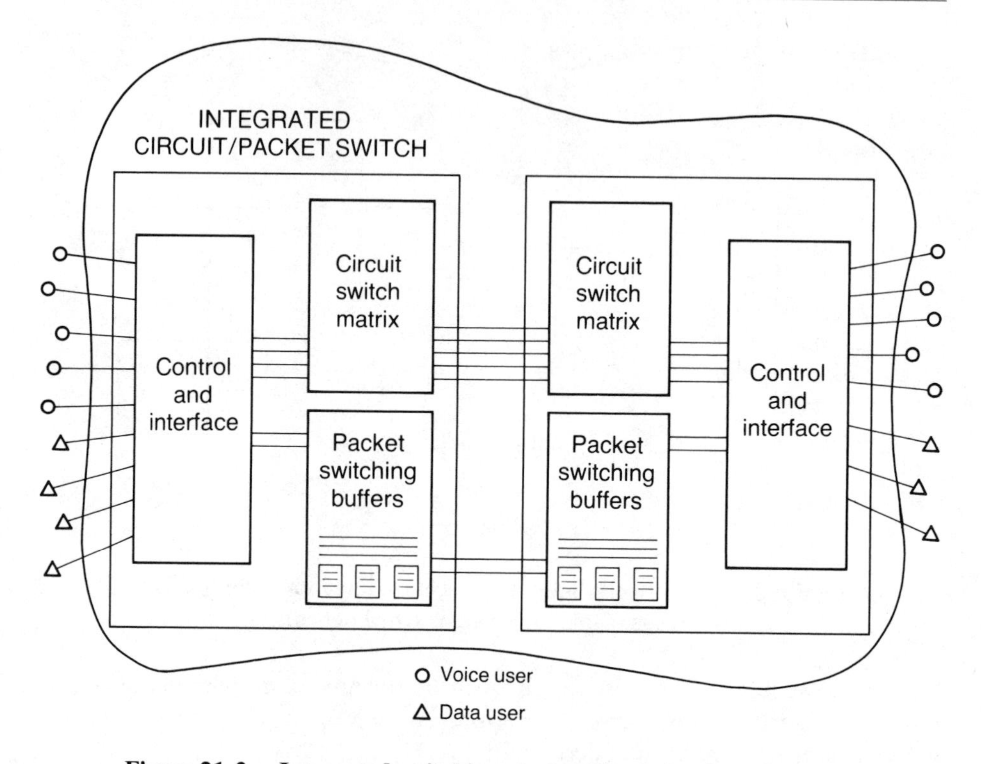

Figure 21-3. Integrated switching via combined circuit and packet matrices with common control.

appropriate matrix. If the information implied a long message between compatible terminals, a circuit switched connection would be established. If not, packet switching would be used. If properly applied, this approach to integration would be very efficient. However, its efficiency is dependent on a signaling technique between the user and the network that would permit the proper choice of facilities to be made in real time by the switches. Such signaling information can be included in the network protocols of computers and **intelligent terminals,** but it is not readily available for terminals designed to operate normally through circuit switched networks.

Integration through Processing and Software

All the approaches we have been considering achieve physical and functional integration of network capabilities through hardware, with the salient feature being the sharing of the network transmission capacity. The most practical approaches, however, utilize sophisticated switching processors and achieve integration through processing and software. This approach to integration can take any of three forms.

The first uses a basically circuit switched technique but adds some aspects of the multiplexing achieved by packet switching. Time assignment speech interpolation (**TASI**) or time assignment data interpolation (**TADI**) are examples of this approach.

A second approach uses a fundamentally packet switched network, following the principles we have developed throughout this book. But by limiting the dynamic routing features and using some additional memory for reference tables, it achieves a basically circuit switched connection through the packet network. Packetized virtual circuits use these principles.

The third approach uses a dynamically processed master frame, with advanced interswitch signaling, to truly integrate the best features of both technologies.

The Basically Circuit Switched Network. The fact that much of the traffic carried through the typical circuit switched network is actually interruptible makes it possible to essentially double the capacity of the transmission plant. The activity on each half of the full-duplex circuits between switches can be sensed and those circuits used to carry only active transmission of information. These principles were applied in the early 1960s, where time assignment speech interpolation (TASI) doubled the capacity of expensive transoceanic underwater cables by assigning physical half-circuits only to currently active talkers. The processing was carried out in special processors, external to the network switches, as illustrated in Figure 21-4.

In Figure 21-5 we see a more flexible implementation using a common switch. When the switch detects a pause in the activity on one of the circuit switched

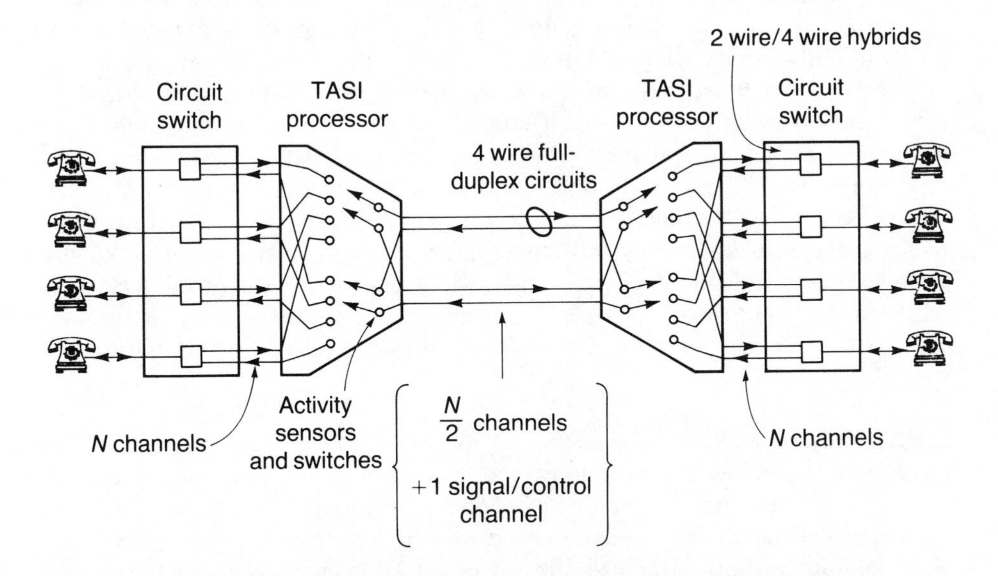

Figure 21-4. Functional implementation of time assignment speech interpolation (TASI).

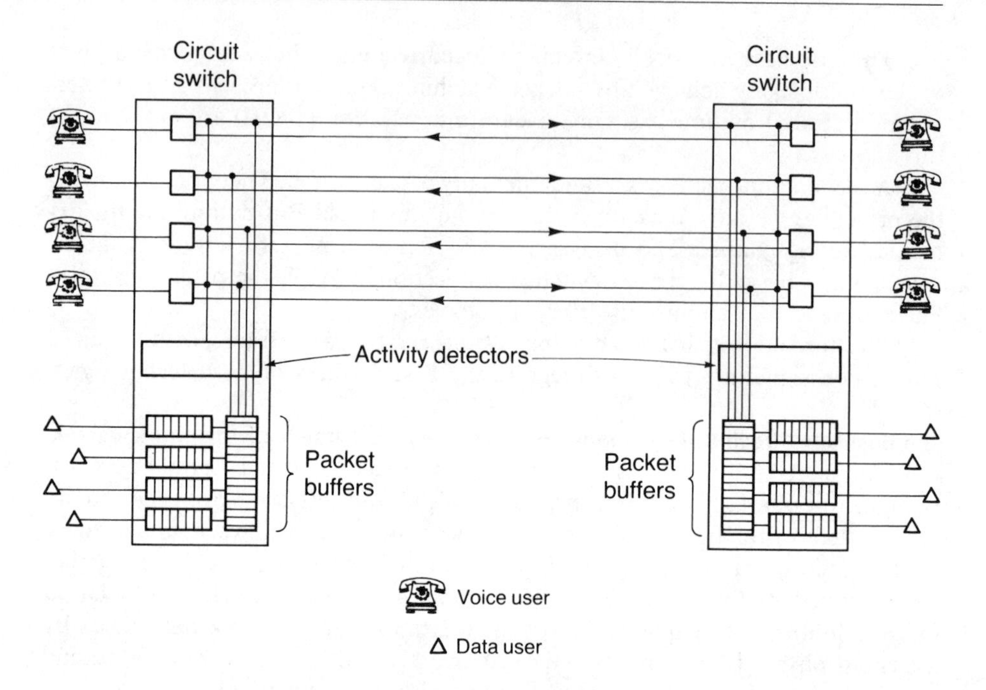

Figure 21-5. Functional implementation of the time assignment data interpolation (TADI)

channels, it inserts a packet of data destined to the next switch. On a 64,000-bits-per-second digital voice channel a 1000-bit packet will fit into a time period of about 15 milliseconds, which is less than the intersyllable gaps in normal speech. Clearly, then, many packets can be interspersed between words and sentences, where the pauses are from one-half second to several seconds. With this time assignment digital interpolation (TADI) approach the network can carry very large amounts of data traffic with little or no increase in transmission capacity. Packets are buffered within the circuit switches. When the activity detectors on the trunks sense an absence of transmission, they signal the switch at the end of the circuit to quiet the circuit to the circuit switched user and meanwhile begin to insert packets onto the circuit. When normal activity resumes on the circuit, the connection is returned to the circuit switched user in, at most, about 15 milliseconds.

The Basically Packet Switched Network. At the other technical extreme a basically packet switched network is used to emulate circuit switched connections. Among the key characteristics of a circuit switched connection is that it follows a fixed path, with constant delay through the network. It has a fixed overhead associated with the initial set-up of the call, but the overhead is independent of the length of the message or transaction. Figure 21-6 shows the typical configuration of a

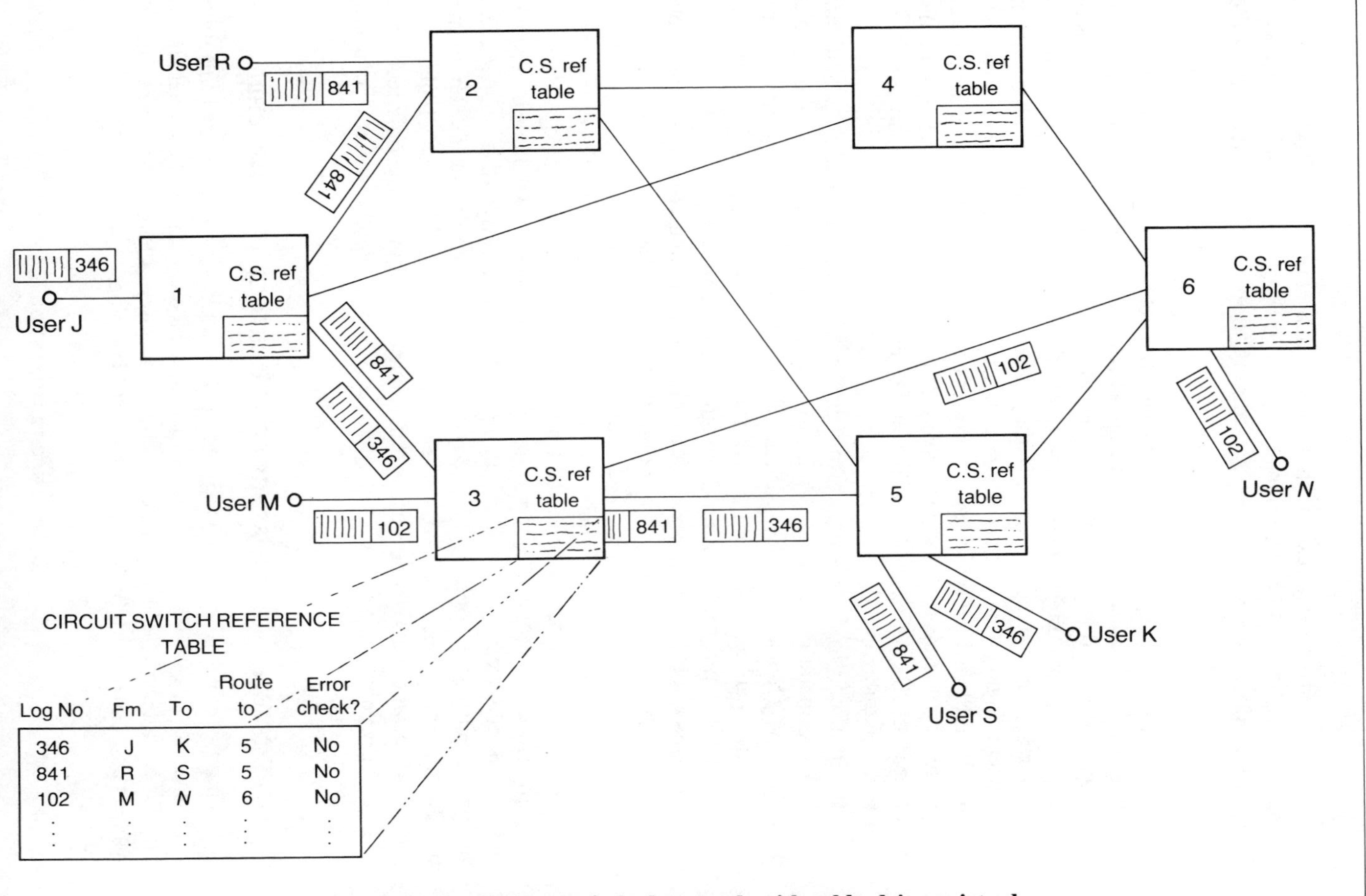

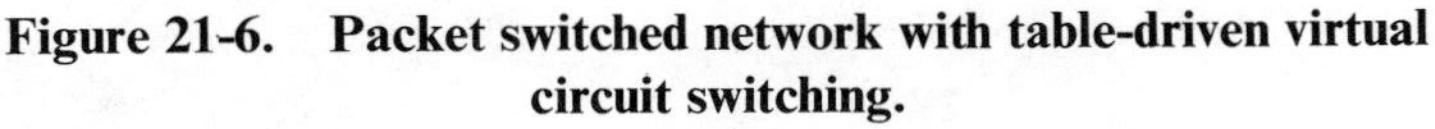
Figure 21-6. Packet switched network with table-driven virtual circuit switching.

packet switched network, except that each switch contains a circuit switch reference table. Short data transactions can proceed through the packet network in the normal fashion with the dynamic routing, flow control, and all the other features we have found in packet switching.

When the properties of a circuit switched connection are required, a call set-up protocol creates a direct route through the network via a set of table entries in each switch along the path. The logical connection through the network is referenced to the tables with a single logical connection number. Transmission proceeds using packet switching principles, except that each packet does not have to carry full overhead, but only the reference number to the table entry in each switch. When a packet arrives at a switch, the reference number is checked against the table, which indicates the proper routing and handling of the packet. Depending on the class of service provided, the packet would normally be put at the head of the queue on the proper outgoing line, thus keeping an essentially constant delay through the connection.

In addition, to fully simulate circuit switching, error checks would not normally be made on the packets, so no retransmission of packets would ever be necessary. Thus a continuous stream of packets, with only a few bits of overhead in each packet, will flow through the network, over a path determined at call set-up time and held constant throughout the connection.

For this approach to operate effectively, the emulated circuit switched connections would require equivalent continuous data rates that are only a small fraction of the overall capacity of the interswitch trunks. Under this condition, additional emulated circuit switched connections could be carried over the trunk, and short, dynamic, packet switched traffic could be accommodated.

The third approach to integrating circuit and packet switching in a single network is achieved by a master framing technique and dynamic management of the capacity allocation between adjacent switches within the master frame. Unlike the approaches we have discussed so far, which, in effect, emulate one technique in a network based on the other, master framing is the only technique that actually achieves integration of circuit and packet switching. Let us consider it now in detail.

THE MASTER FRAME APPROACH TO INTEGRATED SWITCHING

The master frame approach to integrated switching is based on the utilization of a dynamic time-division multiplex structure, which acts as fixed channel allocation for circuit switched traffic but uses any temporary excess capacity to transmit packets associated with bursty and interruptible traffic.

Structure and Operation

The implementation of the dynamic master frame multiplexing depends on very high-speed processing within the network switches and generally high-capacity trunks. The frame length is short relative to the traffic changes in the network, yet it can handle a fairly large number of real and virtual channels within a single frame. The initial structure of the frame approximates that of a standard digital time-division multiplexor, with distinct time slots within the individual frames that can be assigned to particular user pairs for the duration of a call. However, unlike a fixed time-division multiplexor, where unused capacity in the form of vacant time slots is wasted, a large buffer in each switch is used to assemble each frame, so that any unused time slots are recognized and used to carry packet traffic.

Figure 21-7 illustrates the master frame technique with a portion of a switched network, focusing on the trunk between two of the network switches. For our example we will assume a basic trunk rate of 1,544,000 bits per second (which corresponds to the T1 digital multiplex channel rate commonly used in commercial networks) and a master frame time period of 10 milliseconds. With these parameters each frame would consist of 15,440 bits during the 10-millisecond time period.

In the master frame approach the time interval between two successive frames is fixed, as in the example above, at 10 milliseconds. This means that, for a fixed, circuit switched channel, each frame will contain the same number of bits, associated with the data rate of the circuit switched channel and the frame interval. For a 9600-bits-per-second circuit switched data channel, for example, a 10 millisecond frame would always represent precisely 96 bits of data. For a 64,000-bits-per-second digital voice channel, the frame would contain 640 bits associated with each circuit switched channel. Each successive frame would contain the same number of data bits.

The power of the master frame technique is that the fraction of each total frame assigned to circuit switched channels need not be the same in each frame. In the example shown in Figure 21-7 we see a variety of subframes within the master frame structure, beginning immediately after the start of the frame marker (frame synch). If 24 channels were active at the same time, each carrying 64,000-bits-per-second digital voice signals, there would be a total of 15,360 bits of circuit switched information contained in the master frame, leaving only 80 bits of capacity, which is needed for the frame timing and overhead. However, if fewer than 24 channels were active, or if some of the channels were operating at bit rates below 64,000 bits per second, fewer than the maximum number of bits in the frame would be utilized. The frame is built in the buffers associated with each of the switches during the time interval that the preceding frame is being transmitted. As the frame is assembled, the subframes associated with each circuit switched connection are placed in the leading part of the frame. Any residual capacity beyond that needed to support the currently active circuit switched channels is thus readily apparent, from the end of the circuit switched channel subframes up to the end of the master frame.

During the 10-millisecond interval when a frame is being assembled in a switch one or more packets may arrive at the switch for transmission further

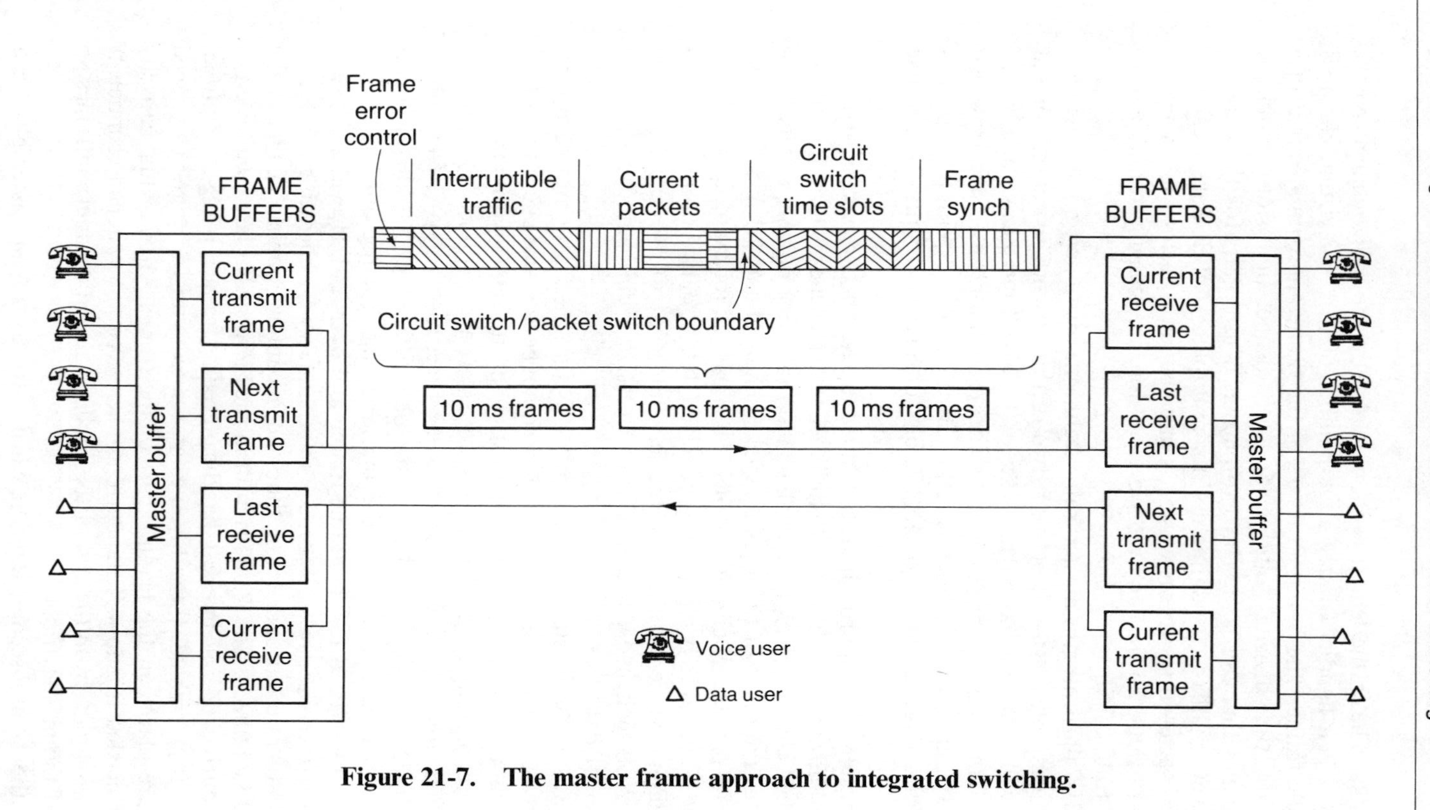

Figure 21-7. The master frame approach to integrated switching.

through the network. Assuming that capacity were to be available at the end of the master frame, the packets would be transmitted at the end of the frame. Furthermore, if additional capacity were still available after all current packets associated with real-time, bursty traffic was cleared, then some long, interruptible traffic would fill in the frame, to fully utilize total transmission capacity. In Figure 21-8 we see a master frame link, supporting twelve 64,000-bits-per-second voice channels, twelve 9600-bits-per-second data channels, and twenty-four 2400-bits-per-second channels, for a total of 9408 bits of circuit switched traffic per frame. The remaining approximately 6000 bits per frame are thus available to carry bursty and interruptible packetized traffic, to the total frame size of 15,440 bits.

Implementation

There are several practical means of implementing a master frame integration of circuit and packet switching.

SENET. One approach views the overall master frame as a large envelope into which data is placed in the form of both dedicated and dynamically assigned slots. This approach, called **SENET** (for *S*lotted *E*nvelope *NET*work), has been studied theoretically by analysis and simulation (see Suggested Readings for details). The analyses and simulations both show that link efficiencies of nearly 100% can be achieved for all three traffic classes. Furthermore, if at least some of the trunk capacity is always reserved to meet the needs of the high-priority bursty traffic, the high levels of trunk efficiency can be achieved while maintaining acceptable average delays for the real-time traffic.

With the SENET technique it is now possible to apply fairly complex processing techniques in order to achieve maximum utilization of the relatively expensive long-haul transmission capacity. Although the SENET switch has not yet seen practical implementation, it is well within the state of the art. A possible functional configuration for such a switch, using parallel, distributed processing for the frame composition/decomposition functions, is shown in Figure 21-9.

PACUIT. A somewhat different, and possibly even more flexible, technique, known as **PACUIT** (for *PAC*ket and cir*CUIT*) switching, has already been implemented in a commercial product line. PACUIT network switching components suitable for a user to develop a private switched network are produced by TRAN Telecommunications Corporation. These switches combine the master frame concept for integrated switching with time-division circuit switching and packet switching.

A typical configuration of PACUIT switches is shown in Figure 21-10, with interswitch trunk circuits operating at 9600 bits per second. Each frame, lasting as much as 0.1 second, is composed of three portions. One portion is reserved for fully allocated circuit switched traffic between pairs of network terminals.

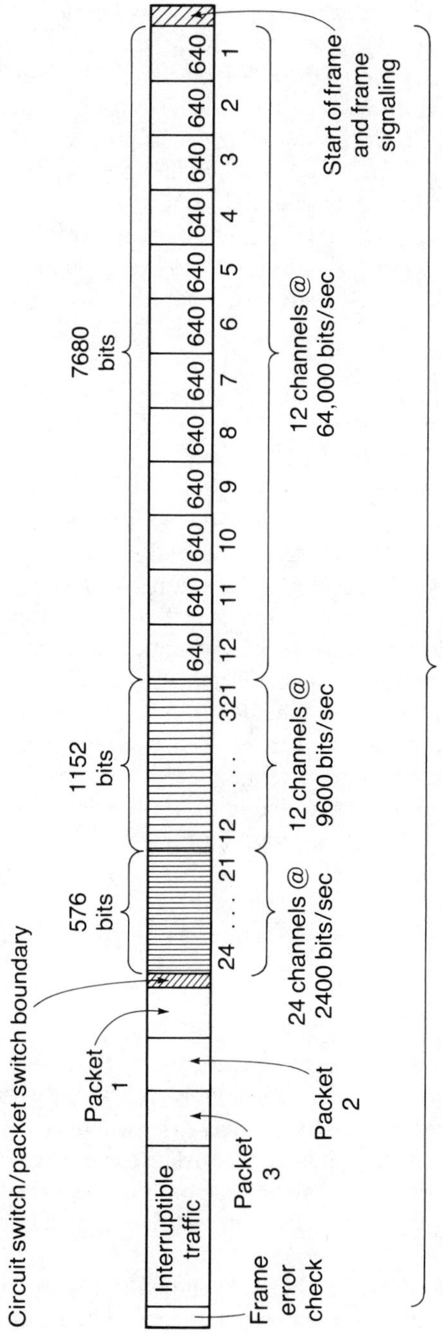

Circuit switch/packet switch boundary

Interruptible traffic

Frame error check

Packet 1

Packet 2

Packet 3

576 bits

24 . . . 21

24 channels @ 2400 bits/sec

1152 bits

12 . . . 321

12 channels @ 9600 bits/sec

7680 bits

640 640 640 640 640 640 640 640 640 640 640 640

12 11 10 9 8 7 6 5 4 3 2 1

12 channels @ 64,000 bits/sec

Start of frame and frame signaling

15,440 bits

Figure 21-8. Example of a master frame structure: mixture of circuit, packet, and interruptible traffic.

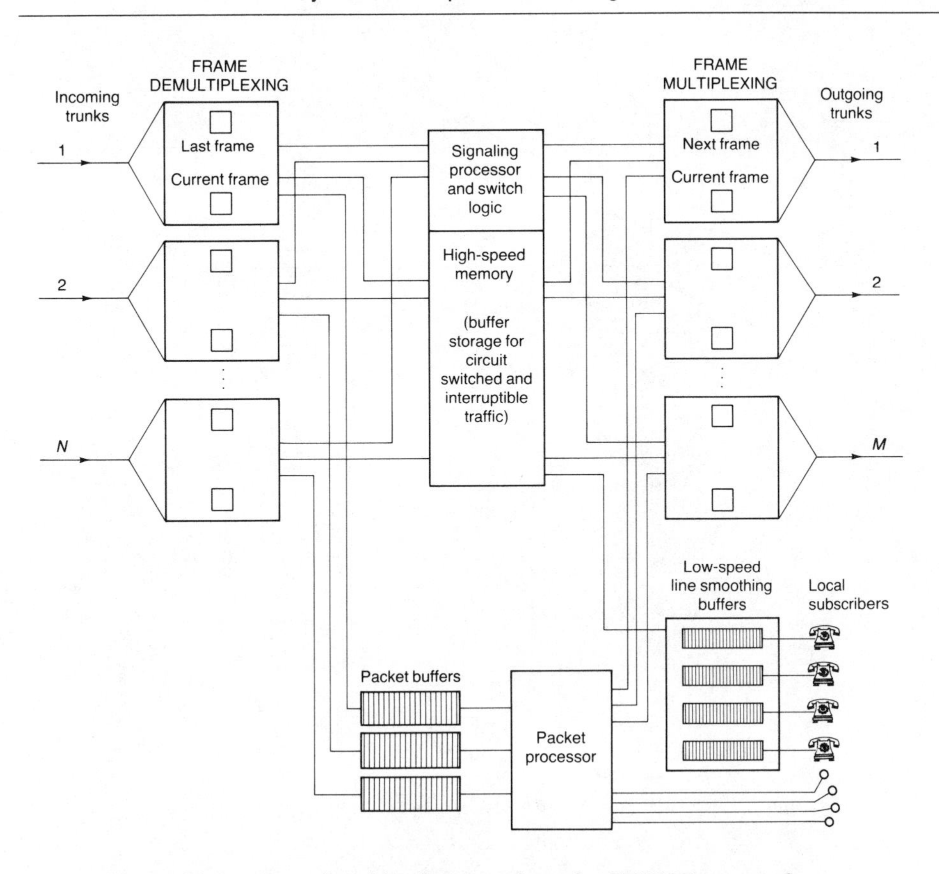

Figure 21-9. Functional implementation of a SENET master frame integrated switch.

Available capacity in each frame that is not assigned to full-period circuits is used to handle bursty traffic.

Most of the network capacity is used not for fully self-contained packets, but for PACUITS. PACUITS are groups of bits or characters going between different user endpoints located at the same switches. That is, data bits or characters transiting the network between the same pair of switches are grouped into a single large packet before the transaction or message is completed. The single packet, carrying data from a number of different users simultaneously, is handled intact between the two switches. PACUITS are assembled at the originating switch and disassembled at the terminating switch. In essence, they are circuit switched between those two switches. PACUITS are not broken down at tandem or intermediate switches, and data can be neither added nor dropped.

The contents of each successive PACUIT flowing between any pair of switches are assembled in a buffer during the transmission interval of the previous frame. At the beginning of the next frame the assembled PACUIT is synchronously

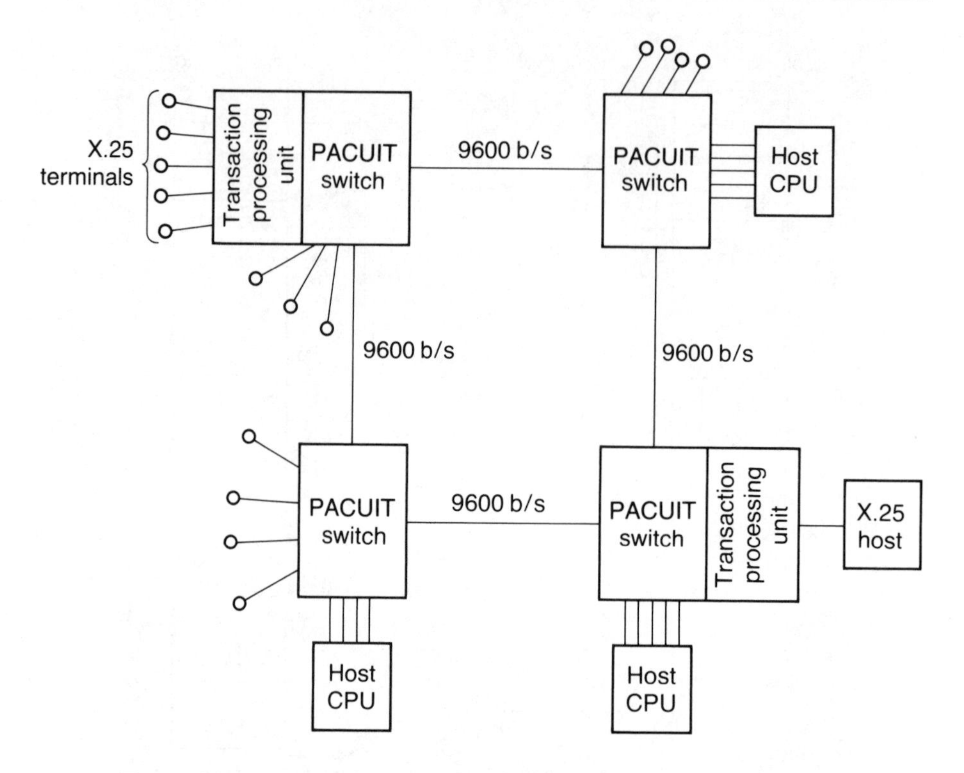

Figure 21-10. A PACUIT network implementation, with packet, circuit, and PACUIT connections.

transmitted through any necessary tandem switches to the destination switch, where the characters and bits within the PACUIT are broken down and delivered to the proper destination line. Signaling overhead in the PACUIT establishment process indicates the proper structure of the frame—that is, the association of the bits in the frame to the proper destination line. However, once established, especially for groups of low-speed users, the frame structure remains relatively constant over a large number of frames, and the resultant overhead is quite small. Finally, any capacity remaining in each frame after the circuit switched time slots and the PACUIT frames are assigned can be used for self-contained packets. The PACUIT switching equipment is configured to accept an additional processor unit dedicated to support the standard X.25 packet switching protocol.

PACUIT switching demonstrates not only the advantages of the integrated approach but the practicality of applying these techniques to relatively small distributed networks. The PACUIT switched approach has much less overhead, and therefore much higher efficiency, than self-contained packet switching, particularly for short messages or low-speed users, where the packet formation period introduces sizable delays to the end-to-end data movement. The PACUIT structure has many similarities to the framing approach used in the nationwide Tymnet Network, as was described in Chapter 18. Nevertheless, for high-speed, high-capacity users, or for users with relatively long messages, which result in essentially full packets, packet switching achieves efficiency essentially as good as, or better than, the PACUIT switched approach.

SUMMARY

This chapter has begun to bring the concepts of packet and circuit switching together. A major incentive for this integration is the wide divergence of operational characteristics of sources requiring end-to-end communication paths. Grouping information sources into three categories—bursty, continuous, and interruptible—helps us to associate the different switching techniques with an entire class, or grouping, of users. Bursty users are a good match for packet switching, and continuous users are a good match for circuit switching. Interruptible users can implement a combination of circuit and packet switching to achieve extremely high utilization and efficiency of the communications circuits. Because some of the user traffic is interruptible, the high efficiencies are achievable without compromising the network responsiveness required by the bursty and continuous users.

Approaches to the integration of circuit and packet switching range from circuit switched networks acting as a transport mechanism for packets, to packet networks that, by additional processing and buffering, can emulate circuit switched connections. Finally, we introduced the master framing approach, where time-division circuit switching and packets with self-contained routing and control information would be carried in a common master frame between switching centers. Two approaches to master frame integration have been extensively investigated: SENET and PACUIT. With slight variations, each technique combines the three classes of traffic, with minimum overhead, to achieve extremely high link efficiencies. The practical implementation of PACUIT switching in relatively small user networks has demonstrated the feasibility of pursuing an integrated approach to communications network technology.

SUGGESTED READINGS

FISCHER, M. J., and HARRIS, T. C. "A Model for Evaluating the Performance of an Integrated Circuit and Packet-Switched Multiplex Structure." *IEEE Transactions on Communications*, vol. COM-24, no. 2 (February 1976), pp. 195–202.

This paper analyzes the SENET type of master framing. A detailed queueing analysis of the traffic flow through such a structure is presented, as are summary results for the channel utilization and delay introduced to the packet and interruptible traffic. An error in the analysis, which was discovered later (corrected in a later paper, listed below), makes the results for some of the cases too optimistic. Nevertheless, the basic principles used in the presentation and analysis are correct.

KEYES, NEIL, and GERLA, MARIO. "Hybrid Packet and Circuit Switching." *Telecommunications*, vol. 12, no. 7 (July 1978), pp. 66–71.

The concepts and operation of PACUIT switching are discussed in this paper. It describes the advantages of the PACUIT approach, which combines packet, circuit, and a hybrid technique in a single dynamic concentrator, multiplexor, and switch. Both the nodal and network architectures are described. A case study is developed to derive and compare the delay characteristics for both packet and PACUIT switching. The paper also presents information about the commercial implementation of the PACUIT switching approach in the M3200 switching systems developed and produced by TRAN Telecommunications Corporation.

ROSS, MYRON J., TABBOT, ARTHUR C., and WAITE, JOHN A. "Design Approaches and Performance Criteria for Integrated Voice/Data Switching." *Proceedings of the IEEE*, vol. 65, no. 9 (September 1977), pp. 1283–1295.

This paper further elaborates on the SENET approach to master frame techniques. Many network details are discussed, including control of the network flow and signaling between the switching nodes. Architectural alternatives are suggested for building a SENET switch with the combined requirements of very high-speed processing and appreciable nodal buffering for the many frames being assembled, disassembled, or transmitted at any given time. Overall system performance in terms of trunk efficiency and packet waiting delays are presented for a variety of cases.

WEINSTEIN, C. J., MALPASS, M. L., and FISCHER, M. J. "Data Traffic Performance of an Integrated Circuit- and Packet-Switched Multiplex Structure." *IEEE Transactions on Communications*, vol. COM-28, no. 6 (June 1980), pp. 873–878.

This paper corrects the error found in Fischer and Harris's paper and presents corrected results. In addition, this paper proposes and analyzes a number of flow-control techniques that can keep the arrival rate of new packets consistent with the dynamic capacity currently available, assuring low-delay for those packets entering the network.

22

Integrated Services Networks

THIS CHAPTER:

will summarize the rapidly evolving technology of voice processing, which provides the possibility of digitizing voice at low data rates.

will describe the general approach of integrating voice and data within a packet switched network.

will summarize the results of a detailed cost/performance study of an integrated voice/data packet switched network.

Though a number of different techniques now exist for integrating circuit and packet switching into a single network structure, there is considerable evidence that technology will send future integrated networks in the direction of packet switching. The driving forces are the rapidly decreasing cost of both switch processing and the conversion of analog signals (voice signals) into digital format. Once digitized, voice signals can be handled in the packet network like any other data signals. Furthermore, since no packets need be transmitted during idle (listening) periods, digitizing speech reduces channel utilization.

VOICE DIGITIZATION

Pulse Code Modulation

Voice signals have been digitized for transmission purposes for many years by a technique known as pulse code modulation (PCM), which results in a digital stream of 64,000 bits per second. This data rate is based on the voice channel bandwidth of 4000 hertz, sampled 8000 times per second. Each sample is assigned one of 256 possible values, represented by an 8-bit binary code. The resultant data stream is thus 8000 samples per second times 8 bits per sample, or 64,000 bits per second.

Digital PCM was originally applied to voice communications in order to multiplex 24 individual voice conversations onto a single pair of wires. The digitization process is totally transparent and undetectable to the users since all the information in the originally transmitted voice signal is captured by the digital conversion process. In addition, the digital signals suffer less degradation during the transmission process since noise cannot accumulate as the signal is repeated over numerous links.

Other Techniques

As the cost of digital processing hardware has been reduced, digitization of voice signals has become cost-effective for many transmission media. More important has been the development of other techniques for digitizing voice signals, which result in a digital data stream at rates considerably less than the 64,000-bits-per-second rate of PCM.

Differential pulse code modulation (**DPCM**) transmits only the difference between the current sample value of the voice signal and the previous sample. This technique can not only reduce the required bit rate but also improve the overall channel fidelity and quality. An extension of the DPCM concept transmits information about the rate at which the value of successive samples changes. Known as continuously variable slope delta modulation (**CVSD**), this technique results in excellent quality voice channel reproduction at data rates of 32,000 bits per second, with good quality possible even at 16,000 bits per second. CVSD analog to digital convertors are readily implemented using a single integrated circuit, at extremely low costs.

To achieve digital rates of 9600 bits per second or less per voice channel, a different type of voice processing is utilized. These techniques capitalize on the structure of the vocal tract and the characteristics of sound formation and articulation. It is thus possible to send a digital data stream containing information about the input voice signal, which is then reconstructed or synthesized at the receiver end. With such advanced processing data rates of 9600, 4800, and even 2400 bits per second can be used to achieve highly intelligible, high-quality speech signal reconstruction. However, considerable processing is involved, and consequently terminal costs, even using advanced microprocessor implementations, are high.

The Future of Voice Digitization

Ultimately, voice digitization will become increasingly widespread in the common carrier networks. As we saw in Chapter 19 (Figure 19-3), as much as 80% of the common carrier networks is likely to be digitized by 1990. Economics is the primary motivation for digitization. Thus new and replacement plant in the commercial networks tends to be digital, with most of that digital plant operating at 64,000 bits per second per voice channel using PCM conversion.

This approach allows continued compatibility to remaining analog plant. At the same time, the ability to restore the digital signal at repeater stations without accumulation of noise produces better transmission quality.

Increasing attention to security and privacy of electronic communications makes digitization highly desirable. Digital data streams can be easily encrypted using commercially available encryption devices at a very low cost per voice channel, especially when such devices are applied in bulk to all the channels on a particular trunk. Voice digitization permits enhanced integration of voice information with computer-based information, as well as the ability to better integrate a wide variety of communications services.

Once digitization has been implemented, further reduction of the digitization rate can lead to significant cost reductions through bandwidth compression of the voice channels. For example, using linear predictive coding (**LPC**) to achieve a voice digitization rate of 4800 bits per second per channel would permit a carrier to transmit more than 12 voice channels in the bandwidth and capacity occupied by a single 64,000-bits-per-second PCM channel.

Another consideration in the conversion of voice signals into digital format is the natural compatibility between the digitized voice stream and other data formats for video, graphics, and general data communications. Because of the word and syllable structure of speech voice signals easily form into packets of 20 milliseconds duration containing between about 50 and 200 bits per packet. This gives a flexible packetizing structure and permits the short packets resulting from the speech digitization process to be readily integrated with the packets resulting from other data information sources.

GENERAL STRUCTURE OF INTEGRATED SERVICES PACKET SWITCHED NETWORKS

Digitized voice can be combined with other digital data services into a common, packet switched network by the proper design of the network interface and flow control protocols. The integrated voice and data network shown in Figure 22-1 is similar to the packet network illustrated earlier in this part, with the addition of voice users. The voice digitization processors may be at the user instrument, at any concentrator in the system (such as an access switch or PBX), or at the input of the packet switch. If the voice processors are placed at concentration points or at the input to the packet switch, their cost can be shared among many more users. This is because the processors can be pooled to handle the number of lines at the switch that are active at any time, and not the total number of end users in the network.

Regardless of the mode of operation of the packet network, be it an X.25-compatible virtual circuit, a pure datagram operation, a special centrally managed packet structure, or any other packet network implementation, two key operational details are necessary. First, the voice packets have to be class marked so that they can be handled as expeditiously as possible. Voice packets need not be

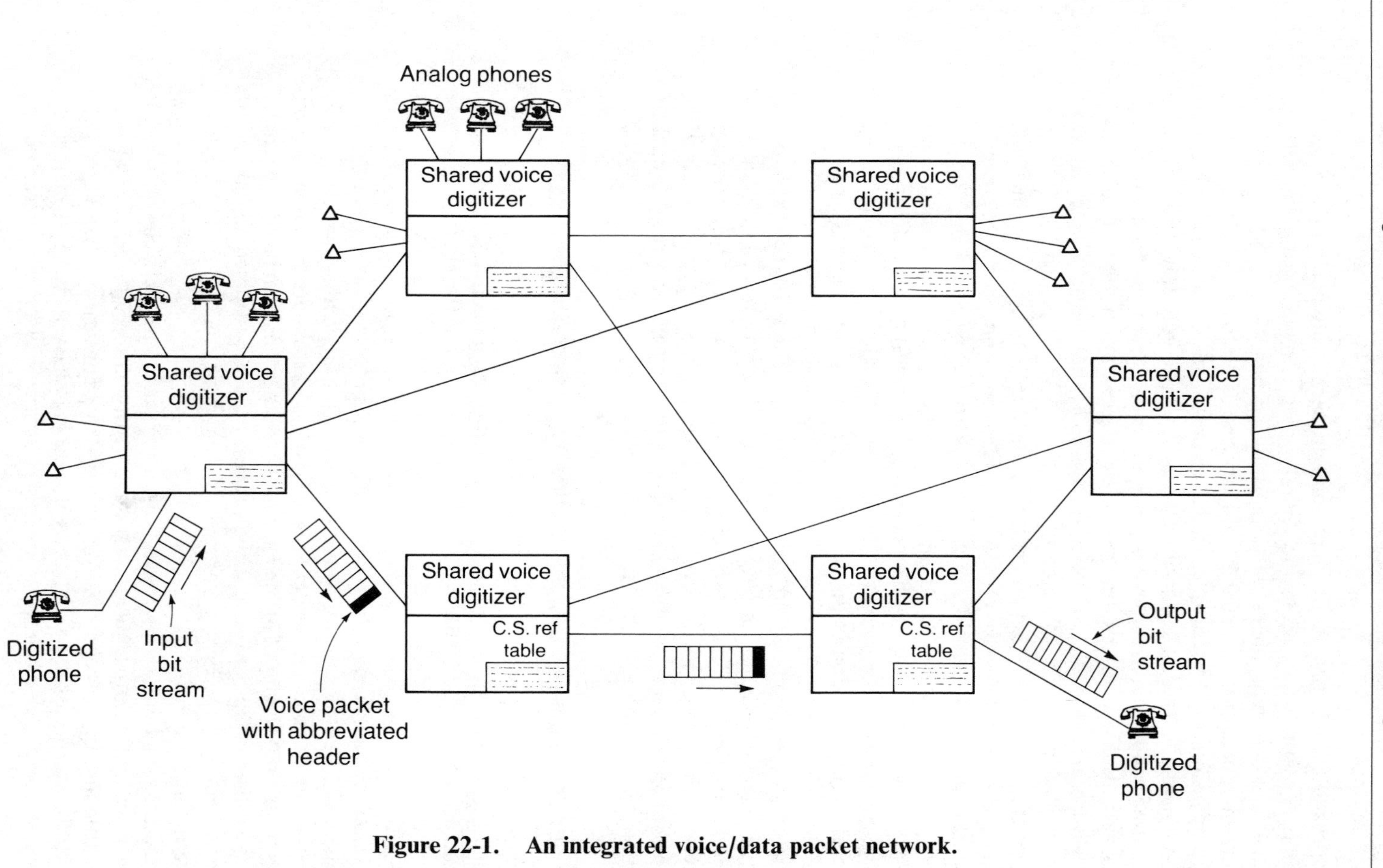

Figure 22-1. An integrated voice/data packet network.

error checked, since there is no time to retransmit errored packets. Packets received with errors will be processed at the output speech synthesizer and will, at worst, result in a short noise burst. Secondly, since voice packets will contain a relatively small number of bits, it is important to minimize their overhead. This is done by establishing a fixed path through the network during the call set-up and transmitting all packets associated with that call over the same path. A table entry in the memory of each switch along the end-to-end path that associates the packets with a logical circuit is needed. In this way the only overhead needed on each packet after the call set-up is the logical circuit reference number. Each switch can then reference this number to the table entry, which instructs the switch on the handling and routing of the packet.

COST/PERFORMANCE ANALYSIS

We have seen that packet switching could offer substantial cost and performance advantages over conventional circuit switching. We can ask how packet switching would compare with either integrated voice and data services or much more rapid, modern, digital circuit switching. Let us look at an analysis of this question.

The Model. The traffic model used for this analysis is drawn from the voice and data networks of the U.S. Department of Defense, with a busy hour traffic load of 2700 Erlangs and 36 Mb/sec of nonvoice data traffic. The nonvoice traffic could be from computers, terminals, facsimile, or any other information sources. Network resources are sized to insure transnetwork delays of less than 1 second for the nonvoice sources, and equivalent blocking for the voice channels of either 1% or 10%. Once a voice logical connection is set up in the packet network, voice packets get priority handling in the queues, so that the network delay for the voice users will be essentially constant over the duration of the voice logical connection.

Results. The overall results of this study are summarized in Figure 22-2. The results are portrayed for (1) traditional circuit switching—that is, circuit switching using the techniques of the nationwide telephone plant; (2) fast circuit switching (with a call set-up time of less than $\frac{1}{4}$ second); (3) "ideal" circuit switching (with zero call set-up time); (4) hybrid switching, using a technique similar to the SENET master frame integrated circuit packet switch; and (5) packet switching, as modified to efficiently handle the voice connections in the packet protocols.

Figure 22-2 shows the total annual cost of the network, with tariffed items and acquired nodal hardware combined with annual operating costs. The cost of the packet switched network is the least for all values of voice digitization rate—significantly lower than any of the circuit switched cases and marginally less than the hybrid technique. Note that the cost of voice digitization devices is not included in the annual costs, but presumably it would affect each of the techniques

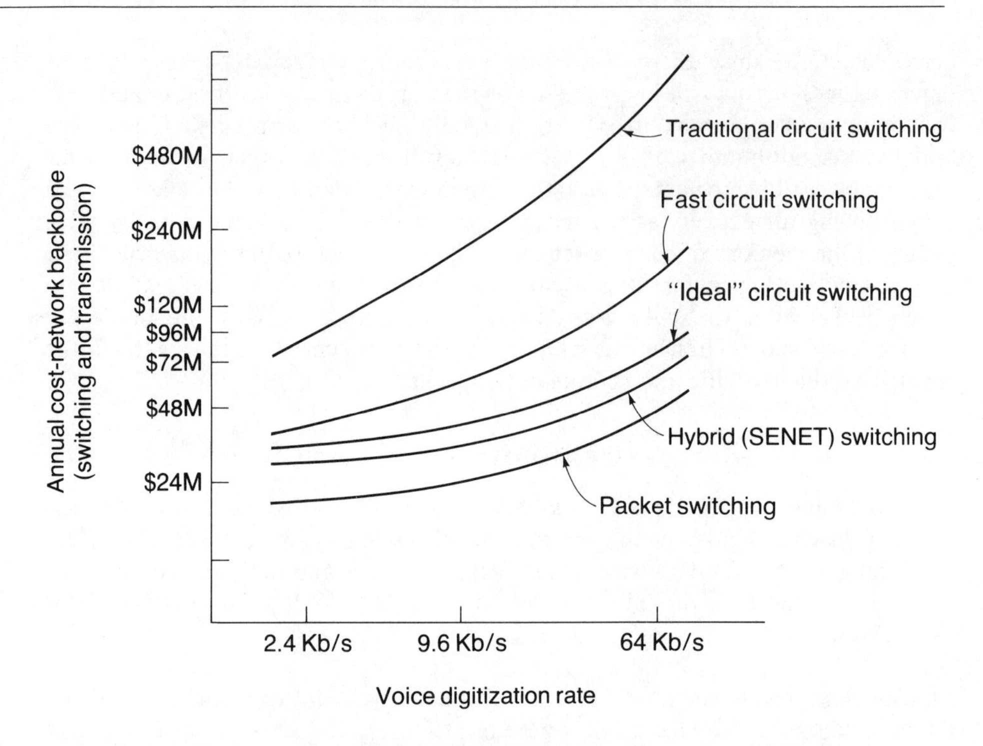

Figure 22-2. Annual network backbone costs for various digitization rates and switching techniques.

in the same way. Since the cost of the digitization devices gets higher as the digitization rate gets lower, the curves would all rise more at the low end than at the high end if the cost of the voice digitization converters were included. However, the relative ranking of the techniques would remain unchanged.

Decisions for the Future

Many assumptions and judgments must go into a tradeoff analysis, and the direction of integrated services packet networks will depend on considerable refinement of those assumptions. However, these results suggest the strong possibility that packet switching networks will prove to be the most efficient and economical way to meet a wide variety information transfer needs.

The results of the analysis are not surprising when we recall the fundamental operation of packet switching. The inherent use of each transmission channel as a pair of bidirectional trunks allows for maximum flow of useful data. When the total demand in the network is sufficiently large so there is data traffic that can be queued for short delay periods, then the idle time in voice conversations can be effective in achieving very high average utilization of the trunks. In smaller networks, or networks with erratic traffic demand, it is likely that packet switching would not show such clear-cut advantages. However, the key point is that, as technology moves the communications art to more digitized services, and as our society moves toward more on-line information-based services, packet switching appears to be the leading candidate to achieve highly efficient, highly cost-effective integrated networks.

SUMMARY

In order to better understand the prospects of integrated services digital networks, which could effectively combine voice and nonvoice services, we first looked at various techniques that could provide large-scale digitization of voice signals. We saw that PCM, CVSD, and LPC would provide a wide range of techniques, varying in channel rate from 64,000 to 2400 bits per second and in cost from a few dollars to as much as $20,000 per voice channel.

The general approach to an integrated services packet switched network would follow the basic packet switching model, modified to minimize the packet overhead structure and to eliminate the error-correction process for voice packets. In this way voice packets would flow through the network in a constant stream, with nearly constant delay and few gaps in the packet flow.

We completed the chapter with a summary analysis showing that an integrated services network appears to be the least expensive of all possible network structures, including ideal circuit switching and hybrid packet and circuit switching. With suitable caveats such results point to the future directions that the progress of packet switching is likely to take.

SUGGESTED READINGS

ARTHURS, E., and STUCK, B. W. "A Theoretical Traffic Performance Analysis of an Integrated Voice-Data Virtual Circuit Packet Switch." *IEEE Transactions on Communications*, vol. COM-27, no. 7 (July 1979), pp. 1104–1111.

This paper provides a theoretical analysis, supported by simulation, of a large, integrated voice/data packet switch. The results show that the number of links required to handle a given traffic load (mixed voice and data) is reduced by at least a factor of two, compared to separate networks, for the same level of performance (blocking for voice, delay for data).

FRANK, HOWARD, and GITMAN, ISRAEL. "Economic Analysis of Integrated Voice and Data Networks: A Case Study." *Proceedings of the IEEE*, vol. 66, no. 11 (November 1978), pp. 1549–1570.

This paper reports on the same study as does the other article by Frank and Gitman but does so in much more technical detail. A total of 15 figures are presented to show the sensitivity of the results to a wide range of assumptions, voice digitization rates, digitizer costs, and switching technologies.

FRANK, HOWARD, and GITMAN, ISRAEL. "Study Shows Packet Switching Best for Voice Traffic, Too." *Data Communications*, vol. 8, no. 3 (March 1979), pp. 43–62.

Figure 22-2 was extracted from the results presented in this paper, which summarizes a major study performed for the U.S. Department of Defense comparing packet switching to other techniques for general applicability to mixed user networks. A full range of network switching technologies is studied. The article summarizes both the technical and the economic parameters considered in the analysis and shows that packet switching provides the most cost-effective topology under a wide range of assumptions.

OCCHIOGROSSO, BENEDICT. "Digitized Voice Comes of Age," parts I and II. *Data Communications*, vol. 7, nos. 3 and 4 (March and April 1978).

This two-part article describes the techniques that are currently available for converting voice signals into digital format and the tradeoffs they entail. The basic technologies are explained, showing the different bit rates, performance characteristics, and costs associated with each. Sources of voice processing hardware are described, both for complete terminals and for specialized microprocessors used in the development of voice processors.

23

The Outlook for Packet Switching—And What to Do About It

We have now had a chance to look at packet switching in all its aspects. We have explored this telecommunications technique particularly as it compares to more conventional techniques. The state of the art of packet switching is already in the third generation, the first generation being marked by experimental packet networks such as the ARPANET and the second by the first commercial packet networks and VANs, such as Telenet. Present systems are evolving toward hierarchical network designs employing hundreds of nodes and a multitude of service features. Large high-capacity switches are handling the bulk of the traffic over long distances, and smaller, local switches are serving the concentrations of users. With the technology moving through three generations in just over a decade, users and developers of packet switching networks are justifiably concerned about what the future holds for packet switching and how they can anticipate those future trends.

The application of digital technology and the introduction of digital telecommunications and information services will continue to accelerate. The mixture of voice, data, graphics, and electronic mail services will place increasing value on the ability to carry mixed media communications within a single network structure. At the same time, digitization of the common carrier voice networks will result in continued decreases in the price per bit of communications capacity, although significant decreases in the cost per voice channel will not be realized until low-rate voice analog-to-digital (A/D) converters are widely used.

The marketplace will also be influenced by the new products and services that are continually being offered. We are seeing broader applications of telecommunications, particularly as an alternative to travel. Teleconferencing, video conferencing, and message-based conferences will reduce the need for many business meetings, while creating a large demand for wider bandwidths. Experiments such as Teletext in Great Britain combine a low-capacity (telephone) inquiry channel with a high-capacity (cable TV) response channel to provide a wide range of information-based consumer services. Such services may largely replace daily newspapers and the "Yellow Pages" as major sources of consumer information.

Almost certainly, these trends will point to the advantages of packet switching in economy, flexibility, and performance. Furthermore, it is likely that processing costs will continue to decrease relative to transmission cost. As a result, packet switching will be increasingly attractive as an integrating medium for telecommunications and teleprocessing services.

However, two countertrends should be recognized. The first is continued improvement of the communications capabilities of satellites. Larger satellites will be launched with extremely large capacities, increased power, and an unrelenting push toward higher frequencies where additional capacity can be deployed, while permitting the use of smaller, less conspicuous, and less expensive earth stations. The overall effect will be reduced cost of communications. Although the many forms of packetized operation over broadcast satellite channels eliminate the need for packet switches per se, they nevertheless underscore the convenience and flexibility of the packet mode of operation.

The second countertrend is a widespread deployment of the video-telephone. In the early 1970s a video-telephone call consumed about 100 times the transmission resources of a voice-only call, and the regulatory and marketing situation would not support that kind of cost differential. By the early 1980s, however, application of advanced video processing and coding techniques could reduce the differential in video-telephone resource utilization to an acceptable level. Consequently, the deployment of video-telephone service may come about through the decade of the 1980s. However, successful implementation of video services in the common carrier networks would be likely to stimulate the rapid increase in total bandwidth and capacity in the networks. This would further reduce the cost per bit of long-haul capacity to the point where the efficient processing and switching of data-only communications is no longer an important factor. On the other hand, the further reduction of transmission costs may ultimately broaden the applicability of packet switching by reducing overhead considerations in long messages and data transfers using packet technology.

At the same time, the cost of carrier-provided transmission facilities is encumbered by a complex legal/regulatory environment in the United States, and by government ownership and regulation in the international and foreign national communications markets. Thus although changes do reflect technological advances, they do so relatively slowly.

What should we do about these trends and rapid advances in the state of the communications art? Flexibility is the key. The ability to reconfigure, alter, modify, and adapt resources must be preserved in any network implementation. As we have seen, users can purchase terminal equipment to use carrier-provided services more efficiently, to form private networks, or to utilize public networks with maximum efficiency at minimum cost. Such acquisitions should be reflected in rapid payback of invested capital. To preserve the required degree of flexibility, any investment in network access and switching equipment, whether by users or by secondary carriers, should have a cost recovery of less than three years. In telecommunications it is not uncommon to recover an investment in user-owned equipment in a year or less. As we have seen throughout this book, packet switching, with its many possible implementations and broad range of applicability, to a large extent provides the cost-effectiveness and flexibility that are needed in the telecommunications networks of the future.

Glossary

Access line. The communications circuit between a user device and the network nodes.

ACK. ACKnowledgement. The control message sent between switches or other intelligent network devices to signal the correct receipt of a block of data.

ACS. *A*dvanced *C*ommunications *S*ervice. A new common user data communications service being developed by AT&T.

Active control. Network control technique that has a frequent and real-time impact on the network.

Adaptive directory routing. Routing technique that changes the path between any two points in a network according to current operating conditions.

ADCCP. *A*dvanced *D*ata *C*ommunications *C*ontrol *P*rocedure. The level 2 procedure that assures accurate transmission of data blocks between network nodes or between user devices and the network.

AFIPS. *A*merican *F*ederation of *I*nformation *P*rocessing *S*ocieties. An organization tying together a number of professional societies interested in computers and automated data processing.

ALL or ALLOCATE. A packet switching function that sets aside specific network resources to assure the completion of a particular call or message.

ALOHA. The broadcast multiple access technique that permits users to transmit packets whenever they desire.

Analog channel. A communications channel that responds linearly to changes in the frequency and amplitude of the information transmitted; will accurately represent the input signal over a specified range of parameters.

ANSI. *A*merican *N*ational *S*tandards *I*nstitute. An organization that develops and distributes standards for a wide range of commercial products.

ARPANET. The computer network developed by and for ARPA (the *A*dvanced *R*esearch *P*rojects *A*gency within the U.S. Department of Defense). It has been the basis for much of the technology of packet switching.

ARQ. *A*utomatic *R*epeat re*Q*uest. A mode of operation where a receiving station requests a retransmission of any data block it perceives to be in error.

ASCII. *A*merican *S*tandard *C*ode for *I*nformation *I*nterchange. The standard code agreed upon by most computer equipment manufacturers for representation of the alphabet, numerals, and a variety of control characters.

Asynchronous. A form of communications where each transmitted character has self-contained beginning and ending indications, so individual characters can be transmitted at arbitrary times.

AUTODIN (or AUTODIN I). *AUTO*matic *DI*gital *N*etwork. The store-and-forward message network used worldwide by the U.S. Department of Defense and various other government users.

AUTODIN II. A complementary system to AUTODIN that uses packet switching to handle computer and ADP information much more efficiently than AUTODIN I.

AUTOVON. *AUTO*matic *VO*ice *N*etwork. The standard analog private voice communications network operated worldwide for the U.S. Department of Defense.

Bit. Contraction of *BI*nary Digi*T*. A single symbol, either a one or a zero, which when used in groups represents the numbers, letters, and other symbols of communications. Generally used in groups of 5, 8, or 16.

Blocking. A phenomenon in a communications network where one user cannot reach another due to any one, or a combination of, network resource limitations.

BPS. *B*its *P*er *S*econd; sometimes written as B/S or b/s. A measure of the speed with which data communications can move over a line. The prefixes K (for thousand) or M (for million) are often used to represent higher speeds.

Buffer. Part of a communications processor or switch used to store information temporarily.

Buffer blocking. Blocking caused by insufficient buffers or by their inefficient use.

Bursty traffic. Communications traffic characterized by short periods of high intensity separated by fairly long intervals of little or no utilization.

Busy hour. The single hour in the day of greatest total communications network usage; often used for sizing the resources of the network.

Byte. A sequence of successive bits, most often a group of eight, handled as a unit in computer manipulation or data transmission.

Call. A complete, two-way interchange of information between two or more parties in a network, extended over a period of time. It will generally consist of a number of sequential messages or transactions passed over the communications circuits in each direction.

Call arrival rate. The number of new call originations that occur over a specified interval of time.

Called party. The destination of a newly originated call in a communications network.

Calling party. The originator of a new call in a communications network.

Call-second. The use of a facility during a call over a period of one second.

Capacity. The ultimate limitation of any resource in a network to hold or move information.

Capture. A phenomenon of communications whereby the stronger of two signals captures the receiver and remains relatively insensitive to interference effects from the weaker signal.

Carterfone decision. A landmark decision in federal communications regulation that permitted the attachment of foreign devices—those not supplied by the franchised carrier—to the telephone network. It opened the way for competition in the communications marketplace.

CCITT. *C*onsultive *C*ommittee for *I*nternational *T*elephone and *T*elegraph. An international advisory committee set up under United Nations sponsorship to recommend standards for international communications.

Centralized control. Control of a network from a single centralized point, which receives status information from various locations and makes and disseminates control decisions.

Centralized topology. A form of data communications network where a central computer is accessed by a large number of remote terminals.

Channel. A single physical communications medium capable of moving intelligence from one point to another. Specific physical and electrical parameters generally define its capacity. See **group, link.**

Channel-mile cost. The portion of a communication tariff that refers to the costs associated with the distance between the circuit endpoints. Also called line-haul cost.

Circuit switching. A form of switched network that provides an end-to-end path between user endpoints under the control of the network switches. Often called channel switching.

Collisions. The condition where two packets are transmitted sufficiently close in time that some portion of each intersects and interferes with the other.

Common carrier. An organization, generally franchised by a governmental body, to provide communications services to the general public.

Common user network. A network, generally provided by a common carrier, whereby public users can gain access to one another.

Concentrator. A device that improves the efficiency of a communications circuit by fitting a number of low-speed inputs into a single, higher speed output that has a lower speed than the sum of the individual input speeds.

Congestion. A network condition that causes information to be delayed or interrupted, even though capacity may be available elsewhere in the network.

Contention packet switching. Implementation of packet switching functions by allowing devices to transmit at will into a commonly available channel or medium. A variety of different protocols may be used to resolve interference (contention) resulting from (nearly) simultaneous transmission.

Continuous traffic. Communications traffic that (nearly) completely fills the resource while it is in progress; for instance, the transmission of a television program. Compare **bursty traffic.**

CRC. *C*yclic *R*edundancy *C*heck. A procedure used to insure the correct transmission of a block of data by performing a known mathematical operation on the data at the transmitter and comparing it with the same operation performed at the receiver.

Cross-office time. In reference to a circuit switch, the time from the receipt of customer-dialed digits until the switch establishes the connection to the next switch or destination party.

CRT. Literally, *C*athode *R*ay *T*ube. Used in a generic sense to refer to data terminals that display transmitted and received information on a televisionlike screen.

CSMA. *C*arrier *S*ense *M*ultiple *A*ccess. A method of contention operation whereby the terminals sense the state of the channel before attempting transmission.

CVSD. *C*ontinuously *V*ariable *S*lope *D*elta *M*odulation. A method for converting analog speech into a digital format by transmitting a signal that is proportional only to the difference between two successive samples of the original analog signal.

Datagram. A mode of packet network operation whereby the contents of a single packet are handled as a distinct entity with no functional connection with the preceding or following packets.

DCE. *D*ata *C*ircuit *E*quipment. The device and connections placed at the interface to a network by the network provider, to which the user's equipment **(DTE)** is connected.

DDD. *D*irect *D*istance *D*ial network. The major nationwide long-distance network, provided mainly by the Bell System, that permits users to complete connections to distant users without the assistance of an operator.

DDS. *D*ataphone *D*igital *S*ervice. A Bell System service that provides for direct connection of digital sources to the communications medium.

Decentralized control. Control of a network from multiple points, using locally known information or information provided by distant points via the network itself.

Dedicated line. A communications circuit between two endpoints that is permanently connected and always available.

Delay. As applied to packet switching, the additional time introduced by the network in delivering a packet's worth of data compared to the time the same information would take on a full-period point-to-point circuit.

Directory routing. Technique for routing information through a network based on directories, or instructions, kept in the memory of each switch.

Distributed control. See **decentralized control.**

Downlink. The transmission path from a satellite to the satellite earth station to which it is transmitting.

DPCM. *D*ifferential *P*ulse *C*ode *M*odulation. A method that permits the conversion of analog information into a digital format, by encoding the difference in amplitude between successive samples.

DSDS. *D*ataphone *S*witched *D*igital *S*ervice. A limited service of the Bell System for high-capacity digital circuits switched among major cities; charged by time and distance of the connections.

DTE. *D*ata *T*erminal *E*quipment. The device, generally belonging to a data communications user, that provides the functional and electrical interface to the communications medium. For instance, a teleprinter, CRT, or computer.

Duplex channel. A communications channel capable of transmitting information in both directions at the same time.

Erlang. A measure of communications traffic intensity representing the full-time use of a communications facility. For example, one erlang represents the traffic that can be carried on a single line used continuously for one hour.

Error detection. The process of using information added to a data transmission to detect the presence of errors in the received information.

ESS. *E*lectronic *S*witching *S*ystem. A generic term for the switching facilities in commercial networks utilizing computerlike processors rather than purely electromechanical switching relays.

Ethernet. A form of contention operation, being commercially deployed by Xerox Corp., used to tie facilities together in local geographic areas.

Fail-soft operation. An operational characteristic whereby failures of individual components only reduce network performance rather than causing loss of service to some users.

Faxpak. A commercial network permitting the interconnection of dissimilar facsimile terminals.

FCC. *F*ederal *C*ommunications *C*ommission. The principal regulatory body in the United States responsible for interstate communications and common carrier services.

FDM. *F*requency *D*ivision *M*ultiplexing. A means whereby a number of separate communications circuits are combined over a common facility.

Flooding. A routing technique whereby copies of the message are transmitted over every possible route to the destination.

Front-end processor. A specialized computer processor, generally used in conjunction with a larger mainframe computer, that interfaces the computer to communications facilities and remote users.

Full-duplex channel. See **duplex channel.**

Full-period. A circuit that is always available to a particular pair of users and is generally paid for on a fixed monthly basis without regard to total usage.

Gateway. A node or switch that permits communication between two dissimilar networks.

Group. A number of communications channels handled as a single entity.

GTE Telenet. The packet switching subsidiary of *G*eneral *T*elephone and *E*lectronics. It provides nationwide common user data communications service via packet switching.

Half-duplex channel. A communications circuit capable of carrying traffic in either direction but only one way at a time.

HDLC. *H*igh-Level *D*ata *L*ink *C*ontrol. A generic name for the digital link control procedure specified by CCITT standards. **ADCCP** is one U.S. implementation.

Header. The initial part of a data block or packet that provides basic information about the handling of the rest of the block.

Hierarchical network. A network composed of various switches with different functions and connectivity operating at various levels of importance in the network.

Hold and forward. Switching technique where each message is held at intermediary switches long enough for them to check the accuracy of the received information before relaying it on to the next switch.

Holding time. The duration of a call in the network, measured from the time connection is established until all requirements for the connection are completed.

Host. An intelligent processor or device, connected to a network, that satisfies the needs of remote users.

IDA. *I*n *D*elivery *A*cknowledgement. A control message used to signal the network that a message is in the process of leaving the network.

IMP. *I*nterface *M*essage *P*rocessor. The name given to the switching nodes in the ARPANET.

Intelligent terminal. A data communications terminal that has sufficient intelligence (processing power) to perform fairly complex interface functions and local formatting and processing.

Interarrival time. A statistical measure of the average time between successive new calls or messages to a network.

Interrupt. A computer operation that temporarily postpones action on a program in progress so that the main processor can service a higher priority function.

Interruptible traffic. Information transfers that, by nature of their content, need not be transmitted in one continuous stream.

ISO. *I*nternational *S*tandards *O*rganization. An international body that standardizes goods and services. ISO works in conjunction with CCITT for standards that impact communications.

Keyboard CRT. A data terminal device that combines a typewriterlike keyboard for data input with a TV-like screen for data output.

Leader. The initial part of a user data block that tells the network the destination and handling of the following data.

Line-haul cost. See **channel-mile cost.**

Link. A physical or electrical connection between two endpoints; for communications purposes may consist of one or more channels.

Logical channel number. A designator of an apparent connection via a packet switched network, by time sharing the channel to the network switch.

Logical multiplexing. The ability of given user to communicate with various network destinations simultaneously, using a single access line, by time sharing the line and using different logical identifiers for each connection in progress.

Loop (local). Telephone terminology which refers to the local connection between a network switch and the subscribers end instrument.

Loop (routing). The undesirable condition in a network where traffic gets routed in a circular path due to an anomaly of the software or address information.

Low duty-cycle. A network user characteristic whereby bursts of transmitted data are separated by relatively long idle periods.

LPC. *L*inear *P*redictive *C*oding. A technique for converting analog speech to digital format that uses information about the vocal tract to synthesize the transmitted speech with a very small number of transmitted bits.

MCI. *M*icrowave *C*ommunications *I*ncorporated. One of the first common carriers licensed to compete with the Bell System for interstate communications services.

M/D/1 queue. Generalized notation of queueing theory, designating a delay process characterized by Poisson (M) arrivals, deterministic (D) message lengths, and a single (1) server.

Message switching. Switching technique where messages are stored in their entirety at each intervening switch.

M/M/1 queue. Generalized notation of queueing theory, designating a delay process characterized by Poisson (M) arrivals, Poisson distributed (M) message lengths, and a single (1) server.

Modem. *MO*dulator-*DEM*odulator. A device that allows digital signals to be transmitted over analog facilities.

MPL. *M*ultischedule *P*rivate *L*ine. A tariff of the Bell System that provides for full-period, point-to-point analog circuits between service locations within the United States.

Multidrop. The data communications analogy of a party line, where various user terminals are connected on a common, shared line.

Multiplexing. See **logical multiplexing.**

Multiplexor. A device that combines a number of low-speed channels into a single higher speed channel.

Multipoint. See **multidrop.**

NCC. *N*etwork *C*ontrol *C*enter. The major control point of a network; collects performance data and issues control commands.

NCP. *N*etwork *C*ontrol *P*rogram. The software that must be executed by a front-end computer in order to interface with a packet switched network in the full packet mode.

Nearest neighbor. Network switches that are directly connected via a single link to a given node.

Node. A point of a network where various links come together; generally containing a switching element used to direct traffic.

Overhead. Information required by a network for its operation, over and above the basic information that is being moved on behalf of the subscribers.

Packet switching. A network technique that divides user messages into relatively short blocks and uses numerous geographically distributed switching nodes, to achieve low end-to-end delay for real-time data traffic.

PACUIT switching. A switching technique, using a transmission frame structure, that combines many of the desirable attributes of packet and circuit switching.

PAD. *P*acket *A*ssembly-*D*isassembly. A packet network–based function that allows terminals with little or no intelligence to interface a packet network.

Passive control. A technique for control over a long-term time frame, based on accumulated statistics on network operation.

PCM. *P*ulse *C*ode *M*odulation. A technique for coding analog signals for transmission on a digital circuit, by sampling the analog signal at regular intervals and converting each sample into a digital codeword.

Persistence. Characteristic of a user who continuously monitors the occupancy of a channel and transmits a packet as soon as he detects that the channel is idle.

Polling. A technique that permits a large number of terminals to share a common channel. A central controller asks each terminal, in turn, to transmit any information it may currently have queued.

POTS. *P*lain *O*ld *T*elephone *S*ervice. The common user voice-based nationwide telephone network.

Protocol. A set of rules and procedures that permit the orderly exchange of information within and across a network.

PVC. *P*ermanent *V*irtual *C*ircuits. A logical connection across a packet switched network that is always in place and available; used to emulate a full-period connection.

Queueing. Any process that combines elements of storage and delay together with a number of servers. The delay experienced by the users of the process can be estimated on the basis of statistical behavior of the various elements involved.

Random routing. Routing technique that moves information through the network in a statistically random manner.

REQALL. *REQ*uest for *ALL*ocation. A control message used in a packet switched network that assigns network resources to the handling of a new call.

Reservation technique. Any of a number of possible packet broadcast methods requiring that users reserve capacity in advance of transmitting their data.

RFNM. *R*equest *F*or *N*ext *M*essage. A network control message that confirms reception of a message and indicates that resources are still assigned for the next message that is part of the same call.

RFNS. *R*equest *F*or *N*ext *S*egment. Similar to **RFNM,** but confirms only a segment of a message, rather than a full message.

RJE. *R*emote *J*ob *E*ntry. A data terminal used to enter complete jobs for processing at a remote computer location.

Robust. A network characteristic indicating the ability to operate nearly normally when certain network elements fail.

Routing. The process of finding a suitable path to move information through the network. See also **adaptive directory routing, directory routing,** and **random routing.**

Routing table. A set of instructions stored at each switch indicating the path to move a given packet to a given destination.

SBS. *S*atellite *B*usiness *S*ystems. A partnership of IBM, COMSAT, and Aetna Insurance to provide private network services via satellite communications facilities.

SDLC. *S*ynchronous *D*ata *L*ink *C*ontrol. IBM's version of the **ADCCP** link control technique.

Segment. A part of an overall information exchange that is transmitted between the user device and the network. It may be the same length as or longer than a packet, depending on the protocol implementation.

SENET Switching. *S*lotted *E*nvelope *NET*work Switching. A master frame technique that combines packet and circuit switched functions in a common network.

Slotted channel. A packet transmission time that has a fixed (rather than random) relationship to allowable transmission times of other packets.

Store and forward. Switching technique where each message is stored in full at each switch it passes through.

SVC. *S*witched *V*irtual *C*ircuits. A logical connection across a packet switched network. It is established on an as-needed basis and can provide connection to any other switched user in the network.

Synchronous. A form of communications where characters or bits are sent in a continuous stream, with the beginning of one contiguous with the end of the preceding one. Separation of one from another requires the receiver to maintain synchronism to a master timing signal.

TADI. *T*ime *A*ssignment *D*igital *I*nterpolation. A digital technique for interleaving data bursts during silent intervals in voice conversations.

Tandem switches. Switches in a network that provide a path between other switching nodes, rather than originating or terminating traffic.

Tariffs. The formalized charges for telecommunications services that are filed and approved by state and federal regulatory organizations.

TASI. *T*ime *A*ssignment *S*peech *I*nterpolation. A technique for carrying a group of voice channels over a physical facility by interleaving conversations in the idle periods of normal voice communications.

TDM. *T*ime *D*ivision *M*ultiplexing. A means whereby a number of separate communications circuits are combined over a common facility by dividing the common facility into discrete time intervals.

Telenet. See **GTE Telenet.**

Telpak. A nationwide AT&T tariff that provided substantial discounts for the lease of 60 or 240 channels in one facility.

Tessellated pattern. A physical arrangement of facilities (such as nodes or earth stations) that follows a regular, repetitive geometric pattern.

Timeout period. The length of time a switch will wait for an expected action (such as an acknowledgement) before it takes unilateral action.

TIP. *T*erminal *I*nterface *P*rocessor. A network switch in the ARPANET that can interface up to 64 user terminals in addition to several computer hosts.

Topology. The physical arrangement of nodes and links to form a network, including the connectivity pattern of the network elements.

Transaction. A computer-based message that represents a complete unidirectional transfer of information between two points on a data network.

Transactional switching. A network switching technique that handles information as discrete entities, rather than providing a fixed end-to-end connection.

Transparent switching. A network switching technique that handles information by providing a logical or physical end-to-end connection.

Trunk. The communications circuit between two network nodes or switches.

Tymnet. A nationwide data service with packet switching-like characteristics, although it is not literally a packet switched network.

Uplink. The transmission path from a satellite earth station to the satellite itself.

VAN. *V*alue-*A*dded *N*etwork. The class of public network that leases facilities in the form of basic transmission from one carrier, adds intelligence, and provides more "valuable" services to end users.

VFCT. *V*oice-*F*requency *C*arrier *T*elegraph. A technique that permits the combination of up to 24 teletype channels over a single voice-frequency channel.

Virtual circuit. A logical connection across a packet switched network that emulates a point-to-point circuit by insuring data integrity, transparency, and data sequence.

WATS. *W*ide *A*rea *T*elephone *S*ervice. A nationwide long distance phone service, where users contract for 10 or 240 hours of use per month, rather than paying for each call individually.

Window. The major element of the flow control mechanism used to prevent the overload of a packet network. The window size indicates the number of packets a given user can have outstanding (unacknowledged) in the network at any given time.

XTEN. *X*erox *TE*lecommunications *N*etwork. A proposed nationwide telecommunications offering, originally developed by XEROX Corp., that combines satellite intercity transmission with local microwave distribution.

X.25. The international standard developed by CCITT that provides the foundation for public packet switching networks.

Index